AF395534

Prestressed Concrete

Charles W. Dolan • H. R. (Trey) Hamilton

Prestressed Concrete

Building, Design, and Construction

 Springer

Charles W. Dolan
University of Wyoming
Laramie, WY, USA

H. R. (Trey) Hamilton
Department of Civil and Coastal Engineering
University of Florida
Gainesville, FL, USA

ISBN 978-3-319-97881-9   ISBN 978-3-319-97882-6   (eBook)
https://doi.org/10.1007/978-3-319-97882-6

Library of Congress Control Number: 2018954932

This Springer imprint is published by the registered company Springer Nature Switzerland AG
The registered company address is: Gewerbestrasse 11, 6330 Cham, Switzerland

Preface

For over a decade we have been teaching Prestressed Concrete using Nilson's *Design of Prestressed Concrete* or Lin and Burn's *Design of Prestressed Concrete Structures*, both of which are over two decades old. These sources are augmented by the *PCI Design Handbook,* the *PTI Design Handbook,* and the prestressed concrete chapters of other texts. None of these are fully satisfactory to present the student or the practicing engineer with the basics and intricacies of prestressed concrete design. Between us, we have nearly 30 years of experience as design engineers and 50 years of experience in teaching and research. We have a unique perspective of having been both practitioners and educators. As such, we have gathered the salient features of the existing literature and combined it with our teaching and practice experience. In this book, we have gathered the fundamental principles coupled with design and construction realities. To this end, the development of the book begins with first principles then proceeds to give results based on those principles, and as interpreted by Building Codes. In some examples, we let the student see that the first trial solution is not acceptable and that multiple trials are needed to obtain a satisfactory solution. By constructing aids in EXCEL or MathCad, students can see the interaction of mechanics in prestressed concrete and how trials are quickly refined. Where appropriate, we discuss prestressing plant and on-site post-tensioning operations that affect the design.

In today's world as a practicing engineer, there is little time for contemplation. The designer must move quickly and efficiently to be productive. This productivity is facilitated greatly by the abundance of software programs that are available for the design engineer or the specialty engineer. While the fundamental purpose of structural analysis software is to implement the mathematical modeling necessary to the design process, today's software has ever increasing features and additions which expedite the design process. We have studiously avoided the details of structural analysis and the software that goes with it. Rather, we have gathered fundamental ideas and practical suggestions that will augment and improve the engineer's use of software and allow the engineer to readily confirm the software output validity.

The book is based on fundamental behavior of pretensioned and post-tensioned concrete and the code interpretation of fundamental behavior. ACI 318-14 is referenced extensively. AASHTO bridge girders are used for many examples because the properties are well established and they provide a greater range of application than just rectangular sections. Reference to the *AASHTO Bridge Design Specification* is included to illustrate that more than one code exists and to illustrate some of the differences between design codes.

The first 13 chapters focus on prestressed concrete and form the basis for a second course in design of concrete structures. Our experience has been that these chapters take one academic quarter or three quarters of a semester. The emphasis is on the mechanics of prestressed concrete and not the analysis needed to obtain structural loads. Today's practice uses finite element or other computer-based programs to determine the member forces. Consequently, we begin by assuming that the loads on the structure are known. Chapter 13 is a look at spliced girder construction. The chapter focuses on the prestressing aspects; however, it simplifies the bridge design aspects to keep the example within limits of classroom presentation.

Chapter 14 on Strut-and-Tie Method and Chap. 15 on Anchorage to Concrete are modified for prestressed concrete from the same chapters in *Design of Concrete Structures, 15th ed.* and are provided by permission of McGraw-Hill Education. The book is based on US customary units as it is tied to the ACI 318-14 *Building Code for Concrete Structures*.

Chapter 16 addresses comprehensive problems where all aspects of prestressed concrete are present. These problems may be used as term project or sequentially through the learning process to envision how a design is developed.

We are publishing the book with Springer because the eBook version is free to students whose libraries subscribe to the Springer library and print copies are available at reasonable cost. We have established an email address, prestressauthors@gmail.com, for communication with faculty members adopting the book and appreciate comments, critiques, or suggestions for new problems.

Lastly, no text is complete without a thorough review and assistance. We are indebted to Dr. Larry Khan at Georgia Tech, Dr. Brandon Ross at Clemson, and Dr. David Mukai at the University of Wyoming for their considerable review of the text and the classroom testing of its contents. Mr. Jacob Montgomery drafted the detailed figures used in the text.

Laramie, WY, USA   Charles W. Dolan
Gainesville, FL, USA   H. R. (Trey) Hamilton

Contents

About the Authors

Charles W. Dolan P.E., S. E., Ph.D., has over 45 years of consulting experience including 25 years of teaching reinforced and prestressed concrete design. His work in reinforced and prestressed concrete has been recognized by his receipt of the ASCE T. Y. Lin award, the ACI Arthur R. Anderson award for excellence in concrete education, and the PCI Martin P. Korn award for meritorious research. He serves on ACI 318 *Building Code for Concrete Structures* and chaired the ASCE/ACI Committee on prestressed concrete. He was the first H. T. Person Chair of Engineering at the University of Wyoming where he focused on undergraduate education and engineering design and was recipient of the University of Wyoming Ellbogen Lifetime teaching award. He is coauthor of the 13th, 14th, and 15th editions of *Design of Concrete Structures*, editor of several volumes addressing the use of FRP for design and to strengthen concrete, author of *The Design Challenge* addressing teaching design to undergraduate students, and author of over 100 technical papers. His design and consulting projects include the monorails at Walt Disney World and the Palm Island Dubai, the overall concept design for the Vancouver Skytrain, the Westin Hotel in Avon, Colorado, the original guideway for the Dallas-Fort Worth Airport, and the guideway for the Detroit Downtown People Mover.

H. R. (Trey) Hamilton P.E., Ph.D., is professor of structural engineering in the Civil and Coastal Engineering Department at the University of Florida, Gainesville, FL. He obtained his bachelor's and master's degrees from the University of Florida and was in private practice as a design engineer for about 7 years prior to obtaining a PhD from the University of Texas at Austin in 1995. He was a faculty member at the University of Wyoming from 1995 to 2001 prior to joining the faculty at the University of Florida. His research work has focused on the development of methods and materials to improve the sustainability of structures, and his professional activities have focused on the implementation of those results in construction and rehabilitation. He is a fellow of both the Post-Tensioning Institute and American Concrete Institute and is past chair of ACI/ASCE 423 Prestressed Concrete committee and the ACI Technical Activities Committee. He has won awards for papers from the American Society of Civil Engineers, ASTM International, the American Composites Manufacturer's Association, the Masonry Society, and International Concrete Repair Institute. Most notable of these is the J. James R. Croes Medal awarded by the American Society of Civil Engineers across all disciplines of Civil Engineering.

Notation[1]

a	Depth of equivalent rectangular stress block, in.
A_{brg}	Bearing area of an anchor head, in.^{2}
A_{cp}	Area enclosed by outside perimeter of concrete cross section, in.^{2}
A_{ct}	Area of that part of cross section between the flexural tension face and centroid of gross section, in.^{2}
A_g	Gross area of the section, in.^{2}
A_l	Area of longitudinal torsion reinforcement, in.^{2}
A_{Nao}	Projected area of a single adhesive anchor, in.^{2}
A_{Nco}	Projected breakout area of a single anchor in tension, in.^{2}
A_{ps}	Area of prestressing reinforcement, in.^{2}
$A_{s,min}$	Minimum area of flexural reinforcement, in.^{2}
A_{se}	Cross-sectional area of anchor, in.^{2}
$A_{se,N}$	Cross-sectional area of anchor in tension, in.^{2}
$A_{se,V}$	Cross-sectional area of anchor in shear, in.^{2}
A_{vf}	Area of shear friction reinforcement, in.^{2}
A_t	Area of transverse torsion reinforcement, in.^{2}
A_{Vco}	Projected breakout area of a single anchor in shear, in.^{2}
b	Width of compression block, in.
b_{eff}	Effective flange width of T-beam based on ACI Building Code requirements, in.
b_{eq}	Equivalent flange width of composite section based on the ratio of modulus of elasticity, in.
b_v	Width of the shear section, in.
b_w	Width of web, in.
c	Distance from extreme compression fiber to neutral axis, in.

---

[1]The following notation is used in this book. In the cases where the units are given for use in unit sensitive equations. In some cases, alternative units are used, and these locations are identified in the text. Certain notations used for derivations are not included in this section but are identified in the text.

$c_{a,min}$	Minimum edge distance, in.
c_{a1}	Distance from the edge of the concrete to the first anchor, in.
c_{a2}	Edge distance perpendicular to c_{a1} or to next interior anchor, in.
c_c	Clear cover over reinforcement, in.
C_c	Creep coefficient
C_{ct}	Creep coefficient at time t where t is in days
cgc	Center of gravity of concrete section
cgs	Center of gravity of prestressing tendon
C_1	Correction factor for relaxation losses due to stress level in strand
c_{na}	Characteristic depth of an adhesive anchor, in.[2]
CR	Load effects due to creep
CR_1	Prestress loss due to creep effects, psi
D	Effect of service dead load
d	Distance from extreme compression fiber to centroid of longitudinal tension reinforcement, in.
d_a	Diameter of an anchor, in.
d_b	Diameter of reinforcing element, in.
d_p	Distance from extreme compression fiber to centroid of prestressing reinforcement, in.
E	Load effects of seismic activity
e	Eccentricity of tendon, in.
e'_N	Eccentricity of tensile forces, in.
e'_V	Eccentricity for shear forces, in.
E_c	Modulus of elasticity of concrete (usually at 28 days unless otherwise specified), psi
E_{ci}	Modulus of elasticity of the concrete at the time of transfer, psi
E_{cp}	Modulus of elasticity of composite topping, psi
e_e	Eccentricity of tendon at end of member, in.
e_h	Distance from the inner surface of the shaft of a J- or L-bolt to the outer tip of the J- or L-bolt, in.
e_p	Eccentricity of the tendon at the critical section, in.
E_{ps}	Modulus of elasticity of the tendon, psi
EQ	Effect of earthquake
E_s	Modulus of elasticity of nonprestressed reinforcement, psi
ES_1	Prestress loss due to elastic shortening effects, psi
F	Effect of loads due to weight or pressures of fluids with well-defined densities
$f'_{c,top}$	Concrete strength of composite topping, psi
f'_{ci}	Specified strength of the concrete at the time of transfer, psi
f_1	First principal stress, psi
f_2	Second principal stress, psi
f_{anc}	Loss of prestress due to anchor seating, psi
f_c	Stress in concrete, psi
f'_c	Specified concrete compressive strength, psi

f_{ce}	Final stress in compression after losses, psi
f_{ci}	Initial stress in compression, psi
f_{cp}	Concrete stress in composite topping, psi
f_{j}	Stress in prestressing reinforcement at jacking, psi
f_{CR}	Loss of prestress due to creep, psi
f_{ES}	Loss of prestress due to elastic shortening, psi
f_{fr}	Loss of prestress due to friction, psi
f_{p}	Stress in prestress reinforcement at intermediate time t, psi
f_{pc}	Compressive stress at neutral axis, psi
f_{pi}	Initial prestress, ksi
f_{ps}	Stress in prestressing reinforcement at nominal flexural strength, psi
f_{pu}	Specified tensile strength of prestressing reinforcement, psi
FR	Effect of friction
f_{r}	Modulus of rupture of concrete, psi
f_{RE}	Loss of prestress due to relaxation, psi
f_{s}	Service stress in reinforcement, psi
f_{se}	Effective stress in prestressing reinforcement, after allowance for all prestress losses, psi
f_{SH}	Loss of prestress due to shrinkage, psi
f_{t}	Concrete tensile strength, psi
f_{loss}	Total loss in stress along a member, psi
f_{te}	Final stress in tension after losses, psi
f_{ti}	Initial stress in tension, psi
f_{uta}	Tensile strength of an anchor, psi
f_{y}	Specified yield strength of nonprestressed reinforcement, psi
f_{ya}	Yield strength of anchor, psi
f_{py}	Specified yield strength of prestressing reinforcement, psi
f_{yt}	Specified yield strength of transverse reinforcement, psi
H	Effect of loads due to weight and pressure of soil, water in soil or other materials
h_{ef}	Effective embedment length of an anchor, in.
I_{c}	Composite section moment of inertia, in.^{4}
I_{ct}	Moment of inertia of cracked transformed section, in.^{4}
I_{cr}	Moment of inertia of a cracked section, in.^{4}
I_{g}	Gross moment of inertia of the section, in.^{4}
I_{ut}	Moment of inertia of uncracked transformed section, in.^{4}
J	Correction factor for effects of creep, shrinkage, and elastic shortening on relaxation
J_{c}	Polar moment of inertia of the critical section of the slab around the centroid c-c, in.^{4}
k	Wobble friction coefficient, lb/in.
k	Depth of service level stress block, in.
k_{c}	Coefficient for calculating concrete breakout in tension and equal to 24 for cast-in anchors, 17 for post-installed anchors

k_{cp} Correction factor for anchor length subjected to pryout
K_{cr} Basic relaxation stress loss by tendon type, psi
L Effect of service live load
l Length of member or tendon, in.
l_d Length to transfer prestress force to the concrete, in.
ℓ_e Load-bearing length of the anchor for shear, in.
l_n Length of clear span measured face-to-face of supports, in.
L_r Effect of live loads on roof
ℓ_{set} Length of tendon in which the prestress forced is affected by anchor set, in.
l_x Tendon length from jacking end to point x in a tendon, in.
M_a Maximum moment in member due to service loads at stage deflection is calculated, in.-lb
M_{cr} Cracking moment, in.-lb
M_{cre} Moment causing flexural cracking at section due to externally applied loads, in.-lb
M_g Moment due to girder weight, lb-in.
M_{int} Internal moment, in.-lb
M_l Unfactored live load moment, in.-lb
M_{max} Maximum factored moment at section due to externally applied loads, in.-lb
M_n Nominal moment strength, in.-lb
M_p Primary moment, in.-lb
M_s Unfactored service load moment, in.-lb
M_{sc} Factored slab moment that is resisted by the column at a joint, in.-lb
M_{sdl} Unfactored superimposed dead load moment, in.-lb
M_T Total moment, in.lb
M_u Factored moment, in.-lb
M_2 Secondary moment, in.-lb
n Modular ratio $= E_{ps}/E_{ci}$
N_b Tensile strength of a single anchor in concrete breakout, lb
N_{ba} Tensile strength of a single adhesive anchor, lb
N_{bag} Tensile strength of an adhesive anchor group, lb
N_{cb} Nominal concrete breakout strength in tension of an individual anchor in tension, lb
N_{cbg} Nominal concrete breakout strength in tension of an anchor group in tension, lb
N_p Basic pullout strength of an anchor, lb
N_{pn} Pullout strength of an anchor in cracked concrete, lb
N_{sa} Nominal steel strength of an anchor in tension, lb
N_{va} Nominal steel strength of an anchor in shear, lb
P Applied load or prestressing force, lb
p_{cp} Outside perimeter of concrete cross section, in.
P_h Perimeter of centerline of outermost closed stirrup, in.
P_i Initial prestress force, lb
P_j Tendon jacking force, lb

P_n	Nominal axial load strength, lb
P_u	Factored axial load, lb
Q	Applied load, lb
Q	Static moment of area above the section under investigation, in.^{3}
q_{ll}	Unfactored live load, psf
q_{sdl}	Unfactored superimposed dead load, psf
R	Radius of curvature, in.
R	Effect of loads due to rain
r	Radius of gyration, in.
RE	Strand relaxation stress, psi
RH	Relative humidity
s	Spacing between elements, in.
S	Effects of loads due to snow
S_b	Bottom section modulus, in.^{3}
S_{bc}	Bottom section modulus of composite section, in.^{3}
S_c	Section modulus at the interface between the precast beam and the composite beam, in.^{3}
SH	Load effects due to shrinkage
SH_l	Prestress losses due to shrinkage effects, psi
S_n	Nominal strength of a member or section
S_t	Section modulus to top of section, in.^{3}
s_w	Center-to-center spacing of beam webs, in.
t	Time of calculated prediction after load is applied, days, hours for relaxation
t	Thickness of an element such as a web, in.
T	Load effect due to temperature variation
T	Tensile force, lb
T	Applied torque, lb-in.
T_{cr}	Torsion that causes cracking, lb-in.
T_n	Nominal torque capacity, in.-lb
T_n	Nominal tension capacity, lb
T_u	Factored torque, lb-in.
T_u	Factored tensile force, lb
TU	Effect of uniform temperature load effects
U	Factored load, sometimes referred to as ultimate load
V	Vertical shear force, lb
V/S	Ratio of volume to surface area of concrete, in.
V_a	Contribution of aggregate interlock to nominal shear strength, lb
V_b	Concrete shear breakout capacity of a single anchor, lb
V_{cb}	Nominal concrete breakout strength in shear of an individual anchor in tension, lb
V_{cbg}	Nominal concrete breakout strength in shear of an anchor group in tension, lb
V_{ci}	Nominal shear strength due to flexure-shear cracking, lb

V_{cp}	Strength of a single anchor in pryout, lb
V_{cpg}	Strength of an anchor group in pryout, lb
v_{cr}	Principal shear stress to cause cracking, psi
V_{cr}	Shear force creating principal shear stress to cause cracking, psi
V_{cw}	Nominal shear strength due to web cracking, lb
V_{cz}	Nominal shear attributed to concrete compression zone, lb
V_d	Nominal shear attributed to dowel action, lb
V_d	Shear force at section due to unfactored dead load, lb
v_h	Horizontal shear stress, psi
V_i	Factored shear force at section due to externally applied loads occurring simultaneously with M_{max}, lb
v_{max}	Principal or maximum shear stress, psi
V_n	Nominal shear strength, lb
V_{nh}	Nominal horizontal shear strength, lb
V_p	Nominal shear attributed to the vertical component of prestress force, lb
V_u	Factored shear, lb
v_u	Factored shear stress, psi
V_{ug}	Factored shear stress due to gravity load, psi
v_{uh}	Factored horizontal shear stress, psi
V_{uh}	Factored horizontal shear, lb
W	Effect of wind load
w	Uniform load, plf
WA	Effect of water loads
W_{cSH}	Unit weight of the concrete, pcf
w_d	Unfactored dead load, plf
w_{eq}	Equivalent uniform load, plf
w_g	Member self-weight, plf
WL	Wind load effects on live load
w_{ll}	Unfactored live load, plf
WS	Load effects due to wind on structure
w_{sdl}	Unfactored superimposed dead load, plf
w_{slab}	Unfactored slab load, plf
x	Length along a member, ft
y	Distance from neutral axis to a point in the section, in.
y	Vertical distance along a member, ft
y_b	Distance from neutral axis to bottom of the section, in.
y_{bt}	Distance from neutral axis to bottom of the composite section, in.
y_c	Distance from neutral axis to the composite interface in the section, in.
y_t	Distance from neutral axis to the top of the section, in.
z	Distance between tension and compression force centroids, in.
α	Angle change of tendon
β_1	Factor relating depth of equivalent rectangular compressive stress block to depth of neutral axis
δ_d	Deflection due to prestress and girder weight, in.

δ_1	Initial deflection due to prestress, in.
δ_o	Deflection due to girder self-weight, in.
Δ_D	Displacement caused by superimposed dead loads, in.
Δ_L	Displacement caused by live loads, in.
Δ_{pi}	Displacement caused by prestressing force, in.
Δ_{sw}	Displacement caused by self-weight, in.
Δ_T	Total displacement, in.
ε	Strain
ε_d	Strain required to bring the bottom strain in the concrete due to prestressing and girder self-weight to zero
ε_f	Strain in prestressing reinforcement due to bending of the member
ε_{in}	Instantaneous elastic strain
ε_{pe}	Strain in prestressing reinforcement after all losses
$\varepsilon_{sh,\,t}$	Unit shrinkage at time t
$\varepsilon_{sh,\,u}$	Total shrinkage after a long time
ε_{sh}	Strain due to shrinkage
ε_{se}	Effective strain in prestressing reinforcement, after allowance for all prestress losses, psi
ε_{cr}	Strain due to creep
γ_v	Factor used to determine the fraction of M_{sc} transferred by eccentricity of shear at slab-column connections
λ	Correction factor for lightweight concrete
λ_a	Correction factor for lightweight concrete used in anchor zones
μ	Coefficient of friction
$\psi_{c,V}$	Correction for cracking for anchors in shear
$\psi_{c,P}$	Correction for cracking for anchors in pullout
$\psi_{ed,N}$	Correction for edge distance for anchors in tension
$\psi_{ed,V}$	Correction for edge distance for anchors in shear
ρ	Reinforcement ratio
ρ_p	Reinforcement ratio of prestressing reinforcement
Θ	Angle to principal stresses, deg.
τ	Shear stress due to torsion, psi
τ_{cr}	Characteristic bond stress in cracked concrete, psi
τ_{uncr}	Characteristic bond stress in uncracked concrete, psi

Chapter 1
Basic Concepts

1.1   Introduction

Most concrete construction in the world is cast-in-place **reinforced concrete**. In reinforced concrete structures, steel reinforcement is placed into the concrete to provide the tensile resistance to flexural loads or to assist the concrete in carrying compressive loads. While a superb building material, reinforced concrete must crack before the steel can significantly contribute to the strength and stiffness. This behavior led engineers to develop **prestressed concrete**, which uses high-strength steel **tendons** that are stretched to apply a compensating compressive load to the concrete prior to the application of the service loads. The prestressing force creates a compressive stress in the concrete to counteract the tensile stresses induced by the service loads. While this prestressing force does not result in an intrinsically higher member strength than reinforced concrete, it does delay cracking so that the service loads are carried primarily by uncracked concrete. This allows for the use of longer spans without increasing the member depth, or shallower members for the same span as can be designed with reinforced concrete.

The concept of **prestressing** involves placing a load on a structure in a direction and magnitude to compensate loads that are applied during the life of the structure. Consider a stack of several wooden blocks. If you attempt to pick up the stack by the first block, the remaining portion of the stack remains unmoved. If you place an elastic band tightly around the stack, you can not only pick up the stack but also you can hold it out as a cantilever beam. The elastic band is in tension and imparts a compression force on the stack. The compressive force overcomes the tensile stress induced by bending. The compressive force created by the elastic bands additionally mobilizes friction between the blocks to provide shear strength. This basic principle of prestressing applies to concrete, timber, and steel only larger and more robust "elastic bands" are used. Concrete, like the stack of blocks, has a low tensile strength. Thus, properly designed prestressed concrete members overcome the low

© Springer Nature Switzerland AG 2019
C. W. Dolan, H. R. Hamilton, *Prestressed Concrete*,
https://doi.org/10.1007/978-3-319-97882-6_1

Fig. 1.1 Long span pretensioned concrete bridge girder

tensile strength of concrete by using compression to prevent cracking under service loads.

The concept of prestressed concrete first appeared in the late 1800s and early 1900s but was generally unsuccessful due to the **loss of prestress force** resulting from volume changes in the concrete due to elastic shortening, shrinkage, and creep of the concrete, and relaxation of the steel. In the 1930s Eugene Freyssinet, a French engineer, was successful in using high-strength reinforcement to successfully overcome the prestress losses (Billington, 1975). When first introduced, prestressed concrete was designed to introduce sufficient axial force in the member to eliminate tensile stresses in the concrete under service load. The lack of tension and cracking gave prestressed concrete the ability to use the full section properties to resist bending and deflection. At the time, prestressed concrete was promoted as an entirely new building material. Freyssinet's definition of prestress was

"To prestress a structure is to artificially create in that structure, either prior to or simultaneously with the application of external loads, such permanent stresses that in combination with the stresses due to external loads, the total stresses remain everywhere, and for all states of load envisioned, within the limits of stress that the material can support indefinitely." (Guyon, 1974)

The benefits of an uncracked section are clear to design engineers. The greater gross section properties allow for design of longer, thinner, and more economical flexural members. Figure 1.1 illustrates a long-span precast-prestressed bridge girder. The girder is shipped by barge to the construction site because the structural depth, span length, and weight exceed allowable limits for highway transport.

Designs that have no tension under normal service load are unlikely to crack. This improves the member stiffness and durability in aggressive environments. Engineers can design prestressed concrete structures to eliminate tension, allow some tension, allow cracking, or just control deflections. The term prestressing applies to any concrete member subjected to some level of initial precompression.

Prestress is a generic term indicating that a preload is applied to the member during construction and prior to the application of external loads. The prestressing force can be applied in one of two methods. One method is to **pretensioned** the member, where the tendon is stressed prior to the placement of the concrete. The **tendon** consists of one or more **prestressing steel** elements such as **wires**, seven-wire **strand**, or high-strength **bars**. Concrete is placed and allowed to harden prior to transfer of the prestress force to the concrete. Transfer of the pretensioning force occurs through bond between the tendon and the concrete. The other method is to **post-tensioned** the member, where the prestressing force is applied to the hardened concrete using a mechanical anchor. The prestressing steel is placed in ducts within the concrete section or external to the concrete section. Post-tensioning tendons may be **bonded**, fully integrated with the concrete to allow for the assumption of plane sections remaining plane, or **unbonded**, able to move relative to the concrete section.

1.2 Loads

Selection of loads on a structure is one of the primary responsibilities of the engineer. The engineer must understand the function and use of the structure and then use engineering judgment for the selection applied loads. Structural loads can be separated into three broad categories: **dead loads**, **live loads**, and **environmental loads**. **Dead loads** deal with the self-weight of the structure plus those portions of the structure that are permanently attached. The self-weight of a concrete structure is often the largest component of the dead load. It can be calculated with a reasonable degree of accuracy using the dimensions of the concrete sections in the structure. A unit weight of 150 pcf is typical for cast-in-place concrete structures. Precast and plant prestressed concrete structures use a unit weight of 160 pcf due to the extra consolidation of the concrete available in the plant operation. The remaining dead load consists of floor systems, ceilings, roofing, curbs, railings, and other permanent attachments.

Live loads represent the loads that are not permanently applied to the structure and are prescribed by the building code. **Environmental loads** are usually external to the structure and include snow, rain, soil pressure, wind, and earthquake. Wind and earthquake loads are dependent on the building system geometry, building framing, site conditions, and hazard occurrence. Definition of these loads requires attention to the geometry and details of the structure and the conditions of the surrounding site.

Service level live loads consist of the occupancy loads in a building, or vehicular loads on bridges. The loads can be fully or partially applied depending on the occupancy. Loads for building structures are typically specified by the **building codes** appropriate for the jurisdiction and application. The *International Building Code (IBC)* (International Building Code, 2015) is a compilation of occupancy requirements, life-safety considerations, and building functional requirements. The *IBC* incorporates by reference independent codes that define the loads on building structures and the required resistance to those loads by building materials such as concrete, steel, wood, and masonry. The American Society of Civil Engineers (ASCE) Structural Engineering Institute (SEI) publication *ASCE/SEI 7 Minimum Loads for Building Structures* (ASCE/SEI 7-16, 2016) defines occupancy and environmental loads in buildings. Table 1.1 contains an abbreviated summary of building loads from *ASCE/SEI 7*. *ASCE/SEI 7* has the responsibility for defining all loads on structures and thus contains wind, rain, snow, and seismic loadings in addition to occupancy loadings. Material building codes such as the American Concrete Institute (ACI) *ACI 318 Building Code for Concrete Structures* (ACI 318, 2014) address the response of structures to the applied loads. While *IBC* and *ASCE/SEI 7* provide guidance for occupancy loads, the materials building codes describe the response to loads. The adoption of a building code is made by the local municipality. As such, local municipalities may have loads or responses that differ from the values given in the national building codes.

The occupancy loads are given as uniformly distributed unit loads. Concentrated loads are specified to account for loads that are temporarily placed in the building and are to be placed at a location to generate the maximum load effect on the member. The specified unit loads are anticipated to be the maximum service loads for the application and are typically higher than the average load in the building. The probability of the maximum live load occurring on all portions of the structure simultaneously is low. Building codes typically prescribe **live load reduction factors** that allow for reduction of the total live load as the area to be supported increases. The details of live load reduction can be found in ASCE/SEI 7.

The American Association of State Highway and Transportation Officials (AASHTO) maintain the *AASHTO LRFD Bridge Design Specifications* (2017), which provides live loads and design criteria for highway bridge structures. The *AASHTO LRFD Specifications* apply to state and federal bridges and by extension to most county and city bridges. Railway bridge design and loads are specified by the American Railway Engineering and Maintenance-of-Way Association in the *Manual of Railway Engineering* (AREMA, 2016).

Load effects resulting from prestressing, thermal changes in the structure, volume changes due to shrinkage and creep of the concrete, and soil settlement are not defined in detail in the building codes. The engineer is required to assess and define the effects of these phenomena. These load effects are important to both member and connection design. The details of the connection can affect the magnitude and direction of the resultant motions and forces.

The compilation of the dead, live, and environmental load effects provides the structural **service load**. The term service load is used because each effect is the best estimate of the maximum load likely to occur during the life of the structure. To provide against structural failure or collapse, the service loads are increased by a **load factor**. The resulting **factored load** represents the maximum probable overload the structure would see in its lifetime. Member design provides a **nominal strength**, the maximum capacity of a member prior to failure, of the structure based on its

Table 1.1 Selected minimum distributed loads from ASCE/SEI 7

Occupancy or use	Live load psf[a,b]	Occupancy or use	Live load psf[a,b]
Assembly areas		Office buildings	
Fixed seats (fastened to floor)	60	File and computer rooms shall be designed for heavier loads based on anticipated occupancy	
Lobbies	100		
Movable seats	100		
Platforms (assembly)	100	Lobbies and first-floor corridors	100
Stage floors	150	Offices	50
Stadiums	60	Corridors above first floor	80
Balconies and Decks		Recreational	
1.5 times live load for area served, not to exceed	100	Bowling alleys, poolrooms	75
		Gymnasiums	100
		Dance halls and ballrooms	100
Catwalks for maintenance access	40	Residential and multifamily houses	
Corridors		Private rooms and corridors serving them	40
First Floor Other floors, same occupancy served except as indicated	100	Public rooms and corridors serving them	100
Dining rooms and restaurants	100	Roofs	
Garages (passenger vehicles only)	40	Flat, pitched, or curved	20
Hospitals		Used as assembly areas	100
Operating rooms, laboratories	60	Schools	
Private rooms	40	Classrooms	40
Wards	40	Corridors above first floor	80
Corridors above first floor	80		
Libraries		Skywalks	250
Reading rooms	60	Stores	
Stack rooms	150	Retail, first floor	100
Corridors above first floor	80	Retail, second floor	75
		Wholesale	125
Manufacturing		First-floor corridors	100
Light	125		
Heavy	250		

Source: Adapted from Table 4.3.1 of ASCE/SEI 7-16

[a]Pounds per square foot

[b]In addition to distributed loads, *ASCE/SEI 7* requires a concentrated load of 2000 pounds for access space, office buildings, and light manufacturing, 3000 pounds for heavy manufacturing and 1000 pounds for hospitals and libraries

sectional properties and materials. The nominal strength is reduced by a **strength reduction factor** to account for variation in sectional and material properties. The **design strength** is the nominal strength times the strength reduction factor and must exceed the factored load for all probable loading conditions.

Service load conditions address the stresses in the steel and concrete in the members, deflections and cracking. In the USA, these limits are defined by *ACI 318 Building Code Requirements for Concrete Structures (ACI 318)* for building structures and by AASHTO for bridge and highway structures. Members are sized to meet these Code prescriptions. Once the design has been adjusted to meet these criteria, the design strength is compared to the factored load. If the design strength is greater than the factored load, the design is satisfactory; otherwise, additional iterations are required to also meet the strength conditions. In practice, the serviceability stress condition of prestressed concrete is designed first, and the strength conditions are checked for compliance.

1.3   Serviceability, Strength, and Structural Safety

Structures must function adequately at service conditions and have sufficient strength to prevent collapse under overload conditions. To meet these objectives the engineer considers the behavior of the structure at several loading stages. Consequently, the engineer examines the capacity of the structure and temporary loads during construction, the completed structure during its service life, and the behavior of the structure under various combinations of factored load combinations.

Considerable research has gone into the definition of load combinations, load factors, and strength reduction factors to be considered and the magnitude of the possible overloads that may occur in a structure to assure structural safety (Winter, 1979; McGregor et al., 1983; McGregor, 1983; Nowak & Szerszen, 2001). This work evolves from reliability theory where statistical studies evaluated the probability of occurrence of various load combinations and the magnitude of the loads associated with these occurrences. A **reliability index** is then established. For example, the reliability index may select a probability of failure of 1 in 10,000 over a 50-year service life. The load factors are then selected to assure that various combinations of loads have the same reliability index. In a parallel effort, the strength of a theoretical section is compared to the variation in strength of the section when tested. Members are investigated for variations in sectional dimensions, placement of reinforcement, and variation of material strengths. These studies led to the definition of a **strength reduction factor** ϕ, the amount that the strength of the section based on the nominal dimensions should be reduced to account for these variations. Reliability indices selected for bridges and buildings are slightly different and, consequently, have different load and resistance factors. The resulting load combinations, load factors, and strength reduction factors are incorporated in the appropriate building codes.

The strength of the structure to resist applied loads is defined symbolically as:

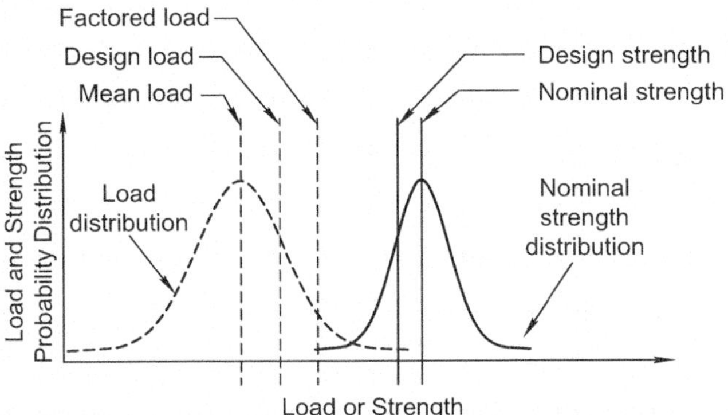

Fig. 1.2 Schematic representation of load and strength distributions

$$\phi S_n \geq U \tag{1.1}$$

where ϕ is the strength reduction factor, S_n is the nominal strength of the member, and U is the factored applied load. Load and resistance factor (LRFD) distributions are shown schematically in Fig. 1.2.

The distribution of loads is shown as a dashed line on the left of the figure and the distribution of nominal strength is shown as a solid line on the right. In a typical structure, the probability distribution of the loads is wider than the strength distribution. This represents the greater variation in loads compared to the higher level of control imposed on the strength. The load distribution indicates the mean load, the design load, which is greater than the mean load, and the factored load, which is the design load times the load factor. The mean nominal strength is multiplied by the strength reduction factor to give the design strength. If the factored load is less than or equal to the design strength, the section carries the applied load. The portion of the curve where the load distribution is greater than the strength distribution represents the conditions where structural failure can occur. Equation (1.1) presents this in general terms, but load and resistance factors are extended to all member loading effects. These loading effects include bending, shear, axial load, and torsion. Each condition is then presented as:

$$\phi M_n \geq M_u$$
$$\phi V_n \geq V_u$$
$$\phi P_n \geq P_u \tag{1.2}$$
$$\phi T_n \geq T_u$$

In Eq. (1.2) the subscript n refers to the nominal strength in moment, shear, axial compression, and torsion respectively and the subscript u refers to the factored load combination, sometimes referred to as the "ultimate" load, creating the largest effect in the member. This relationship is required for every section of a member.

Load and resistance factors are independent functions. Consider a bridge designed for a standard truck load. Equation (1.1) says only that the bridge should be strong enough to safely support the truck. A larger load than a standard truck can cross the bridge providing the design is examined for the heavier loading and accommodations are made to distribute the load on the structure. Such engineering allowed for a retired space shuttle to cross a bridge on its journey to a museum in California (USA Today, 2012). The space shuttle weighs 4.4 million pounds while the standard truck used to design the bridge nominally weighs 72,000 pounds.

A basic principle in building codes is that the structure displays **ductile** behavior. Ductile behavior in reinforced concrete structures implies that the reinforcement yields prior to secondary crushing of the concrete. This behavior is associated with substantial deflections and cracking, thereby warning the occupants of an overload and possible collapse. Conversely, **brittle** behavior occurs when the concrete crushes, often suddenly, prior to yielding of the reinforcement. The ACI Building Code achieves this by limiting the **net tensile strain**, the strain in the reinforcement farthest from the compression face, in the member. Ductile members have a higher strength reduction factor than brittle members to encourage ductile failure modes to occur in overload conditions. AASHTO limits the reinforcement in the section by restricting the **reinforcement ratio** ρ the area of reinforcement A_{ps} divided by the area of concrete A_{ps}/bd.

1.3.1   ACI Provisions

The provisions of the *ACI 318 Building Code for Concrete Structures* are used as the basis for development of analysis and design methodologies for building systems.

Table 1.2 ACI 318-14 Load Combinations

Load combination	Notation
$U = 1.4(D + F)$	D = Effect of dead load
$U = 1.2(D + F + T) + 1.6(L + H) + 0.5$ $(L_r$ or S or $R)$ Often only $U = 1.2D + 1.6\,L$	E = Effect of seismic activity F = Effect of weight or pressures of fluids with well-defined densities
$U = 1.2D + 1.6(L_r$ or S or $R) + (1.0\,L$ or $0.8\,W)$	H = Effect of weight and pressure of soil, water in soil or other materials
$U = 1.2D + 1.6\,W + 1.0\,L + 0.5(L_r$ or S or $R)$	L = Effect of live load
$U = 1.2D + 1.0E + 1.0\,L + 0.2S$	L_r = Effect of live loads on roof
$U = 0.9D + 1.6\,W + 1.6H$	R = Effect of loads due to rain
$U = 0.9D + 1.0E + 1.6H$	S = Effect of loads due to snow
	T = Effect of loads due to thermal changes
	W = Effect of wind loads

Table 1.3 ACI 318-14 strength reduction factors

Condition	Strength reduction factor, ϕ
Tension controlled section	0.90
Compression controlled section with spiral reinforcement	0.70
Other compression controlled sections	0.65
Shear and torsion	0.75
Bearing on concrete	0.65
Post-tensioned anchor zones	0.85
Strut and Tie Models	0.75
Flexural sections in pretensioned members where strand embedment is less than the transfer length	0.75
Plain concrete	0.60

Table 1.2 summarizes typical load combinations for building structures and the ACI load factors, and Table 1.3 summarizes the strength reduction factors based on calibration efforts. For many members, the factored load combination that directly affects section design is U = 1.2D + 1.6L. This derives from the second entry in Table 1.2 in which other effects in the load combination often are not present.

The ACI Building Code provisions include the requirement that ductility is provided by requiring that a **tension-controlled section** have a net tensile strain equal to or greater than 0.005 and that a **compression-controlled section** have a net tensile strain less than or equal to 0.002. The strength reduction factor varies linearly between 0.9 and 0.7 or 0.65 depending on the detailing of the ties used to provide compression confinement. Table 1.2 uses the phrase "effect of" to describe the loadings. The "effect of" implies that the external loading, say wind, produces "effects" such as axial loads, moments, shears, and torques in the structural members.

1.3.2   AASHTO Provisions

The AASHTO *LRFD Bridge Design Specification* (2017) applies to bridge and highway structures. Both load and resistance factors vary from those in the ACI 318. The differences are due in part to the fundamental difference in the use of the structure. The AASHTO Design Specification addresses variation in live load, fatigue, and environmental issues related to water crossings and aggressive environments such as road salt that many building designs do not encounter. The AASHTO strength reduction factors for prestressed concrete reflect the inspection requirements for bridges. An understrength prestressed member is not allowed to be placed in a bridge, and thus a strength reduction factor of 1.0 is selected. AASHTO load combinations and load factors are summarized in Table 1.4 and Strength reduction factors are given in Table 1.5.

Table 1.4 Sample AASHTO load combinations and load factors

Load combination strength[a,b,c]	Notation
$U_1 = 1.25D + 1.75(L + I) + 1.0(WA + FR) + 1.2 (TU + CR + SH)$	D = Effect of dead load CR = Effect of creep
$U_2 = 1.25D + 1.35(L + I) + 1.4 WA + 1.0 FR$	EQ = Effect of earthquake
$U_3 = 1.25D + 1.35(L + I) + 1.0 (WA + FR) + 1.4 WS + 1.2 (TU + CR + SH)$	FR = Effect of friction
$U_3 = 1.25D + 1.0 (WA + FR) + 1.0 EQ$	H = Effect of weight and pressure of soil, water in soil or other materials
Load combination service	L = Effect of live load
$S_1 = 1.0D + 1.0(L + I) + 1.0 WA + 0.3 WS + 0.3WL + 1.0 (TU + CR + SH)$	SH = Effect of shrinkage TU = Effect of uniform temperature variation
$S_2 = 1.0D + 1.3(L + I) + 1.0(WA + FR) + 1.0 (TU + CR + SH)$	WA = Effect of water
	WL = Effect of wind on live load
	WS = Effect of wind on structure

[a]The AASHTO load factors for both strength and service loads are given in tabular format that require the engineer to select the appropriate components. This table is a partial summary of the conditions to demonstrate the combinations considered for the reliability calibration
[b]Many of the AASHTO requirements have a heavy and light condition. For example, the TU + CR + SH combination is 0.5 if the effect decreases the member loading, or 1.2 if the effect increases the member loading
[c]AASHTO has multiple strength and serviceability conditions requiring checking. Not all combinations from the AASHTO Specification are included in this table

Table 1.5 AASHTO strength reduction factors

Condition	Strength reduction factor, ϕ
Flexure and tension of reinforced member	0.90
Flexure and tension of prestressed member	1.00
Axial loads with spiral reinforcement	0.70
Shear and torsion	0.90
Bearing on concrete	0.70
Post-tensioned anchor zones	0.80
Strut and tie models-compression	0.70

AASHTO bridge loadings include a uniform load in conjunction with concentrated loads that represent the axles on a truck. The concentrated loads are positioned on the structural model to result in the largest shear or bending stresses. Alternatively, AASHTO allows a train of standard trucks to be placed on the structural model to create maximum shear and bending envelopes. Two other differences in design philosophy are in the definition of maximum reinforcement and shear design. AASHTO provisions use maximum reinforcement ratios to limit the reinforcement in a member instead of net tensile strain. AASHTO shear design uses a compression field theory approach but does allow for an ACI concrete contribution approach.

1.4   Structural Integrity and Sustainability

Structural integrity addresses the behavior of the structure to prevent or limit progressive collapse or disproportionate collapse of a structure when subjected to extreme loadings. The terms **disproportionate collapse** or **progressive collapse** refers to the failure of a primary member leading to either failure of adjoining members or complete collapse. The ACI Building Code incorporates structural integrity by prescriptive requirements. **Prescriptive requirements** specify reinforcement placement required to meet the building code objectives, for example, placing continuous reinforcement in a perimeter beam. Post-tensioned structures require a minimum number of slab tendons to be placed between column vertical reinforcement. Continuous perimeter reinforcement or tendons through column cores form a catenary support system should an intermediate column be lost. The catenary supports the structure above even if there is excessive deflection and cracking of the affected beam and slab. These details suggest that structural integrity is best ensured by providing redundant load paths through the structure.

Sustainability is defined as a requirement of our generation to manage resources such that the average quality of life that we ensure ourselves can potentially be shared by all future generations (World Commission, 1987). Prestressed concrete, using high performance reinforcement and higher strength concrete, implies that strength conditions can be satisfied with less material. As such, prestressed concrete supports overall sustainability objectives. Sustainability also considers the resilience of the structure. Prestressed concrete structures are often candidates for repurposing due to their adaptability to new uses.

1.5   Serviceability and Stress Control by Prestressing

The primary advantage of prestressing is to improve the serviceability response of a concrete structure. This includes improvements in short-term deflection, long-term deflection, and cracking. To understand how prestressing improves serviceability behavior, it is instructive to compare the flexural behavior of a nonprestressed and prestressed concrete beam.

1.5.1   Comparison Between Nonprestressed and Prestressed Concrete Beams

Service level stresses in prestressed concrete are assumed to be linearly elastic. Throughout this book sketches of the stress distribution are provided. Figure 1.3 is a sketch of the strain and stress distribution over the height of the beam section. The compressive stresses are on top and the corresponding strain and stress sketches are

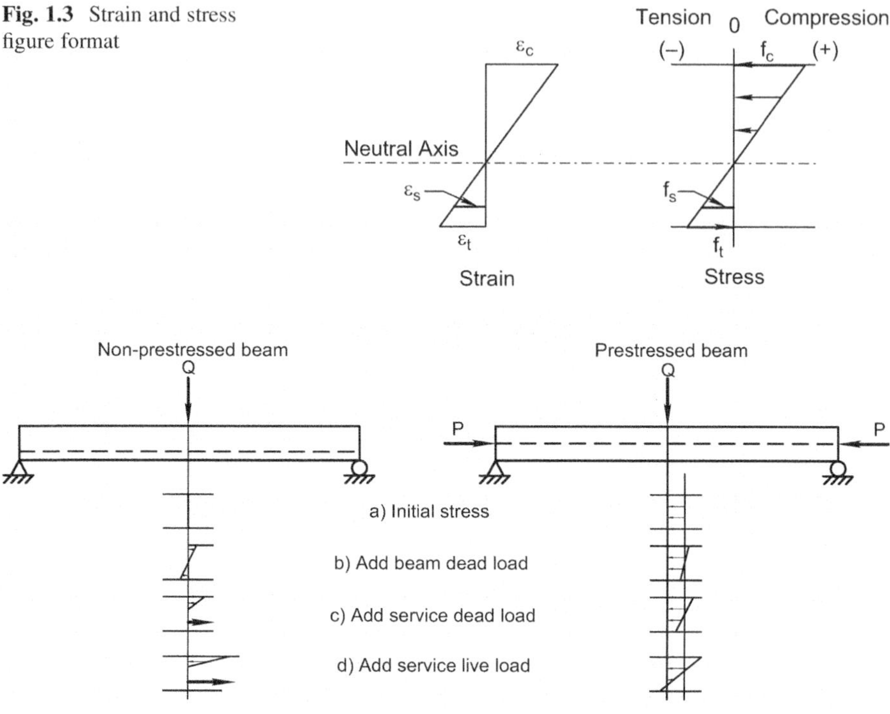

Fig. 1.3 Strain and stress figure format

Fig. 1.4 Behavior comparison of nonprestressed and prestressed concrete beams

to the right. This strain and stress convention, with compressive stresses and strains indicated as positive, is used throughout.

Figure 1.4 compares the midspan stresses in a rectangular nonprestressed concrete beam with a similar beam prestressed with a straight tendon located at the centroid of the section. Both beams are subjected to identical concentrated loads Q at midspan. For the purposes of this comparison, it is assumed that the compressive stresses due to the prestress are within all allowable limits and that the prestressed concrete beam has no tensile stress under full service load. Figure 1.4a indicates the initial stress in the beam at midspan ignoring the beam self-weight. The nonprestressed concrete beam has no initial stress while the prestressed beam has a uniform compressive stress equal to the prestressing force P divided by the gross area of the section A. When the beam is first removed from the formwork, the self-weight of the beam is applied to the member. Figure 1.4b shows the cumulative effect of the initial stress plus the tensile and compressive stress from the beam self-weight. The nonprestressed concrete beam is assumed to remain uncracked under self-weight so a tensile stress remains at the bottom of the section. The prestressed beam correspondingly shows an increase in compressive stress in the top and a decrease in compressive stress at the bottom for the same loading. The full service dead load is applied and the resulting cumulative stresses are seen in Fig. 1.4c. For this comparison, the service dead load is sufficient to exceed the tensile capacity of the concrete in the nonprestressed concrete beam. The concrete cracks, the compressive stress at the top of the beam increases, the tensile stresses are engaged by the reinforcement, and the neutral axis of the section moves upward. The prestressed concrete beam exhibits an increase in compressive stress and an additional reduction of compressive stress on the bottom of the beam. Lastly, the full service live load is applied. In Fig. 1.4d the nonprestressed concrete beam has an increase in the compressive stress, in the tensile stress in the reinforcement, and the neutral axis again moves upward. The compressive stress is shown as linear providing the service load remains within the linear range of the concrete stress–strain behavior. The prestressed beam top compressive stress increases and the bottom stress goes to zero.

Under full service load, the nonprestressed concrete beam has tension cracks extending up to the neutral axis and the calculation of beam deflection is based on cracked section properties. The prestressed beam has no cracks and the deflection is calculated based on full section properties. Thus, for the same loading, the prestressed beam has less deflection and cracking than the corresponding nonprestressed concrete beam.

1.5.2   Stress Control Using Prestressing

In the preceding comparison of nonprestressed and prestressed concrete beams, the prestressing force is shown to be beneficial in controlling the service stresses in the beam. An examination of the design suggests that a tendon at the centroid of the section is not the most effective method of applying the prestressing force. Figure 1.5a shows a beam with a rectangular cross section and the prestressing force applied at the section centroid. The prestressing force creates an uniform compressive stress with an average prestress at the beam centroid of magnitude f_c. For the purposes of the example, the self-weight of the beam is neglected. Upon application of the load Q, the maximum compressive stress is $2f_c$ and the tensile stress is zero. If we now move the prestressing force downward to 2/3 of the height of the beam h, the prestressing creates a linearly varying stress distribution with the maximum compression stress on the bottom of the section of $2f_c$ and zero stress at the top of the section, Fig. 1.5b. The applied load is increased to $2Q$ creating a flexural stress of $2f_c$. The final stress in the beam is $2f_c$ in compression and zero in tension. Thus, by adjusting the location of the prestressing force, the applied load is doubled without exceeding the stress limits of the beam or increasing the prestressing force.

Next we examine the beam stresses due to the eccentric tendon from the viewpoint of deflections. If applied loads are ignored, the prestressing force in Fig. 1.5b generates an upward deformation in the beam. This upward deformation is called **camber**. The stresses at the end of the beam are the same as the midspan stresses prior to the application of the load. This end moment increases the camber in the beam. Serviceability conditions or the strength of the concrete at the time the

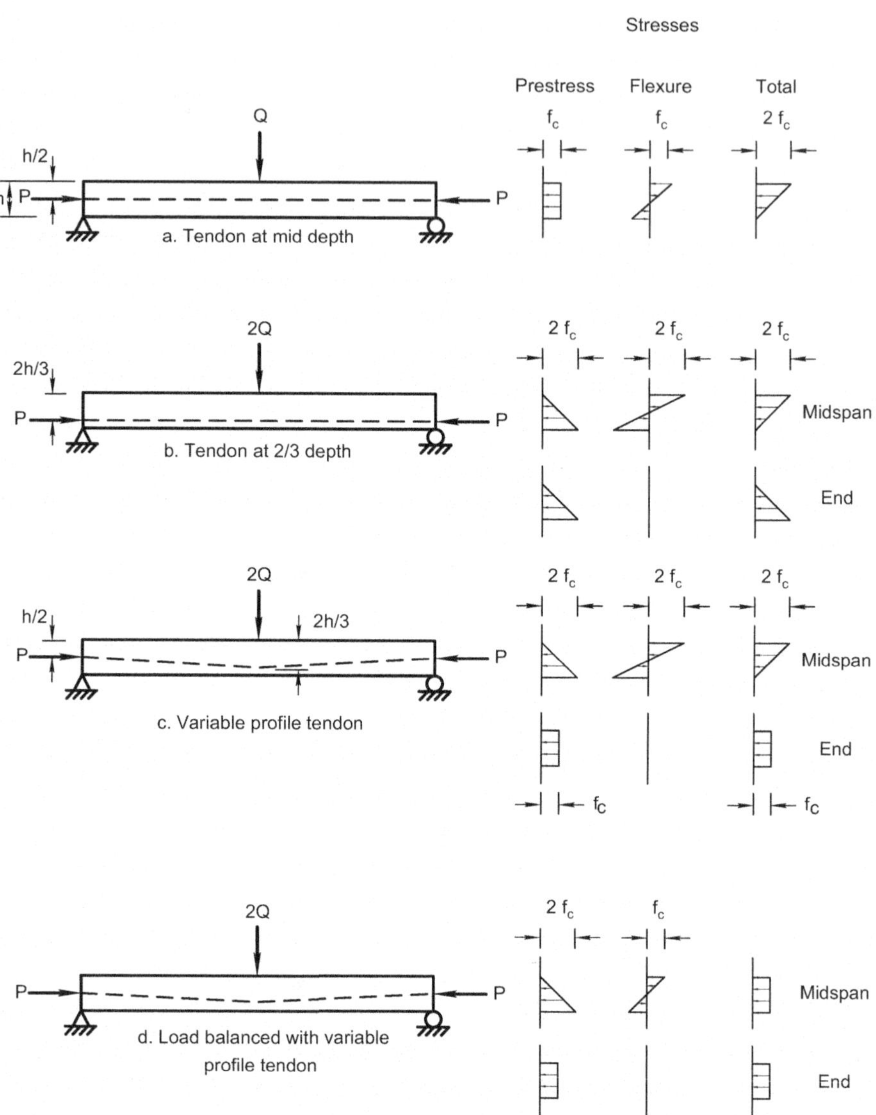

Fig. 1.5 Tendon location and stress distribution

prestress force is transferred to the concrete can make this an undesirable condition. To overcome this condition, it may be desirable to use a variable eccentricity tendon.

Figure 1.5c indicates the same beam with a tendon having variable eccentricity. The tendon is located at the centroid of the section at the end of the beam and at $2/3h$ at midspan. The midspan stresses are identical to the condition in Fig. 1.5b; and the end stresses remain uniform like Fig. 1.5a.

If, instead of a load of $2Q$, a load Q is placed on the beam, the flexural stress is f_c and the final midspan stress is uniform and equal to f_c, Fig. 1.5d. In this condition, the entire beam section has a uniform stress distribution. The implication of this stress distribution is that the beam deflection under a load of Q is zero, neither up nor down; however, the beam is still subject to axial shortening. Adjusting the prestress force and eccentricity to create this condition is called **load balancing** and is explored further in Sect. 1.6. Increasing the applied load to a magnitude of $2Q$ gives the same midspan stress as Fig. 1.5b.

1.5.3   Serviceability and Partial Prestress

In the comparison with nonprestressed concrete beams, prestressed beams can be designed to carry the entire service load with zero tension. Historically this was called **full prestressing** and various design criteria allowed no tensile stresses under service load (Leonhardt, 1964). As the use of prestressing became more common, engineers began allowing some tension stress to form in the concrete. In other situations, prestress was reduced to avoid excessive deflections and some cracking of the section was allowed under full service load. These applications were called **partial prestressing**; however, the description and use of partial prestress remained poorly defined and often depended on the engineer's selection of service criteria. In recognition of the range of prestressing options available to the engineer, the 2002 ACI 318 Building Code identified three separate classifications for prestressed concrete structures. The classes are tied to the maximum tensile stress in the member under full service load and are designated uncracked—Class U, transitional— Class T, and cracked—Class C. The stresses associated with these classes are given in Table 1.6. The definition of these classes eliminates the ambiguity created by the term partial prestress. Serviceability checks for Class U members use the uncracked section while Class C members use the cracked sections. Class T members require close attention to the loading sequences to assure use of the proper section properties.

Table 1.6 ACI 318 tensile stresses for classification of prestressed concrete

Class	Condition	Tensile stress
U	Uncracked[a]	$f_t \leq 7.5\sqrt{f'_c}$
T	Transitional	$7.5\sqrt{f'_c} < f_t < 12\sqrt{f'_c}$
C	Cracked	$f_t \leq 12\sqrt{f'_c}$

[a]Two-way prestressed concrete slabs systems are designed as Class U with a maximum tensile stress of $6\sqrt{f'_c}$

1.6   Equivalent Loads and Load Balancing

In the discussion of stress control, Fig. 1.5d indicates that it is possible to design a tendon configuration to counteract the applied load. Properly configuring the tendon profile allows the prestress force to generate an **equivalent load** that has the same effect as the applied load but in the opposite direction. Creating equivalent loads provides a powerful design tool. For example, if the equivalent load is exactly equal and opposite to the applied load, then the service load is **load balanced**. As seen in Fig. 1.5d, the stresses in the beam are uniform along the member and there is no net moment, shear, or deflection in the section. While this is an idealized condition, it is possible to balance portions of the load in the structural analysis, and design only the portion of the load that is unbalanced. This concept considerably simplifies the design and the detailing of many structures and provides a easy and convenient method to check the results of more complex structural analyses.

The number of theoretical load balancing schemes is large, and geometric and construction considerations result in a smaller set of practical solutions. Figure 1.6 illustrates several common load balancing configurations. At service load, the beam behaves elastically, and therefore, superposition of the load balancing effects is possible.

Consider first a straight tendon with an end eccentricity e_e from the centroid of the section, Fig. 1.6a. The tendon exerts an axial force on the beam equal to the prestressing force. In addition, the tendon creates an end moment equal to the prestressing force times the end eccentricity or Pe_e. Thus, the beam is subjected to a constant moment over the entire length. If the end eccentricity is reduced to zero, then the beam is subjected to axial force and there is no moment resulting from the prestress.

Next, consider the prestressed beam with a tendon depressed to a midspan eccentricity e, Fig. 1.6b. In this case, the end eccentricity is zero. If the tendon has an end eccentricity, the effects of the end eccentricity can be superimposed on the results of the deflected tendon. The tendon intersects the end of the beam at an angle of $\alpha = \tan^{-1}\left(\frac{e}{L/2}\right)$ relative to the beam centerline. The deformed geometry causes the prestressing force in Fig. 1.6a to be concurrent with the tendon geometry. Thus, the axial force on the beam is $P\cos(\alpha)$. In practice, the angle of the tendons is small, and the axial force used for calculation of prestressing forces is approximated as P as shown in Fig. 1.6b. Equilibrium of the inclined prestressing force generates two vertical forces: $2Pe/l$ applied to each end of the inclined tendon. The force is doubled at midspan, creating an internal concentrated force of $4Pe/l$.

The internal moment at any location along the simple span beam is equal to the prestressing force times the eccentricity from the beam centerline. Thus, at midspan the internal moment is Pe. If we consider that a concentrated load Q is placed at the midspan of the beam, the moment from the external force would be $QL/4$. Equilibrating the internal and external moments and solving for the required prestress force for a given eccentricity gives:

$$P = \frac{Ql}{4e} \tag{1.3}$$

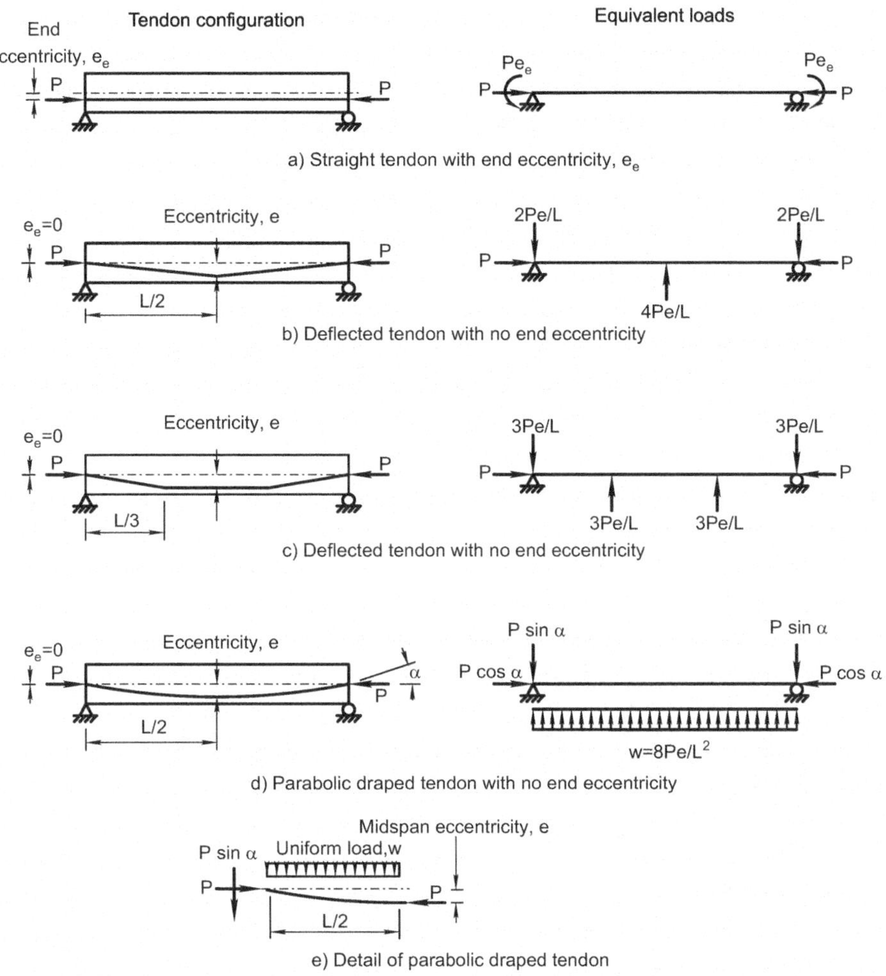

Fig. 1.6 Tendon configurations for load balancing. Additional load balance information is in Appendix A

This is an example of load balancing. In this example, the eccentricity was selected as a known value. This follows from the geometric constrains to the beam. For most beams, the tendon is inside the beam with sufficient cover to protect the steel. Thus, the maximum eccentricity of the tendon is selected based on the beam geometry and the required prestress force is minimized, or the beam section depth is determined based on the eccentricity required to balance the applied load.

If the tendon is deflected in two locations as in Fig. 1.6c, the equivalent load counteracts two concentrated applied loads. The resulting trapezoidal moment diagram closely replicates the parabolic moment diagram for a uniformly distributed load. Thus, a tendon with two deflection points is often used to partially compensate distributed loads on a simple span beam. This is not a true load balancing but is convenient for control of both end stresses and applied loads and simultaneously attainable in the prestress plant. The tendon profile leads to three critical service level stress locations in the beam: the beam end, midspan, and the tendon deflection point.

To exactly balance a uniform load, the tendon requires a parabolic drape. Figure 1.6d indicates the details of such a tendon. In Fig. 1.6d the horizontal force is shown as $P \cos(\alpha)$ even though conventional practice would use a value equal to the prestress force, P. The end shear is $P \sin(\alpha)$, where the angle α is slope of the tendon at the end of the beam. A free body diagram of half the tendon provides the geometry of the tendon, Fig. 1.6e. The left end has an axial load and shear. By symmetry, the prestressing force is horizontal at the beam midspan and the shear is zero. This configuration allows computation of the equivalent load. Summing forces about the left-hand support gives

$$Pe = w\frac{l}{2}\frac{l}{4} \tag{1.4}$$

solving for the required prestress force gives the required prestress

$$P = \frac{wl^2}{8e} \tag{1.5}$$

In general, the equivalent loads from a tendon are result of the geometry and placement of the tendon. Each angular change in the tendon creates a lateral force equal to the tendon force times the sine of the angle change.

Example 1.1: Determine Prestress to Balance Load
A prestressed concrete beam, Fig. 1.7, is 30 ft long and has a rectangular cross section of 10 in. wide and 20 in. deep. A concentrated service load of $Q = 22$ kips is placed at midspan. (A) Find the required prestressing force to balance the service load using a tendon with an end eccentricity of zero. (B) Determine the prestress force if the end eccentricity of the tendon is such that no tension is allowed at the end of the beam. The self-weight of the beam can be ignored.

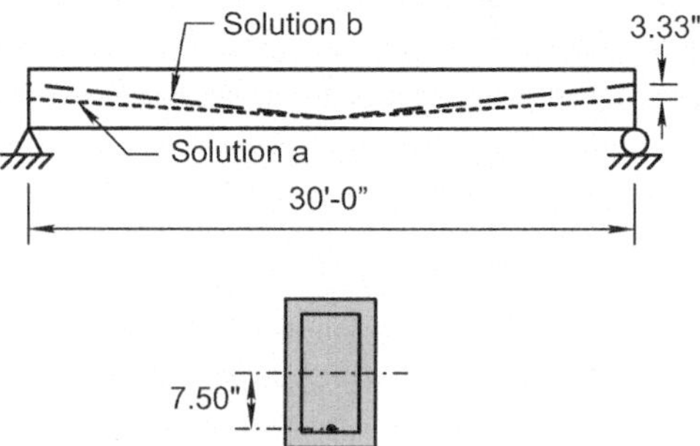

Fig. 1.7 Beam details for Example 1.1

Fig. 1.8 Beam details for Example 1.2

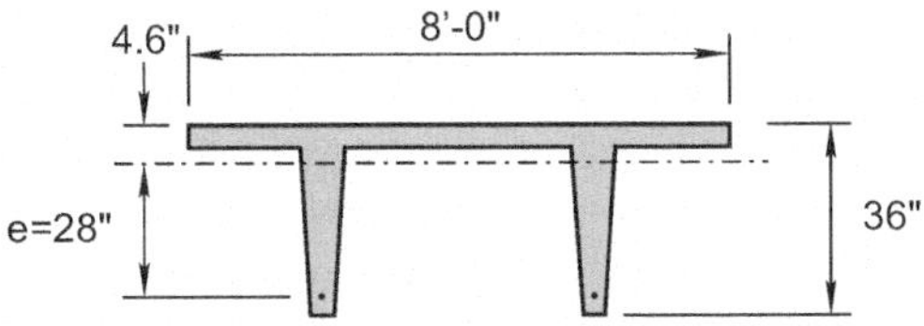

Solution: Allowing for 1.5 in. cover and No. 4 stirrups, the lowest the tendon can be placed in the beam at midspan is: $h/2 = 10$ in. $- 1.5$ in. cover $- 0.5$ in. for No. 4 bar $- 0.5$ in. allowance for the tendon radius $= 7.5$ in. This is the largest eccentricity that is possible and corresponds to the lowest prestressing force. (A) Computing the prestress force from Eq. (1.3) gives $P = QL/4e = 22$ kip $\cdot$ 30 ft $\cdot$ 12 in./ft/(4 $\cdot$ 7.5 in.) $= 275$ kips. (B) If the end eccentricity is raised above the neutral axis to a point such that there is no tension in the bottom of the beam, the end eccentricity e_e is $h/6 = 20$ in./6 $= 3.33$ in. The total eccentricity in the tendon at midspan is then 3.33 in. $+ 7.5$ in. $= 10.83$ in. The required prestress force is then P $= 22$ kip $\cdot$ 30 ft $\cdot$ 12 in./ft/(4 $\cdot$ 10.83 in.) $= 183$ kip.

Comment: By raising the tendon 3.33 in. at the end of the beam, the required prestress force is reduced by 33%.

Example 1.2: Determine Prestress to Balance Load

A 40 ft long double-T beam is 36 in. deep with a cross-sectional area of 210 in.^{2} and a centroid 4.6 in. from the top surface. It carries a superimposed dead load of 1230 lb/ft and a service live load of 2160 lb/ft. Determine the prestressing force required to balance the self-weight of the beam plus the superimposed dead load plus one half of the live load if the tendon eccentricity is 28 in., Fig. 1.8.

Solution: The self-weight of the beam is $A_g \cdot w_g = 210$ in.^{2} $\cdot$ 1/144 in.^{2}/ft^2 $\cdot$ 160 lb/ft$^3 = 230$ plf. The total distributed load to be balanced on the beam is $w_g + w_d + w_s = 230$ plf $+ 1230$ plf $+ 2160/2$ plf $= 2540$ plf. From Eq. (1.5), the required prestress force is $P = wL^2/8e = 2540$ plf $\cdot$ (40 ft)^2 12 in./ft/ (8 $\cdot$ 28 in.) $= 217.7$ kips.

Comment: A draped tendon must be post-tensioned to achieve the desired profile. A tendon with two hold-down points as in Fig. 1.6c would require the same prestress force.

1.7   Prestressing Concrete

Successful design and construction of prestressed concrete structures requires an understanding the theoretical calculations, how the members are fabricated, and on-site construction. In his book *Prestressed Concrete: Design and Construction* (Leonhardt, 1964), Dr. Fritz Leonhardt, an early adopter of prestressed concrete, posited ten commandments for prestressed concrete. In the late 1950s and early 1960s there were virtually no codes for prestressed concrete. These early adopters worked diligently to understand the behavior and performance of their designs. With slight modification, Leonhardt's guidance is as valuable today as when first presented.

Ten Commandments for the prestressed concrete engineer
(Adapted from Leonhardt, 1964)

In the design office

1. Prestressing means compressing the concrete. Compression can only take place where shortening is possible. Make sure that your structure can shorten in the direction of prestressing!
2. Any change in tendon direction produces "radial" forces when the tendon is tensioned. Changes in the direction of the centroidal axis of the member are associated with "unbalanced forces," likewise acting transversely to the general direction of the member. Remember to take these forces into account in the calculations and structural design.
3. The high permissible compressive stresses must not be fully used regardless of the circumstances! Choose cross-sectional dimension of the concrete, especially at the tendons, in such a way that the member can be properly concreted—otherwise the laborers on the job will not be able to place and consolidate the concrete correctly, which is essential to prestressed concrete construction.
4. Avoid tensile stresses under dead load and do not trust the tensile strength of concrete.
5. Provide non-tensioned reinforcement preferably in a direction transverse to the prestressing direction and, more particularly, in those regions where the prestressing forces are transmitted to the concrete.

On the construction site

6. Prestressing steel is a superior material to ordinary reinforcing steel and is sensitive to rusting, notches, kinks, and heat. Treat it with proper care. Position the tendons accurately, securely, and immovably held in the lateral direction, otherwise friction will take its toll.
7. Plan your concreting program in such a way that the concrete everywhere be properly consolidated, and deflections of scaffolding will not cause cracking of the young concrete. Carry out the concreting with the greatest possible care, as defects in the concreting are liable to cause trouble during the tensioning of the tendons.
8. Before tensioning, check that the structure can move and shorten freely in the direction of tensioning. Make it a rule to always cover high pressure hydraulic lines and never stand in the line of the tendon during stressing.
9. Tension the tendons in long members at an early stage, but at first only apply part of the prestress to produce a moderate compressive stress, which prevents cracking in the concrete due to shrinkage and temperature. Do not apply the full prestress force until the concrete has developed sufficient strength. The highest stresses in the concrete usually occur during the tensioning of the tendons. When tensioning, always check the tendon elongation and the jacking force. Keep careful records of the tensioning operations.
10. Do no start grouting of tendons until you have checked that the ducts are free from obstructions and water. Perform the grouting strictly in accordance with the relevant directives and specifications.

Prestressed concrete deals with large forces in the tendons and more slender members than nonprestressed concrete structures. Leonhardt's "Commandments" illustrate the most important aspect of prestressed concrete. The successful engineer must understand behavior, design, and construction of prestressed concrete. Failure to include these aspects compromises the project.

1.7.1   Pretensioning and Plant Operations

Pretensioned members are usually fabricated in a prestressing plant. The precast pieces are then transported to the site for erection and installation into the final structure. The most common method of pretensioning is a **longline** stressing bed, Figs. 1.9 and 1.10. The longline bed consists of end bulkheads that serve as reaction points for the stressing forces. The tendon can be individual strands or groups of strands. The tendon can run straight through the form or it can be deflected. **Hold down points** allow the tendons to be depressed or **harped** to form the permanent deflection points in the tendon, Fig. 1.10. The tendons are stressed individually or as a group, Fig. 1.11. A longline bed can contain multiple forms so several pieces can be fabricated at the same time.

Prestressing plants represent a considerable investment in formwork, stressing beds, and curing equipment. Consequently, there is a production emphasis on the daily reuse of the equipment. Common pieces such as **double-T beams, bridge girders** and **hollowcore planks** are produced on a daily cycle, Fig. 1.12. Once the tendons are stressed, the concrete is placed in the forms. The concrete typically hardens in 14–18 h, then the tendons are detensioned and the piece removed from the

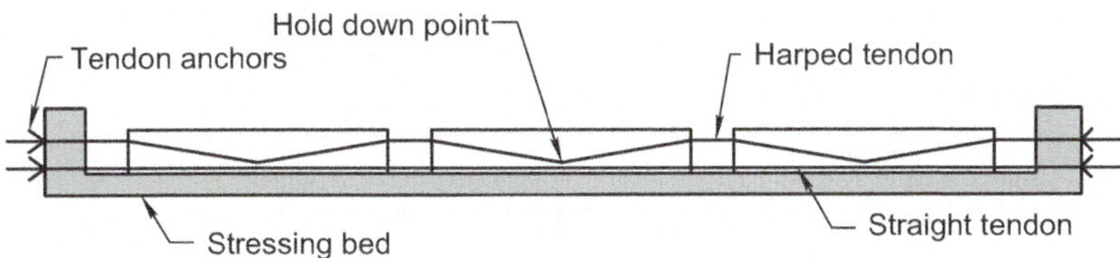

Fig. 1.9 Longline pretensioning bed

Fig. 1.10 Longline prestressing bed; stressing abutment and jacks are on the left and the form to be placed in the line is on the right

Fig. 1.11 Tendons stressing in a longline operation

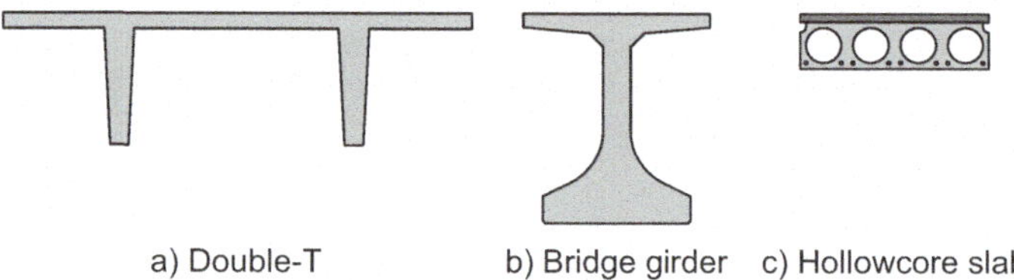

Fig. 1.12 Typical pretensioned concrete sections

form. A typical plant operation begins with detensioning at 4–5 AM followed by cleaning the forms, installing reinforcement and tendons, stressing and finally concrete placement at about 11 AM. The work is rotated through the various longline operations at the plant. Following casting, the crews prepare the reinforcement cages and tendons for the next day production. Prestressing plants often use accelerated curing in the form of steam or electrical heat. Type III **high early cement** and insulated forms further accelerate initial curing. High range water reducers or self-consolidating concrete are commonly used in pretensioning operations. These mixtures reduce the need for extensive vibration to consolidate the concrete and reduce noise in plant operations.

Quality control methods are imposed on both tendon stressing and concrete curing. Tendon stressing is typically conducted by stressing to a specified force level measured by the pressure gages on the hydraulic jacks used to stress the tendons. The elongation of the tendon is recorded and elongations that vary by more than 5% for pretensioned members and 7% for post-tensioned members from the theoretical calculation require correction in accordance with 26.10.2 of ACI 318-14. One of the major sources of error in post-tensioning is misreading either the gage or the elongation. Strands ends are cut off immediately following transfer or field stressing, so field confirmation of such an error is difficult.

Pretensioned concrete is monitored by either by placing test cylinders directly on the forms to be cured or by using **Surecure**™ cylinder molds. Surecure™ cylinder molds contain internal heating elements and are synchronized with the form curing using thermocouple monitors. Thus, the cylinder is cured as closely as possible to the

Fig. 1.13 Reinforcement cage and internal ducts for post-tensioned system

prestressed member. Detensioning the tendons is determined by testing the cylinders. The test cylinders must reach the specified **transfer strength** f'_{ci} prior to detensioning to ensure proper stress transfer. Field post-tensioned concrete is monitored by laboratory and field cured cylinders. The engineer specifies which method is to be used on the project.

1.7.2 Post-tensioning Operations

Post-tensioning applies the prestressing force to the hardened concrete. The post-tensioning can be either **internal** or **external**. Internal post-tensioning places the tendon in hollow voids created by casting **ducts** into the concrete, Fig. 1.13. A common form of internal post-tensioning is flat slab construction for office building, Fig. 1.14.

Figure 1.14 is indicative of unbonded monostrand construction. A **monostrand** tendon consists of a single strand fabricated with a plastic sheath, which in turn is filled with protective coating. The protective coating both provides corrosion protection and reduces friction during stressing. Unstressed tendons are shown extending from the top floor in the oval in Fig. 1.14, lower floors are stressed, strand tails cutoff, and anchor pockets grouted. Shoring is used for temporary support of upper floors while concrete hardens.

External post-tensioning tendons are secured by anchors at the end of the member and at specified points on the structure. The tendons are passed through **deviators** to change the tendon geometry. A common application of external post-tensioning is on the interior of hollow post-tensioned girders. Figure 1.15 shows a section and elevation of a hollow box girder. Internal tendons are in ducts cast into the walls of the box section. The external tendons are placed in the void space inside the hollow box. Tendons are anchored in the end block of the beam and pass through deviators cast into the bottom slab of the girder.

Fig. 1.14 Post-tensioned flat slab system

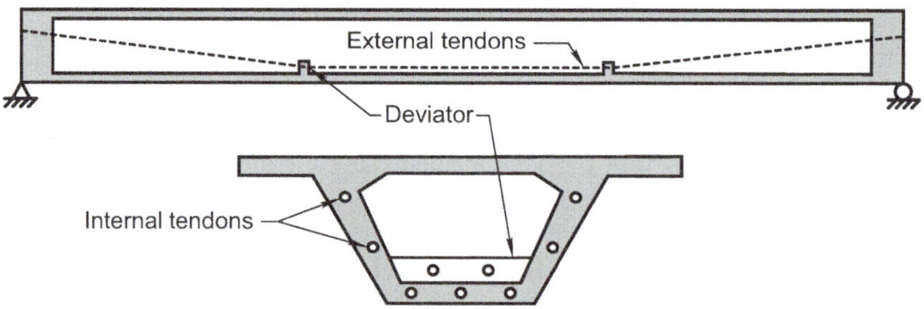

Fig. 1.15 Internal and external duct system

1.7.3 Precast Concrete

Unlike cast-in-place concrete **precast concrete** is fabricated separately from the primary structure then installed into the structure. Precast members may be reinforced or prestressed concrete. The architectural panel shown in Fig. 1.16 is an example of a nonprestressed precast concrete architectural element. The precast member can become part of the structural load carrying system, an architectural element, or a non-structural portion of the building envelope. Structural precast

Fig. 1.16 Precast concrete architectural panel

concrete elements include floor, column, wall, fascia, or roof components. Each element is designed to carry its appropriate design load and to connect to the building structure. Connections, handling and installation of precast concrete are incorporated into the element and offer the engineer a full spectrum of design options.

The ability to manufacture concrete elements away from the construction site offers potential savings in construction time and economy. This is especially true for crowded construction sites or projects with very tight construction schedules. Precasting can begin as the site work and foundation construction is initiated. Fully hardened concrete members then are available for incorporation into the structural system. Elements are delivered to a remote construction site as needed and placed directly in their final location, Fig. 1.17. Members may be certified for strength, dimensional tolerance, and finish prior to shipping. Plant fabrication quality control is often superior to onsite inspection. Fabrication plants work 12 months of the year while on site construction is impacted by adverse weather.

1.8 Loss of Prestress

The discussion so far has examined prestressed concrete as having a constant prestressing force. The reference to prestress losses in Freyssinet's early work suggests that the prestress force varies over time. The loss of prestress is discussed in detail in Chap. 4. The concept of prestress loss is critical to understanding the

Fig. 1.17 Erection of precast prestressed elements

long-term behavior of prestressed concrete. Limits on tendon and concrete stresses at the time of initial prestressing restrict the total amount of prestress that can be applied to a structure. Losses in prestress force due to volume changes in the concrete and relaxation of the steel have a pronounced effect on the final service stresses that are allowed on the structure.

1.9 Supplemental Reading

The following books and articles are provided for the reader to gain an understanding of structural engineering, prestressed concrete design and behavior, and some of the individuals making contributions to the development of prestressed and precast concrete.

- "The wisdom of the structure" by Halvard W. Birkeland *ACI Journal* April 1978, pg. 105–111, provides an interesting philosophical look at structural and material behavior.
- "Rethinking Bridge Design: a new configuration" by Man-Chung Tang, *Civil Engineering* July 2007, pg. 38–45, gives the insight of the chairman of T. Y. Lin international.
- "Analysis software analyzed" *Engineering News Record*, April 12, 1984. While dated the article summarized that analysis software was plus or minus 60% of the load 95% of the time.

- *Structures, or why things don't fall down*, J. E. Gordon, DeCapo Press, New York, NY, 1978, 395 pg. provides interesting insights to structural systems and behavior.
- "What structural engineers know" by Jon A. Schmidt, *Structure Magazine*, March 2008 pg. 9, provides an oversight of how engineering knowledge is organized.
- "The engineer's calculations" by Robert Mote, *Structures Magazine*, September 2009, Pg. 14–16, explores the advances in engineering calculations and their underlying assumptions.
- *The new Science of Strong Materials or why you don't fall through the floor*, J. E. Gordon, Princeton University Press, 1986, 278 pg. gives an introduction to materials and their behavior.
- "The Hyatt Regency decision: one view" by Robert A. Rubin and Lisa A. Banick,. *Journal of Performance of Constructed Facilities*, 1987, *1*(3), 161–167, gives an assessment of negligence and lack of oversight leading to the collapse.
- The following articles from the PCI Journal examine the development and future of prestressed concrete in the USA.

  – Arthur R. Anderson

  "An Adventure in Prestressed Concrete" in several parts. *PCI Journal*,— beginning July–August, 1979 24(4) pp. 116–139; through Vol *24*, No. 6, pp. 76–93.

  – Norm Scott

  "Precast Prestressed Concrete Beyond the Year 2000 in the United States," July–August, 1994 *PCI Journal 39* (4) pp. 42–53.
  "Reflections on the Early Precast/Prestressed Concrete Industry in America" *PCI Journal*, 2004, *49*(2), pp. 20–33.

  – Ted Gutt

  "Reflections on the beginnings of prestressed concrete in America— Prestressed concrete developments in the Western United States." *PCI Journal*, 1979 *24*(2), pp. 15–36.

  – George C. Hanson

  "Prestressed Concrete in Colorado", *PCI Journal*, May–June, 1979, *24* (3) pp. 15–39.

- The following three articles examine a major collapse and the recommendations that followed:

  – "Flawed Connection detail triggered fatal L'Ambiance Plaza collapse", *Engineering News Record*, October 29, 1987, pg. 10–18

- "After L'Ambiance Plaza", K. A. Godfrey, *Civil Engineering*, January 1988, pg. 36–39.
- "What Happened at L'Ambiance Plaza?", David R. Wonder, *Civil Engineering*, October 1988, pg. 68–71.

- "Lessons from nonfailures," Daniel A. Cuoco, *Civil Engineering*, October 1985, pg. 58 discussed over reliance on computer output.

Problems

In addition to detailed problems at the end of each chapter, Chap. 16 contains a series of comprehensive problems. These problems can be assigned as term projects or portions as individual problems.

1.1. A double-T beam has a cross-sectional area of 978 in.^{2}, is 8 ft wide and spans 48 ft. Determine the service loads per foot of length if the beam is to be used to support a library reading room.

1.2. The end 12 ft same double-T beam from problem 1.1 supports a corridor on an upper floor of the building. Immediately adjacent to the corridor is a partition wall with a sustained dead weight of 80 lb per linear foot. Compute the service and factored loads (based on ACI load factors) on the beam and draw the shear and moment diagrams for the factored loads.

1.3. A high-rise apartment complex has a floor plan with columns placed 25 ft on center. The floor is a post-tensioned flat slab is 10 in. thick with 10 psf of supplemental service dead load. Determine the total service and factored load in pounds per square foot for a typical upper story floor.

1.4. Determine the equivalent loads and internal moments for the beam shown in Fig. 1.18.

1.5. Determine the equivalent loads and internal moments for the beam shown in Fig. 1.19. The tendon is concurrent with the centroid of the section. Comment on your result.

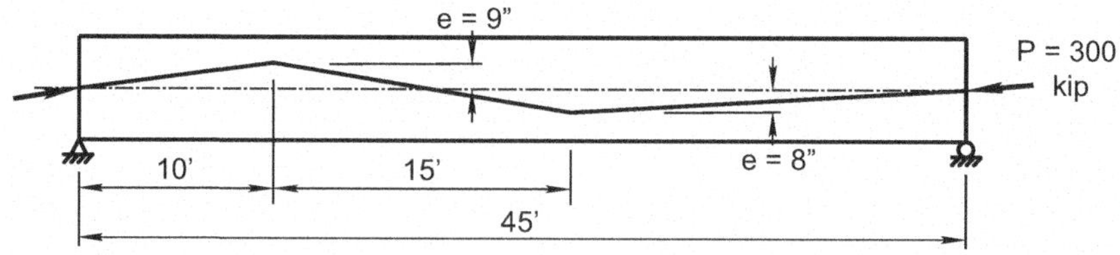

Fig. 1.18 Stressing configuration for problem 1.4

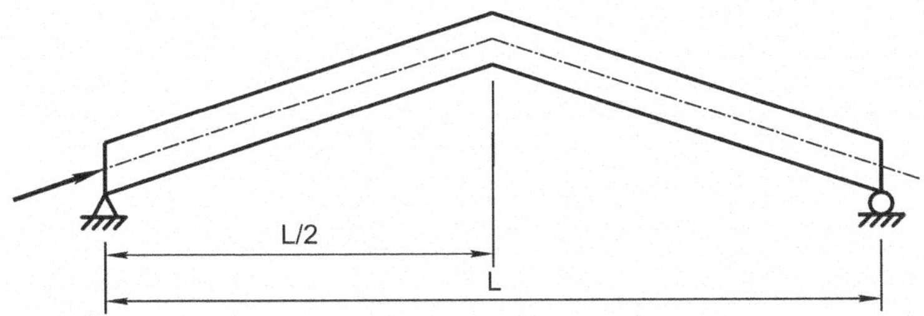

Fig. 1.19 Stressing configuration for problem 1.5

1.6. Determine the prestress force to balance the beam self-weight, superimposed dead load and 30% of the live load. The beam is 40 ft long beam with a self-weight of 270 plf, a superimposed service dead weight of 370 plf and the service live load is 620 plf. The maximum allowable eccentricity is 15.5 in.

1.7. A beam has a tendon deflected to an eccentricity e at the third point of the span length, L. Compare the internal moment of equivalent load created by the tendon with the moment diagram for a uniformly distributed service load, w. What prestress force is needed to balance the applied load? Comment on the assumption made to solve this problem and the final deflection of the beam under full service load.

References

AASHTO LRFD Bridge Design Specifications, Customary U.S. Units 8th ed. (2017). American Institute of State Highway and Transportation Officials (AASHTO), Washington, DC.

AREMA Manual of Railway Engineering and Maintenance-of-Way. (2016). American Railway Engineering and Maintenance-of-Way Association, Landover, MD.

ASCE/SEI 7 Minimum Loads for Buildings and Other Structures. (2016). American Society of Civil Engineers (ASCE), Reston, VA.

Billington, D. P. (1975). Historical perspective on prestressed concrete. *PCI Journal, 21*(5), 48–71.

Building Code for Concrete Structures (ACI 318-14) and Commentary for Building Code for Concrete Structures (ACI 318R-14). (2014). American Concrete Institute, Farmington Hills, MI.

Guyon, Y. (1974). *Limit-state design of prestressed concrete* (Vol. 1 and 2). Halsted Press.

International Building Code. (2015). *International Code Council.* VA: Falls Church.

Leonhardt, F. (1964). *Prestressed concrete: Design and construction.* W. Ernst. (in German).

MacGregor, J. G. (1983). Load and resistance factors for concrete design. *Journal ACI, 80*(4), 279–287.

MacGregor, J. G. (1976). Safety and limit states design for reinforced concrete. *Canadian Journal of Civil Engineering, 3*(4), 484–513.

MacGregor, J. G., Mirza, S. A., & Ellingwood, B. (1983). Statistical analysis of resistance of reinforced and prestressed concrete members. *Journal ACI, 80*(3), 167–176.

Nowak, A. S., & Szerszen, M. M. (2001). *Reliability-based calibration of structural concrete* (Report UMCEE 01-04). Department of Civil and Environmental Engineering, University of Michigan, Ann Arbor, MI.

USAToday.com. (2012). *Moving the space shuttle endeavor*. Retrieved Aug 1, 2018, from https://www.usatoday.com/picture-gallery/tech/2012/10/14/moving-the-space-shuttle-endeavour/1626769/.

Winter, G. (1979). Safety and serviceability provisions of the ACI building code. Concrete design: US and European Practices, ACI Special Publication SP-59, American Concrete Institute, Farmington Hills, MI, pp. 35–49.

World Commission on Environment and Development's (the Brundtland Commission report Our Common Future), reflections. (1987). Oxford University Press.

Chapter 2
Prestressed Concrete Applications

2.1 Introduction

Prestressed concrete is adaptable to a wide variety of structural systems. These include pretensioned and post-tensioned structures, both cast-in-place and precast, and other prestressed elements in conjunction with normally reinforced concrete. Case studies are presented in this chapter that represent both traditional construction and unique engineering projects that incorporate the advantages and design considerations associated with prestressed concrete. The case studies explore some of the details that make precast and prestressed concrete particularly attractive to the application. The design concepts and techniques needed to complete the case studies are developed in later chapters. While there is no general classification for precast and prestressed concrete, it is useful to group certain elements and structures together to explain how prestressed and precast concrete is designed and constructed. Prestressed and precast concrete may be considered in four broad categories: **Standardized Elements**, **Fixed Cross Section Elements**, **Fully Engineered Elements** and **Precast Nonprestressed Elements**. While there is some overlap, each group has its own unique characteristics.

The role of the engineer varies with the type and complexity of the structural system being constructed. Indeed, multiple engineers may be involved in some aspect of the design, fabrication, and construction of the project. In general, the **design engineer** who is typically the **licensed design professional** or **engineer of record** is responsible for the overall design. The unique characteristics of prestressed concrete often require the additional services of a **specialty engineer**. The specialty engineers can either provide consulting services to or be employed by a precast plant or contractor. Specialty engineers can also be associated with post-tensioning companies either as an employee or consultant. In either case, the specialty engineer takes the concept prepared by the licensed design professional and prepares final detailed design calculations as well as developing fabrication or construction details necessary to complete the project. **Engineer** is used throughout this book to indicate

© Springer Nature Switzerland AG 2019
C. W. Dolan, H. R. Hamilton, *Prestressed Concrete*,
https://doi.org/10.1007/978-3-319-97882-6_2

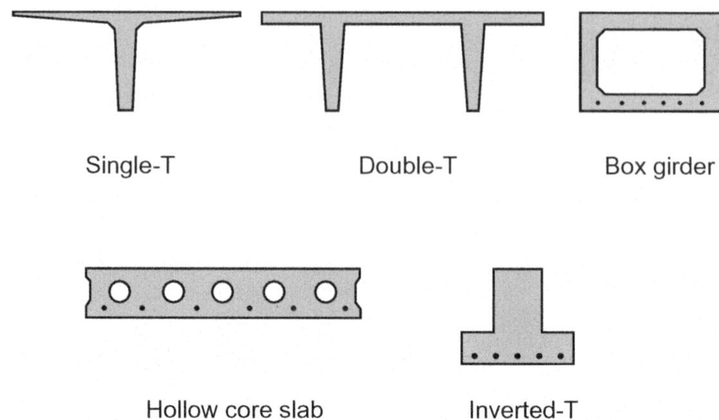

Fig. 2.1 Typical standardized sections

the individual responsible for the details of prestressed concrete design, production, or construction; this could mean either the specialty engineer or licensed design professional depending upon the application.

2.2 Standardized Precast Prestressed Elements

Pretensioned concrete beams and slabs are typically constructed in reusable steel forms in a precast plant. Although a modest amount of custom formwork is used at precast plants, improved quality and reduced costs are realized only when **standardized** elements are used. They consist of standard sections such as single-T and double-T beams, box girders, hollowcore slabs, inverted T-beams, and bridge girders (Fig. 2.1). The capital investment required to construct and equip a precast plant includes the concrete mixing equipment, forms, stressing beds, curing systems, and heavy lifting equipment. To obtain a return on this investment, the forms and stressing facilities must be in constant use. Efficiencies in production allow the precast pieces to be fabricated on a routine and daily basis. The cost efficiencies of this type of fabrication allow architects and engineers to select the sections for a wide number of uses and be sure of availability and competitive cost. Hollowcore planks, single-T, and double-T beams are used for as floor elements in building construction, Figs. 2.2 and 2.3. Inverted-T beams support double-T and hollowcore elements. These elements are commonly used in combination in office space, bridges, and parking garages, Fig. 2.4.

Standardized elements are creatively incorporated in building structures. For example, entire buildings have been constructed of double-T sections as is discussed in the commercial building case study. Double-T beams and box girders are used for short-span low-volume bridge girders. For example, following the flood in the Big Thompson Canyon in Colorado, double-T bridges were installed to replace the

Fig. 2.2 Single-T floor beam before topping and cast-in-place beam

Fig. 2.3 Double-T floor element with suspended ceiling removed

original structures, Fig. 2.5. The double-T bridges allowed a standard design to be developed and installed in multiple locations in the canyon. This solution accelerated the reconstruction effort.

Engineer's Role Standardized elements The design engineer typically selects one of these standardized elements from references such as the *PCI Design Handbook* (2017) or the *Manual for the Design of Hollow Core Sections* (1998). The design engineer may also contact precast plants located near the project to determine availability of sections. The section type and the design loads are provided to the precast plant. Final detailed design engineering is completed by the precast plant or

Fig. 2.4 Precast concrete panel for parking garage

Fig. 2.5 Precast concrete double-T specialty bridge

their specialty engineer in the form of shop drawings. This process allows the design engineer the efficiency of selecting desired shapes for their function and allows the plant to select the appropriate number of strands, strand configurations, harping locations, and other details to maximize the performance of the plant operations to meet the project objectives.

Fig. 2.6 Flat plate system with banded tendons (Photos courtesy of VSL)

Fig. 2.7 Flat plate post-tensioned slab construction

2.3 Fixed Cross Section Elements

The design engineer is required to determine the prestressing forces and tendon locations in fixed cross section situations. Two common fixed section design conditions are post-tensioned **beams** and **slabs** for building or parking garage construction, and **girders** for bridge construction. Other applications of fixed section elements include structures such as water tanks and post-tensioned slabs on-ground.

Flat plate and flat slab floor systems are ideally suited for the use of post-tensioning tendons, Figs. 2.6 and 2.7. Another popular system is one-way slab and beam floor systems that are cast-in-place, Figs. 2.8, 2.9, 2.10, and 2.11. The design engineer specifies a tendon profile geometry and an average effective post-tensioning force necessary to satisfy the design requirements. The specialty engineer for the post-tensioning company then takes this requirement and produces a detailed design with tendon sizes and spacing along with anchorage and splice locations. Pour strips and other detailing requirements necessary to isolate the post-tensioned element from other elements in the structure should be detailed by the engineer. Selection of the tendon location is determined by the thickness of the slab. The

Fig. 2.8 One-way beam and slab system showing tendon passing through column at the top of the section and coiled slab tendon ready to be placed (Photo courtesy of VSL)

Fig. 2.9 One-way beam and slab system. Bundled tendons are seen at the beam bottom. Single slab tendons are on top of beam (Photo courtesy of VSL)

Fig. 2.10 One-way slab and beam floor system. Slab tendons placed parallel with the slab span (Courtesy of VSL)

Fig. 2.11 One-way beam
end anchorage detail.
(Courtesy of VSL)

maximum tendon eccentricity available to the engineer is determined by minimum cover requirements for corrosion and fire protection over the top and bottom of the tendon. Therefore, in these applications, the section shape does not vary, but rather the design is controlled by the selection of the prestressed force and tendon spacing.

Another popular fixed section are two-way slab systems used as podium slabs. Podium slabs are typically a single-story post-tensioned concrete floor system

Fig. 2.12 Spliced girder bridge (Courtesy of Freyssinet Inc.)

supported by columns that support a lighter superstructure above, which is usually wood or metal stud walls with a light floor system. These are popular for use in residential construction where the upper stories serve as the living areas and the area below the podium slab serves as parking. The podium slab is usually designed as a separate structure from that of the wood or metal stud superstructures. The individual structures may have two separate structural engineers.

Fig. 2.13 Boggy Creek Road interchange at State Road 417 and Orlando International Airport's South Access Road

Spliced bridge girders are an example of design to a fixed section using partially standardized precast, pretensioned elements that are also post-tensioned during the final stage of assembly, Fig. 2.12. State departments of transportation and AASHTO specify standard beam sections. Precast plants have forms for bridge girder sections used in their market area. The section selection is dependent on the state practice and is further influenced by the distance that the girders are shipped. The variation and the magnitude of loads, load placement on the bridge, the girder spacing in the bridge, and the bridge deck design, preclude defining standard prestressing tendon forces and locations. Details in Appendix A indicate the possible strand locations in a section. The design engineer selects from several choices regarding the layout and loading of the bridge prior to design of the prestressing force and location. Unlike standardized products, the design engineer specifies all details of the bridge girder.

Another example of spliced segmented precast construction using standardized shapes involves the use of plant-produced horizontally curved, precast concrete

Fig. 2.14 Liquified Natural Gas tank showing circumferential post-tensioning tendons to ensure tank wall integrity under cryogenic conditions (Photo courtesy of Freyssinet Inc.)

U-girders (Hamilton & Dolan, 2016). These U-girders use standardized shapes and geometry along with post-tensioning to facilitate design and construction efficiency. One example of this approach is shown in Fig. 2.13.

Walls and tanks are a condition where the tendon location and force are determined within a fixed rectangular section, Fig. 2.14. The case study on tanks in Sect. 2.6 indicates how standardized elements, precast element and post-tensioning can be combined to create unique structures.

Engineer's Role with fixed section elements The engineer's role with fixed cross section element structures varies with the client and project. Some examples include:

- Building design engineers specify the desired final prestress force. The contractor or post-tensioning company specialty engineer completes the design by determining the tendon spacing and stressing forces. The design engineer then approves the contractor's shop drawing submittal.
- Building design engineers specify the final prestress force, tendon location, and hardware detailing. Post-tensioning company engineer develops the tendon layout, anchorage location, and stressing sequence.
- Bridge design engineers prepare the complete beam design, including detailed determination of prestress forces, tendon location, and construction sequence.
- Projects such as tanks are often procured on a design–build basis. The contractor and either the contractor's in-house engineering staff or a consulting engineer prepares the design to meet project requirements and the contractor's preferred construction practice.

Fig. 2.15 Parkland Hospital, Dallas, Texas. Seven stories are supported by girders with 62-ft cantilevers and 120-ft spans over an opening. (Courtesy of VSL)

2.4 Fully Engineered Elements

Fully engineered elements require detailed engineering continuously during design and construction. Examples of fully engineered structures include segmental bridges, specialty transit structures, tanks, towers, stadiums, floating facilities, and unusual building construction. Design of these structures requires considerable engineering effort and often includes on-site inspection. The complexity of these structures necessitates the engineer have a fundamental understanding of structural behavior, loads, prestressing effects, and material behavior. Collaboration of efforts among engineers, precast plants, and general contractors is required.

Engineer's role in fully engineered elements Fully engineered elements require the engineer to define the loads, structural system, concrete section, prestressing force, tendon location, and details (Figs. 2.15, 2.16, 2.17, and 2.18).

Fig. 2.16 Construction of Ironton Russell Bridge over the Ohio River. Longitudinal and transverse post-tensioning was used in the deck, which was cast-in-place using a form traveler. (Courtesy of VSL)

Fig. 2.17 St. Anthony Falls Bridge over the Mississippi in Minneapolis. (Courtesy of DYWIDAG-Systems International USA Inc.) Each bridge has a main span of 154 m that consists of precast concrete box girder segments supported by eight 21 m high piers. The end spans are 108 m long each cast-in-place, post-tensioned concrete box girders built on false work which seamlessly blends into the precast main span sections

2.5 Precast Nonprestressed Elements

The standard elements discussed in Sect. 2.2 are pretensioned concrete. In addition to pretensioned concrete, there are important nonprestressed precast elements. The decision to separate prestressed and precast elements into two groups is arbitrary.

Fig. 2.18 Woodrow Wilson Bridge replacement across the Potomac River near Washington, DC. (Courtesy of DYWIDAG-Systems International USA Inc.)

The major difference in grouping is that pretensioned elements require significant plant capitalization and stressing beds. Precast pieces can be fabricated on the jobsite or in a facility without stressing beds and other equipment associated with a plant operation. Tilt-up walls are an example of on-site precasting. If a small amount of prestressing is required for delivery, erection or final loads, it is provided in the form of single-strand post-tensioned tendons.

Two examples of precast nonprestressed elements are **architectural precast panels** and **tilt-up** construction. Architectural precast panels can be used either as structural elements or the exterior finish of buildings, Figs. 2.1, 2.4, and 2.6. The architectural panel finish can include color, texture, or simulated alternative materials such as a brick or stone (Fig. 2.19). Dyes or colorants are used in these special concrete mixtures. The architectural surfaces are made in small quantities and placed only on the outermost one to 1–1/2 in. of the precast piece. The backing concrete would be normal concrete to reduce costs. Textures are fabricated by sandblasting, retardants that are power washed off, Fig. 2.19, or liners in the form to develop more complex surface features like Fig. 2.20.

Tilt-up construction is a specialized form of precast construction where wall elements are fabricated on-site in a horizontal position. The floor of the structure is cast first. Edge forms are then laid out on the floor and the floor surface becomes the bottom of the wall form. The wall elements, complete with block-outs for windows and electrical or mechanical inserts, are then cast and allowed to cure in-situ. After the concrete has cured, the entire wall panel is lifted into a vertical position (ACI 551.2R, 2015). The tilt-up panel is temporarily braced against wind loads, Fig. 2.21. Connections between wall elements and roof elements provide stability. The roof diaphragm carries lateral loads to the end panels, which act as shear walls.

Fig. 2.19 Architectural wall panel finishing

Fig. 2.20 Architectural panel finish simulating sandstone rock

Tilt-up construction is commonly used for commercial structures such as warehouses, and industrial facilities. While some architectural finish is possible, the most economical tilt-up construction uses a plain or painted concrete finish. Tilt-up elements require two design considerations in addition to the design for vertical and lateral loads. These conditions are determination of the lifting positions and associated lifting hardware and the temporary bracing systems. The temporary bracing prevents damage under wind loads and is designed for a 6 month return period rather than the full 50 or 100-year return period (ACI 551.1R, 2014; Shah, 1995).

Fig. 2.21   Tilt-up wall panel construction

Engineer's Role Precast concrete   The design engineer is typically responsible for all design elements in precast pieces. The specialty engineer is responsible for the lifting details and temporary bracing as part of the construction effort.

2.6   Case Studies

The following case studies summarize illustrated examples of precast and prestressed concrete, methods of construction, and integration of prestressing into the structures. Critical design parameters leading to the overall design solution are identified. The design contributions to successful project completion are explored. Several of the projects are older structures. They are selected because references are available to allow the reader to examine the design and construction of the project in additional detail.

2.6.1   Commercial Precast Concrete Building

Standardized Precast Prestressed Elements

Figure 2.22 illustrates a commercial building that uses double-T beams for both roof elements and vertically for wall elements. This type of structure was common in the late 1970s, when engineers used alternative applications of standard precast elements. Precast plants place a bearing plate on the top surface of the double-T web to

Fig. 2.22 Commercial building using double-T elements for roofs and walls

serve as bearing for the roof beams. The roof beams provided a clear span working area to satisfy the functional requirements of the facility. The high ceiling allows vehicles to enter the building. A corbel cast on the top surface of the double-T beam at an intermediate height permits a second story for storage and inventory. The sales floor consists of double-T roofs and glass curtain walls.

The construction of similar buildings demonstrates the flexibility of using precast and prestressed concrete elements. Newer construction is more sophisticated but is developed using the same design principles.

The double-T beams are used as floor elements and spaces between the webs house utilities in the ceiling space. The use of precast concrete for building systems is gaining wider acceptance due to it sustainable nature and its energy conservation characteristics (Shaw et al., 1994).

Parking garages are another common use of standardized precast elements. The parking structure in Cheyenne, Wyoming illustrates the use of double T-beams for floor members, inverted-T beams spanning between precast concrete columns, Fig. 2.23, and precast wall panels to replicate the city streetscape, Fig. 2.5.

2.6.2 Solleks River Bridge

Standardized Precast Prestressed Elements and Fixed Cross Section Elements

The Solleks River Bridge was designed for the Washington State Department of Natural Resources and is intended as an access road to the remote logging areas in the Olympic National Forest, Fig. 2.24. This project illustrates several features of precast and prestressed concrete. The logging trucks are overweight vehicles with loads up to 75 tons. These loads are greater than those prescribed by the AASHTO Bridge Standard or standard building codes. The engineers worked with the client to develop project specific design criteria. The bridge crosses, a salmon spawning river, which required coordination with the Department of Natural Resources to maintain the overall water quality during construction. The remote location of the project,

Fig. 2.23 Parking garage using standardized prestressed elements

Fig. 2.24 Solleks River Bridge construction

55 miles from the nearest town, meant that delivery of ready mix concrete and construction materials be minimized to control overall project quality and costs (Casad & Birkeland, 1970).

The project solution used precast concrete struts anchored into the steep hillside of the river gorge. The bridge was designed as a three-span structure with pinned supports at the abutments and the top of the struts. The 15 ft wide bridge uses three girders to carry the heavy loads. The beam section is a bulb-T girder with end spans of 75 ft long and a center span 90 ft long. The ability to adapt the tendon design, beam length, and beam spacing to the heavy load conditions while still using a standard section was an attractive feature of the prestressed bridge solution. The foundation for the struts was excavated during times of minimum fish migration. The rock was fractured and inferior quality, so the abutments were post-tensioned into the rock and then shotcrete applied to control erosion. Once the foundation cured, the struts were fastened to the abutments.

Precast beams were temporarily anchored to abutments and placed on the inclined struts by means of pin anchorage assemblies cast into the beams and the strut ends. The final girder was lowered into place between the two and spans. Forms were fastened to the flanges of the bulb T beams and cast-in-place concrete was placed to complete the bridge deck and curbs. Reinforcing steel in the deck and a welded positive moment connection makes the structure continuous for live load. A complete description of the project, including the loads and structural analysis of the bridge and struts can be found in White et al. (1974).

2.6.3 Precast Concrete Water Storage Tanks

Fixed Cross Section Elements

The most common fixed cross section elements are building slabs. An examination of a reservoir structure offers a look at another application for a fixed section. There are three different methods of constructing prestressed concrete tanks: precast-prestressed walls, precast internally post-tensioned walls, and post-tensioned wire wrapped walls. Precast-prestressed concrete elements for the walls are fastened together with mechanical connections to provide the resistance to circumferential forces. Precast post-tensioned tanks use precast wall elements and internal circumferential post-tensioning. Wire wrapped post-tensioned tanks use cast-in-place walls with vertical reinforcement. Circumferential prestressing is applied to cast-in-place walls by a large wrapping machine. The wire wrapping requires specialty equipment not available to the general contractor and is done by specialty contractors.

Design by any of these methods addresses three distinct conditions peculiar to tanks. These conditions are vertical bending of the wall, circumferential forces due hydrostatic pressure, and seismic forces. If the wall is free to move radially when the tank is filled, there would be no vertical bending in the wall. In a tank design, the base of the wall is restrained by a ring girder, which prevents the base of the wall from moving radially and creates bending in the wall above the girder. The top of wall can be tied to the tank cover, creating an additional restraint and compatibility induced bending moments. The vertical bending of the wall can be further complicated on sites where the tank is partially buried or the soil slopes and a portion of the wall is subjected to earth pressure.

Circumferential forces develop due to the hydrostatic pressure of the fluid. Forces increase with the depth of the tank; however, they diminish near the base of the wall due to the restraint of the foundation ring girder. Seismic loads result from both the inertial forces of the ground motion and the sloshing of the fluid. In extreme cases, the fluid motion creates substantial uplift on the roof of the structure. The engineer decides whether the walls are secured to the foundation, if the walls are fastened to

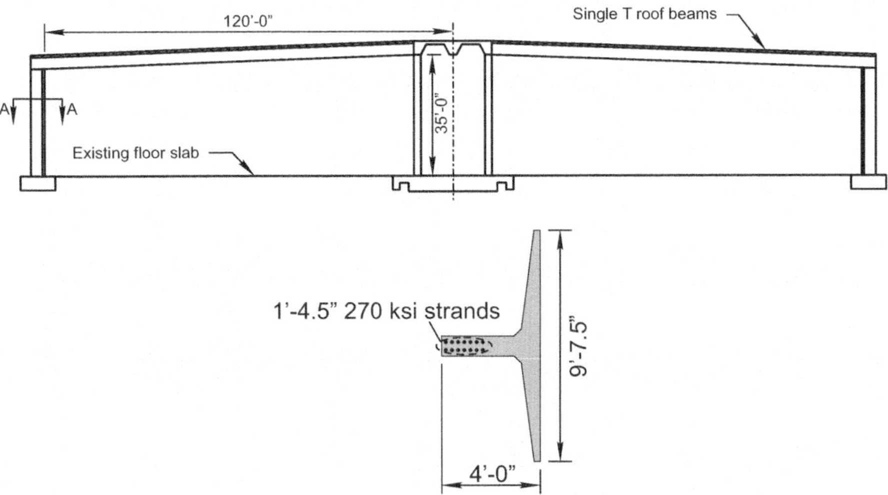

Fig. 2.25   Lincoln Heights reservoir schematic. Adapted from Lennen et al. (1996)

the roof or if fluid can spill over. More details on the design of concrete tanks can be found in Imper et al. (1983), ACI 350.3 (2001), and Hu and Hendrickson (1999).

The Lincoln Heights reservoir in Spokane Washington is an example of a ten-million-gallon precast tank design. Single-T beams were installed vertically as wall elements on a nonprestressed concrete foundation beam. The prestressing in the web of the beam provided the bending resistance Fig. 2.25. Mechanical fasteners between the flange tips provided the circumferential strength. Since the walls are not post-tensioned circumferentially, a flexible water restraining membrane provided a water tight structure. Pie-shaped single-T beams were used for the roof of the tank. The site sloped and the wall panels were designed for the active earth pressure in the backfill areas (Lennen et al., 1996).

The Riverton Heights Reservoir designed for the City of Seattle has a 50-million-gallon capacity, Fig. 2.26. It consists of a 169 ft radius post-tensioned foundation ring beam, a series of precast wall elements, and precast concrete roof panels (Birkeland, 1981). A walls design criterion includes a 300-psi minimum compressive stress for water tightness when the tank is at full capacity.

A post-tensioned ring beam forms the wall the foundation. The precast wall panels containing the internal post-tensioning ducts centered in the wall are erected on a waterproof elastomeric bearing placed on the foundation beam. Once the walls were erected ducts are connected and a cast-in-place joint was placed between the panel edges. Duct spacing is determined by the total prestress force to overcome the circumferential forces plus the dynamic pressure from seismic events. 12 ½ in. diameter seven wire prestressing strands constituted a single tendon. The frictional losses resulting from stressing the tendons around the perimeter of the tanks were sufficiently large that intermediate anchor locations were required. The anchors

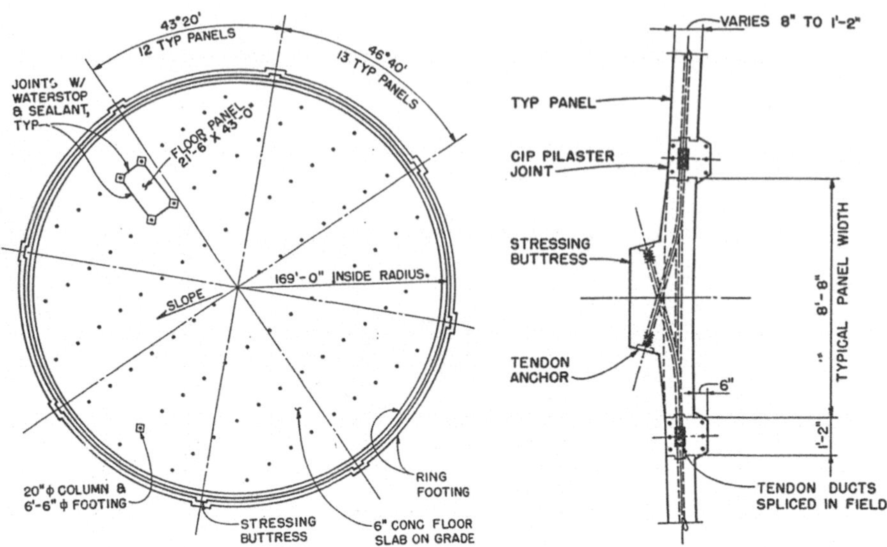

Fig. 2.26 Details of Riverton Heights reservoir. Courtesy of Berger/ABAM Engineers

allow the tendons to be stressed from one end, and if the losses were too great, to be restressed from the anchorage end. The tendons crossed at the anchor zone to assure the full compressive force across the joint, Fig. 2.27.

2.6.4   *Montreal Olympic Stadium*

Fully Engineered Structure

The Montreal Quebec Olympic Stadium is a major post-tensioned structure and is typical of many large stadium projects such as a scalloped dome at the University of Virginia (Berger, 1966), the Seattle Kingdome (Christiansen, 1976), the Iowa Unidome (Geiger & Dick, 1976), and rehabilitation of the stadium at the University of Oregon (Korkosz et al., 2004). The Montreal stadium was constructed for the 1976 Olympics and as major league baseball facility. The basic stadium was to have an open roof to satisfy International Olympic Committee rules for outdoor sports records. Baseball required a covered roof for play in inclement weather. The design solution was to construct an open-air stadium and a tower to support a fabric cover for indoor events. Standard building codes are typically insufficient for the design of stadium structures. Environmental forces, such as wind, require special attention (Irwin & Wardlaw, 1979). Even with the additional attention to engineering details, exceptional loads can create problems as seen when the stadium roof ripped during a heavy snowstorm in 1999 (Anon, 1999).

Fig. 2.27 Riverton Heights reservoir construction. Courtesy of Berger/ABAM Engineers

Fig. 2.28 Montreal Olympic Stadium initial construction

The basic stadium structure is a cast-in-place post-tensioned concrete frame with precast concrete seating units, Fig. 2.28. By using reusable forms and repeatable shapes, the frames were cast sequentially to gain maximum reuse of the forms. Segmental post-tensioning allowed each subsequent casting to be secured to the existing structure. Two types of post-tensioning were used in the project. Vertical post-tensioning secures the frame to the foundation. Horizontal post-tensioning is used in the beam elements. As the frames cantilever over the field, sequential post-tensioning allowed the total prestress force at the base of the frame to increase and the prestress force at the tip of the frame to be minimized. The weight of the frame and the precast concrete seats is load balanced by the post-tensioning forces.

The Montreal stadium tower supporting the fabric roof cantilevers from the foundation out over the top of the playing field. The tower is segmental cast-in-place and post-tensioned construction to compensate for the cantilever tower loads and final roof load.

2.6.5 *Sydney Opera House*

Fully Engineered Structure

Few structures in the world are as readily recognizable as the Sydney Harbour Opera House, Fig. 2.29. The Sydney Harbour Opera House resulted from a design competition. The competition sponsors realized that the location for the Opera House was the prime real-estate location for the entire city. Consequently, Sydney wanted a landmark structure. The winning design was based on a small number of architectural sketches. The sketches were to represent a series of sails in the Sydney Harbour. While the winning design achieved the objectives that the competition sponsors envisioned, it proved difficult to fabricate and expensive to construct (Arup & Zunz, 1971).

The completed opera house is a series of carefully matched precast concrete panels on a post-tensioned concrete frame. After considerable discussion, the surfaces of the Opera House were designed to have a spherical shape likened to an orange peel. By using spherical sections, a small number of forms were required to

Fig. 2.29 Sydney Harbour Opera House

fabricate the roof elements. The major design issue became one of the defining the interfaces between the panels so that each one would fit the complex geometry when delivered to the site.

2.6.6   Disney World Monorail

Fully Engineered Structure

The Walt Disney World Monorail was selected as one of the outstanding engineering projects in the first 50 years of the Prestressed Concrete Institute. The monorail uses precast prestressed concrete beams as the guideway structure and precast columns for support, Fig. 2.30. The design criteria for the monorail included several unique demands. Each beam must meet exactly the geometry of the route to provide guidance for the train. The two tracks operate independently from each other because approaching trains either create large end moments at the column top or a pinned connection is needed to release the end rotations. Continuous structures with fixed joints provide superior ride quality compared to simple span structures but continuity requires resolution of volume changes due to thermal creep and shrinkage effects (Dolan & Mast, 1972).

The typical structure is a six-span continuous beamway. The center column in the six-span structure has a larger cross section to absorb the majority of longitudinal

Fig. 2.30 Monorail structure

Fig. 2.31 Precast pier being erected

forces generated by acceleration and deceleration of the train. The end column section is thinner to flex and allow for temperature, creep, and shrinkage effects. The column top is fabricated from a steel plate, Fig. 2.31. The plate is designed to provide longitudinal and vertical bending resistance while being torsionally flexible. The torsional flexibility allows the column top to act as a pinned connection to uncouple the motions of approaching trains.

The prestress design for dead load anticipated the prestressing losses in the beam prior to erection on the site. The prestressing force balanced the self-weight of the beams, allowing them to be erected with neither camber nor sag. The beams are fabricated as either straight or curved members. Straight beams were pretensioned and had a hollow void in the section to reduce weight. Curved beams were fabricated in a specially designed form, which allowed each beam to be built to its required geometric profile. The curved beams have hollow cavities formed with Styrofoam blocks, Fig. 2.32. Beams were precast then post-tensioned at the plant for dead load.

Once the beams were erected, the six-span structure was post-tensioned together, Fig. 2.33. The haunched shape of the beam facilitated both the plant and field post-tensioning. Varying the concrete center of gravity allowed the field post-tensioning to run straight and yet be at the bottom of the section at midspan and the top of the section at the support. In areas where only uniform cross sections are allowed, the beams used draped tendons to balance the dead load and straight tendons for field post-tensioning.

Fig. 2.32 Reinforcement cage, parabolic duct and Styrofoam core

Fig. 2.33 Post-tensioning for continuity

2.6.7   Floating Concrete Structures

Fully Engineered Structure

Archimedes' principle states that if you can displace a sufficient volume of water, any structure will float. The principle holds true for concrete structures as well as steel vessels. There is a variety of floating concrete structures beginning with the concrete ship program of WWI. More recently, concrete structures have been used for **floating bridges** in Washington State and British Columbia, Canada, **floating piers** in Alaska and numerous different **platforms** for energy extraction (Gerwick, 1976).

The LPG plant is an example of the application of concrete for floating structures. The hull is designed as a series of precast concrete panels post-tensioned together to form a monolithic structure, Fig. 2.34 (Anderson, 1976, 1977). The facility consists

Fig. 2.34 LPG platform and hull under construction

of 12 large insulated steel LPG storage tanks and a LPG processing and liquefaction plant housed in concrete hull that displaces over 65,000 tons. Six of the tanks are stored below the deck and an additional six tanks are mounted above deck. Curved hull elements were used on the bottom of the platform to match the shape of the tanks and provide support for the saddles holding the storage tanks. The curved sections provide an arching effect to reduce bending stresses resulting from the hydrostatic water pressure. External side walls are designed for hydrostatic pressure. Internal walls are designed for possible hull breaches and for overall structural stiffness.

Loadings on ocean structures are beyond the limits of standard building codes. In addition to the hydrostatic pressure, the structural design considers the differential and fatigue loadings due to loading and unloading tanks and sea action. The sea action is manifest as a wave moves along the length of the hull. If the valley of the wave is amidships, the hull acts like a simply supported beam. As the crest of the wave moves to amidships, the hull behaves more like a double ended cantilever. The post-tensioned wall and hull panels are designed for these differential bending combinations.

2.6.8 Segmental and Cable Stayed Bridges

Fully Engineered Structures

Balanced cantilever and **cable-stayed** bridges represent major advances in bridge design. Balanced cantilever bridges initiate the construction at a central pier and construction takes place as each segment is added. Incremental post-tensioning provides the support as the length of the cantilever increases. Cable stayed bridges begin at a support tower. Each segment is added and tied back to the central tower with a stay. The vertical reaction due to the weight of the segment creates a compression force in the section, thereby reducing the required post-tensioning. These structures require close coordination between the engineer and the contractor.

Balanced cantilever bridges are constructed in one of two different methods. They can be segmentally precast and erected segmentally or segmentally cast-in-place

Fig. 2.35 Balanced cantilever bridge construction. Photo courtesy Roger Hauser

(Mathivat, 1983). Figure 2.35 illustrates precast balanced segmental construction. Precast segmental bridges fabricate each segment near the construction site and deliver it to the bridge. Cranes or launching trusses, mounted on the erected portion of the bridge, allow each segment to be moved and positioned at the end of the cantilever (Palmer, 1988). The segments are typically **match cast**; that is, each cast segment is placed at the end of the form for the adjoining segment to assure an intimate connection in the field. The design of balanced cantilever bridges requires attention to the deflection of the bridge during construction to assure that the adjacent balanced cantilevers fit when they meet. The deflections are controlled by monitoring both the concrete delivered to the site, the deflection of each segment and the post tensioning force used to connect each segment. The post-tensioning extends from the center column for each segment. As each segment of the bridge extends out from the center, the total amount of post-tensioning over the support is increased. Once the closure segment is installed, the final bridge is post-tensioned so live load is carried on a continuous structure. Cast-in-place segmental bridges move the form piecemeal out from the support column and follow a similar post-tensioning approach.

Cable-stayed bridges use precast concrete bridge segments similar to balanced cantilever construction, Fig. 2.36 (Kumarasena et al., 2003; Muller and Barker 1985; Walther et al., 1988; Grant, 1979). The cable-stayed bridge becomes more

Fig. 2.36 Cable Stayed Bridge tower under construction on the Bangkok Industrial Ring Road. Photo courtesy Norconsult AG

structurally indeterminate as each segment is added. Like balanced cantilever, the cable stayed bridge is sensitive to deflections during construction. Post-tensioning each segment is accomplished by the compressive force generated from the stay cable and internal post-tensioning to assist the construction. Cable stayed bridges can be constructed in a balanced fashion, like the balanced cantilever, or from one side if the back stays can be adequately anchored, Fig. 2.36.

2.6.9 Slabs-on-Ground

Fully Engineered Structures

Slabs-on-ground account for nearly 60% of the unbonded post-tensioning construction by weight in the United States. Design of slabs-on-ground are not covered by the ACI Building Code. Slabs-on-ground for residential housing are addressed in the Post-Tensioning Institute report DC10.5-12 (2012) and referenced in the International Building Code. Other slab-on-ground applications are addressed in ACI Committee 360.

An innovative application of slab-on-ground design and construction is the Sky View Parc in Flushing, New York. The client was looking to construct two tennis courts and a multisport court on top of a seven-story parking garage. The parking

Fig. 2.37 Slab-on-ground design used for tennis courts on a parking garage. Example and photo courtesy of the Post-Tensioning Institute, Farmington Hills, MI. (http://www.post-tensioning.org/sog-case-studies.php retrieved 02/12/2018)

garage designers did not consider that this area was going to be tennis courts, resulting in the slope of the deck being incorrect. Weight restrictions on the existing structure needed to be maintained, thus eliminating many construction alternatives. An elevated deck was constructed using ridged foam insulation boards ranging in thickness between 2 in. and 2 ft. Electrical conduits were run underneath the elevated platform for the tennis court lighting and outlets. A 4-in. thick post-tensioned slab-on-ground was installed on top of the elevated deck. The slab was designed to have a residual prestress in the center of 120 psi. After the slab was completed, custom fence post brackets were fabricated and installed on the edge of the post-tensioned slab and final court finish applied (Fig. 2.37).

References

ACI 350.3-01. (2001). *Seismic Design of Liquid-Containing Concrete Structures 350.3-01, and Commentary (350.3R-01) Reported by ACI Committee 530*, American Concrete Institute, Farmington Hills, MI, 53p.

ACI 551.1R-14. (2014). *Guide to Tilt-Up Construction Reported by ACI Committee 551*, American Concrete Institute, Farmington Hills, MI, 42p.

ACI 551.2R-15. (2015). *Guide for the Design of Tilt-Up Concrete Panels, Reported by ACI Committee 551*, American Concrete Institute, Farmington Hills, MI, 72p.

Anderson, A. R. (1976). Concrete vessel is star of first totally offshore LPG facility. *Civil Engineering (New York), 46*(4), 58–60.

Anderson, A. R. (1977). World's largest prestressed LPG floating vessel. *Journal of the Prestressed Concrete Institute, 22*(1), 12–31.

Anon. (1999). New fabric roof at Montreal Olympic Stadium tears under heavy snow. *ENR, 242*(6), 19.

Arup, O. N., & Zunz, G. J. (1971). Sydney opera house. *Civil Engineering, 41*(12), 50–54.

Berger, H. (1966). Scalloped prestressed dome from prestressed elements. *ACI Proceedings, 63*(3), 313–323.

Birkeland, C. (1981). Riverton-heights Reservoir Seattle, Washington. *Journal Prestressed Concrete Institute, 26*(3), 16–28.

Casad, D. D., & Birkeland, H. W. (1970). Bridge features precast girders and struts. *Civil Engineering, 40*(7), 42–44.

Christiansen, J. V. (1976). King County Multipurpose Domed Stadium. *Chemical Engineering Science, v 2, IASS (Int. Assoc. of Shell and Space Struct.) World Congr. on Space Enclosures (WCOSE-76)*, p 1049–1061.

Dolan, C. W., & Mast, R. F. (1972). Walt Disney World Monorail Designed for Smooth Riding. *Civil Engineering (ASCE)*, 4p.

Geiger, D. H., & Dick, J. S. (1976). Design, fabrication and erection of unidome stadium. *Journal of the Prestressed Concrete Institute, 21*(6), 94–107.

Gerwick, B. C. (1976). Current trends in concrete sea structures. *PCI Journal, 21*(5), 176–190.

Grant, A. (1979). Pasco-Kennewick Intercity Bridge. *Journal of the Prestressed Concrete Institute, 24*(3), 90–109.

Hamilton, H. R., & Dolan, C. W. (2016). Prestressed concrete—The innovator's industry. *Concrete International, 38*(10), 28–33.

Hu, D. P., & Hendrickson, B. (1999). Seismic design of prestressed concrete tanks. *Technical Council on Lifeline Earthquake Engineering Monograph, 16*, 572–581.

Imper, R. R., Arafat, M. Z., Birkeland, C. J., Carpenter, J. E., Jorgensen, I. F., Koestring, E., Kulka, F., Kurtz, J. K., Mallet, J., Mujumdar, V., Sanderson, K., Stackpole, J. H., Tadros, M. K., & Verma, S. P. (1983). State-of-the-art of precast prestressed concrete tank construction. *Journal of the Prestressed Concrete Institute, 28*(4), 36–83.

Irwin, H. P. A. H., & Wardlaw, R. L. (1979). Wind tunnel investigation of a retractable fabric roof for the Montreal Olympic Stadium. *National Research Council of Canada, Quarterly Bulletin of the Division of Mechanical Engineering, 3*, 19–33.

Korkosz, W. J., Haris, A. A. K., & Andrews, D. (2004). Precast concrete transforms the University of Oregon's Autzen Stadium. *PCI Journal, 49*(3), 44–54.

Kumarasena, S., McCabe, R., Zoli, T., & Pate, D. (2003). Zakim—Bunker Hill Bridge, Boston, Massachusetts. *Structural Engineering International: Journal of the International Association for Bridge and Structural Engineering (IABSE), 13*(2), 90–94.

Lennen, R., Miller, G., & Prussack, C. (1996). Precast prestressed concrete—Solution of Choice for Lincoln Heights Water Tanks. *PCI Journal, 41*(1), 20–33.

Manual for the Design of Hollow Core Slabs. (1998). Precast/Prestressed Concrete Institute, Chicago, IL, 95p.

Mathivat, J. (1983). *The cantilever construction of prestressed concrete bridges* (352p). New York: Wiley.

Muller, J. M., & Barker, J. M. (1985). Design and construction of Linn Cove Viaduct. *Journal of the Prestressed Concrete Institute, 30*(5), 38–53.

Palmer, W. D. (1988). Concrete in the Canyon. *Concrete International: Design and Construction, 10*(2), 19–23.

PCI Design Handbook, 8th Ed. (2017). Precast and Prestressed Concrete Institute, Chicago, IL.

PTI DC10.5-12. (2012). *Standard Requirements for Design and Analysis of Shallot Post-tensioned Concrete Foundations on Expansive Soils*, Post-tensioning Institute, Farmington Hills, MI, pp. 52.

Shah, N. K. (1995). Tilt-up Construction in Two Charlestons. *Concrete International, 17*(7), 45–47.

Shaw, M. R., Treadaway, K. W., & Willis, S. T. P. (1994). Effective use of building mass. *Renewable Energy, 5*(5–8), 1028–1038.

Walther, R., Houriet, B., Isler, W., & Moia, P. (1988). *Cable Stayed Bridges* (196p). London: Thomas Telford.

White, R. W., Gergely, P., & Sexsmith, R. (1974). *Structural engineering* (Vol. 1, pp. 3–18). New York: Wiley.

Chapter 3
Materials

3.1 Introduction

The primary material considerations for prestressed concrete structures are the mechanical properties and durability of concrete, prestressed and nonprestressed reinforcement, and anchorage devices. The first half of the century saw the understanding of these properties develop significantly. As discussed in Chap. 1, the development of high-strength steel was essential to the development of prestressed concrete. Without the ability to stretch the prestressing steel to offset the time-dependent effects of concrete creep and shrinkage, prestressed concrete is not practical. By the 1950s, steel manufacturing techniques had developed sufficiently that stress-relieved steel wire with high strength and ductility was produced at a cost that was economically suitable for use in producing prestressed concrete structures. Since the inception of prestressed concrete, prestressing strand has further advanced with the advent of low-relaxation prestressing strand, which is the most commonly used prestressing steel in the world today (Mindess et al, 2003; Neville, 2012).

Prior to about 1960, concrete was essentially composed of portland cement, water, and fine and coarse aggregates, which would produce modest strength gains with time. It was quickly discovered that increased production rates for both cast-in-place and precast concrete required higher compressive strength earlier in the project. This was necessary so that the large prestressing force could be applied to the member soon after casting to allow the next member to be produced. In the late 1960s, the first plasticizing admixtures became available. These admixtures evolved into today's **high range water reducing (HRWR)** admixtures. HRWR admixtures reduce the mixture viscosity allowing concrete mixtures to use lower **water-cementitious material ratio** (w/cm) while maintaining the same workability. The lower w/cm also improved the early strength gain. This combined with improved portland cement production techniques led to greatly improved production rates, which, in turn, has improved the economic viability of prestressed concrete.

© Springer Nature Switzerland AG 2019

C. W. Dolan, H. R. Hamilton, *Prestressed Concrete*,

https://doi.org/10.1007/978-3-319-97882-6_3

More recent developments of admixture chemistry have resulted in flowable **self-consolidating concrete mixtures**. SCC allows concrete placement into tighter form locations, reduces segregation, and reduces the vibration effort and noise in a precast plant, which has improved production rates, plant efficiency, and safety, while reducing energy consumption.

3.2 Specified Mechanical Properties

The design of prestressed concrete requires knowledge of the mechanical properties of the materials used to construct the member. Those mechanical properties are estimated during the design process, specified in the contract documents, and then confirmed with factory certifications or field testing during construction. The ACI 318-14 Building Code covers concrete properties in Chap. 19 and steel properties in Chap. 20. The following notation describes specific mechanical properties commonly used in the design of prestressed concrete:

E_c = modulus of elasticity of concrete
E_s = modulus of elasticity of nonprestressed reinforcement
E_{ps} = modulus of elasticity of prestressing reinforcement
E_{ci} = modulus of elasticity of concrete at time of initial prestress
f'_c = *specified* compressive strength of concrete usually at an age of 28 days
f'_{ci} = *specified* compressive strength of concrete at time of initial prestress
f_{pu} = *specified* **tensile strength** of prestressing reinforcement
f_{py} = *specified* **yield strength** of prestressing reinforcement
f_r = modulus of rupture of concrete
f_y = *specified* **yield strength** of nonprestressed reinforcement
f_{yt} = *specified* **yield strength** of transverse reinforcement

The definitions described as "specified" indicate that the engineer's design requires this property and acceptable methods to achieve these properties is the responsibility of the contractor. The engineer can further specify other standards to refine or restrict the selection of materials. For steel properties, ACI 318 Building Code does not describe how to make the steel, but rather makes use of ASTM International standard specifications. This approach provides a consistent and technically current method of describing the material. These standard specifications are established through ASTM's consensus process, which must satisfy certain procedures and regulations. Stakeholders with an interest in the material, such as owners, building officials, engineers, contractors, producers, among others participate in their development and use. ASTM standards are not law, but rather are used in the contract documents for the procurement of materials and legally binding if incorporated into local building codes.

Specifying concrete is not currently possible with a single ASTM standard because concrete is composed of materials that are locally or regionally available and are quite variable. Consequently, concrete mixtures must be designed and tested

Table 3.1 Mechanical properties used to design prestressed concrete

| Mechanical properties | | | Described |
Specified	Derived	Design	in Chapter
f'_c	E_c, E_s, E_{ps}	Prestress losses	4
f'_c	f_r, E_s, E_{ps}	Service stresses	5
$f'_c f_y f_{ps}$	–	Flexure strength	5, 6
$f'_c f_y f_{yt}$	–	Shear and torsion strength	7
f'_c	E_c, E_s, E_{ps}	Deflection and camber	8

in accordance with ACI 318 Building Code (2014) and ASTM standards to ensure adequate strength and serviceability in the completed structure. Table 3.1 provides a summary of both the specified and derived mechanical properties and thier use in the design of prestressed concrete. For example, stresses and moduli of elasticity are calculated from the specified strength values as described in the following sections.

The sections that follow provide an overview of material properties suitable for most prestressed concrete design projects, for calculation of prestress losses, and for calculation of member deflections.

3.3 Concrete Mechanical Properties

Several concrete properties are required for prestressed concrete design. These include tensile strength, time related properties of elastic modulus, shrinkage and creep. These properties are derived from the concrete specified compressive strength.

3.3.1 Compressive Strength and Ductility

Most practical concrete properties are based on the compressive strength of the concrete. Prestressed concrete members have two compressive strength requirements. The compressive strength at transfer f'_{ci} is the minimum strength required to resist the initial prestress force to the member. The **specified compressive strength** f'_c is used in calculating other serviceability and strength conditions of the section. The transfer strength is usually specified at 16–18 h after casting for pretensioned members and 2–7 days for post-tensioned members. The specified compressive strength is typically specified at 28 days. For prestressed plant operations, the transfer strength ranges between 2500 psi and 5000 psi, and higher strengths are used regionally. The corresponding design strength range is 4500–10,000 psi. Cast-in-place post-tensioned concrete specifies 28-day strengths between 4000 and 8000 psi.

Two methods are used to monitor and obtain the transfer strength: accelerated curing and cure time. Plant prestressed elements transfer the prestress at 16–18 h to reuse the forms on a 1-day cycle. To obtain these short cure times, the concrete mixtures may use **Type III—high early** strength cement, or may contain admixtures to accelerate initial curing, or can use steam or heat to further accelerate curing. To avoid unnecessary testing and to assist in production planning, transfer strength can be monitored using the Sure-Cure© system. Concrete in both the prestressed member and test cylinders are instrumented with thermocouples. The cylinders are placed in an environmental chamber which is controlled to cure the cylinder at the same conditions as the member. **Maturity methods** augment cure time to estimate additional cure time if initial cylinder tests are below the specified strength.

Proportions of concrete components in precast concrete batch plants are usually computer controlled. Using computer quality control procedures, coefficients of variation of the cylinder strength below 10% are attainable. This level of quality control allows plants to make more efficient use of constituent materials to achieve the statistical validation of the mixture design in an economical manner.

Concrete in cast-in-place slab construction is usually cured for 24–72 h prior to post-tensioning and admixtures may be used to accelerate strength gain. Slabs are shored until the prestress is applied. Low initial prestress levels typically required in slab construction facilitate transfer at lower compressive strengths without accelerated curing. Shoring remains in place for construction of three to four stories above the current slab to reduce early-age overload.

The expanded use of chemical admixtures led to changes in long-term concrete strength. Prior to the introduction of HRWR admixtures, the engineer could count on continued cement hydration to increase the concrete compressive strength. Present-day mixtures can be fine-tuned to satisfy specified strengths with little or no strength gain beyond 28 days. This practice results in little or no residual strength beyond the specified strength, Fig. 3.1.

Not all admixtures create the same effects and the engineer rarely has detailed involvement in the prestressed concrete mixture design. For fully engineered projects, however, awareness of the variation possible in material properties due to the mixture is important for understanding prestressed concrete behavior.

Figure 3.2 provides representative stress–strain curves for concrete tested at 28 days. Two observations from the stress–strain relationships are important. First, the concrete modulus of elasticity increases as the strength increases. The

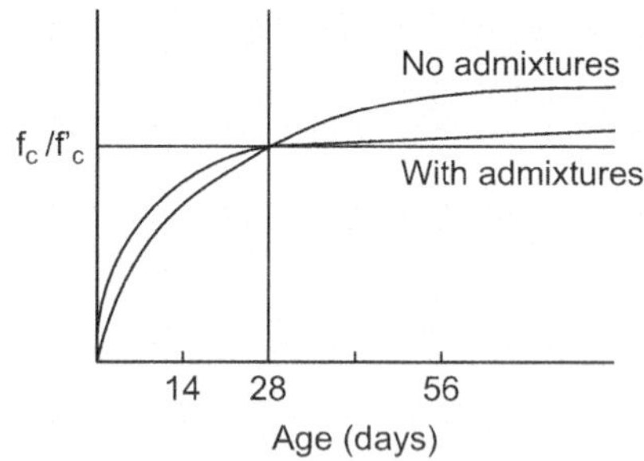

Fig. 3.1 Strength gain versus time for various concrete mixtures (From Dolan et al., 1993)

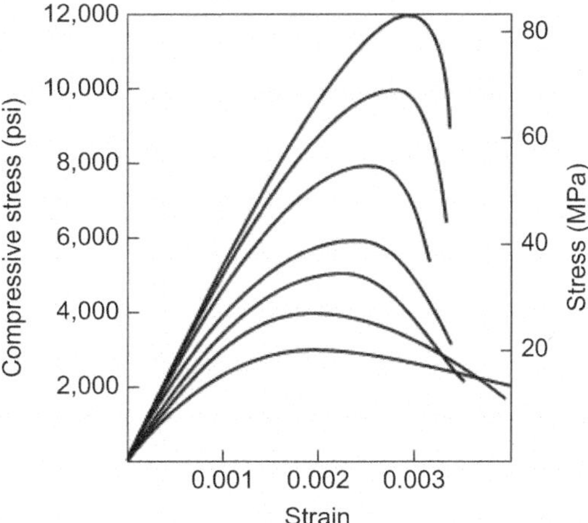

Fig. 3.2 Characteristic concrete stress–strain curves (Adapted from Darwin et al., 2015)

approximation of a linear relationship between stress and strain is valid over a greater portion of the stress curve as the strength increases.

Second, the ductility, as measured by the stress–strain ability beyond the maximum stress, decreases with higher strength. The maximum strain limit of 0.003, as assumed in the ACI 318 Building Code, remains valid. Use of the Whitney equivalent stress block continues to be applicable for all bending behavior and is the basis for calculating nominal bending strength. Bae and Bayrak (2013) suggest that the use of the ACI equivalent rectangular stress block may be slightly unconservative for axially loaded members using high-strength concrete.

3.3.2   Tensile Strength

Tensile stress is used to both classify prestressed members and to determine section properties. The tensile strength of concrete is established by one of two methods: split cylinder test or flexural beam test. ASTM C496/496M (2017) provides the test method to determine the split cylinder tensile strength. The split cylinder test places nearly the entire cross section of the cylinder in tension, f_t, Fig. 3.3. While considerable scatter occurs in the data, the tensile strength derived from the spit cylinder test for normalweight and lightweight concrete as well as the ACI 318 Building Code value are given in Table 3.1. The symbol λ is used in the ACI Building Code to adjust the tensile strength for the unit weight of the concrete where $\lambda = 1.0$ for normalweight concrete, 0.85 for "sand lightweight" and 0.75 for "all-lightweight" concrete. Correlation of λ to the unit weight of the concrete mixture provides an alternative method of defining the effects of lightweight concrete.

Concrete tensile strength is also measured by a four-point beam test and is referred to as the concrete flexural strength or **modulus of rupture f_r**, Fig. 3.3 (ASTM C496/496M, 2017). Because the stress distribution in the flexural test varies

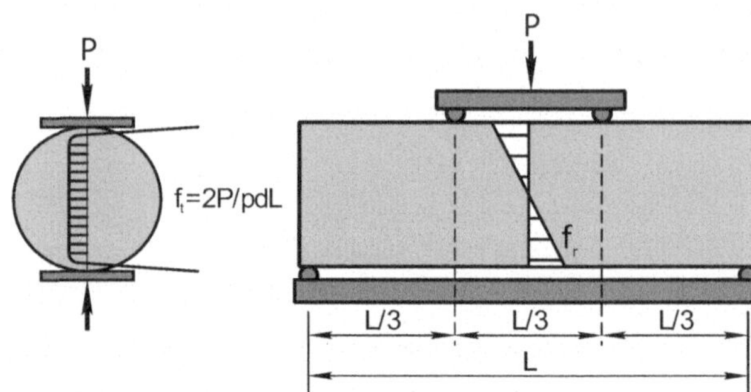

Fig. 3.3 Schematic of split cylinder and flexure test

Table 3.2 Range of concrete tensile strength

Test method	ACI Building Code value, psi	Range in normalweight concrete, psi	Range in lightweight concrete, psi
Split-cylinder, f_{ct}	$6\lambda\sqrt{f'_c}$	6 to $8\sqrt{f'_c}$	4 to $6\sqrt{f'_c}$
Modulus of rupture, f_r	$7.5\lambda\sqrt{f'_c}$	8 to $12\sqrt{f'_c}$	6 to $8\sqrt{f'_c}$

linearly across the beam section, only a small portion of the specimen is under maximum tensile stress. As a result, the modulus of rupture is typically greater than the split cylinder tensile strength (ASTM C78/C78M, 2015) for any given concrete mixture.

Neither split cylinder nor modulus of rupture test is typically conducted to determine the tensile strength for design purposes. Rather, the tensile strength is derived from the specified compressive strength by correlating the cylinder compressive strength with the tensile strength results from a variety of concrete mixtures into the equations shown in Table 3.2.

The modulus of rupture strength f_r is used to classify prestressed concrete members. A member is classified as uncracked, Class U, if the maximum tensile stress is less than the modulus of rupture, a cracked member, Class C, if the tensile stress is greater than $12\sqrt{f'_c}$ or a transitional element, Class T, if the stresses are between these two limits.

3.3.3 Elastic Modulus

The **modulus of elasticity E_c** of concrete is based on the initial slope of the stress–strain curve of 6 in. diameter by 12 in. long cylinders loaded in compression. Research indicates that modulus of elasticity tests contain considerable scatter. Equations given in the ACI 318 Building Code correlate the modulus of elasticity to the unit weight of the concrete in pounds per cubic foot w and f'_c. Equation (3.1) provides a value for E_c for concrete with unit weights between 90 and 165 pcf.

$$E_c = 33w^{1.5}\sqrt{f'_c}\,(\text{psi}) \tag{3.1}$$

For normalweight concrete of 145 pcf, Eq. (3.1) can be taken as:

$$E_c = 57,000\sqrt{f'_c} \tag{3.2}$$

Equation (3.1) is used for concrete of any age with the value of f'_c equal to that of the concrete at the specified age. Thus, the modulus of elasticity at transfer E_{ci} uses f_{ci} in lieu of f_c in Eq. (3.1). Therefore, the equation is valid for use for calculations at the transfer strength and for calculations at strengths used to evaluate service load behavior. The ACI formulation of modulus of elasticity is commonly used wherever the ACI Building Code is the primary design reference. Other formulations for the modulus of elasticity are available (ACI 423.10 2016). Statistical studies suggest that the variation in modulus of elasticity test data do not justify more refined equations.

3.3.4 Shrinkage

Shrinkage is the change in volume of the portland cement paste due to both loss of water and hydration of the portland cement. Drying shrinkage occurs as water migrates out of the cement paste as curing progresses. Curing involves a chemical reaction between the cement and the water. The final volume of hydrated cement paste is less than the volume of the two initial constituents; water and cement. Consequently, shrinkage occurs over the life of the structure, with most of the shrinkage occurring in the first year.

Shrinkage can be partially reclaimed if the concrete is immersed in water. Because the concrete has undergone additional hydration between the initial moisture loss and the time of rehydration, shrinkage strain ε_{sh} is never fully recovered.

Shrinkage occurs in the cement paste, not the aggregate. The magnitude of shrinkage in any concrete mixture is a function of the initial *w/cm* material ratio and the mixture constituents. A mixture with a high coarse aggregate content has less shrinkage than a mixture with less coarse aggregate. Values of final shrinkage $\varepsilon_{sh,\,u}$ for ordinary concretes with aggregates such as granites and some limestones are generally on the order of 400×10^{-6} to 800×10^{-6} in./in. depending on the initial water content, ambient temperature, and humidity conditions. Highly absorptive aggregates with low moduli of elasticity, such as some sandstones and slates, result in shrinkage values two times more than those obtained with less absorptive materials. Some lightweight aggregates, in view of their higher porosity and corresponding higher absorption capacity, result in much larger shrinkage values than concrete with normalweight aggregates.

Estimating long-term shrinkage for time-dependent losses of prestress is complicated by these variables. Long-term studies indicate that for moist cured concrete the shrinkage can be satisfactorily predicted by Eq. (3.3) at any time after the first 7 days (Branson, 1977).

$$\varepsilon_{sh,t} = \frac{t}{35 + t}\,\varepsilon_{sh,u} \tag{3.3}$$

where $\varepsilon_{sh,t}$ is the unit shrinkage at time t in days, and $\varepsilon_{sh,u}$ is the total shrinkage after a long time. Eq. (3.3) is for "normal" conditions, that is, adequate aggregates, humidity more than 40%, and average thickness of members of about 6 in. ACI 209 (2008) provides a detailed discussion of shrinkage.

The Precast/Prestressed Concrete Institute (PCI) addresses the magnitude of final shrinkage in a different format (PCI, 2017). Total baseline shrinkage is selected as 8.2×10^{-6} and then adjusted for volume-to-surface ratio V/S and relative humidity RH, Eq. (3.4).

$$\varepsilon_{sh,u} = \left(8.2 \times 10^{-6}\right)\left(1 - 0.06\frac{V}{S}\right)\left(100 - RH\right) \tag{3.4}$$

Example 3.1: Shrinkage Calculation Using PCI Approach
For a rectangular beam section 24 in. deep and 16 in. wide in a relative humidity of 25%, the corresponding magnitude of final shrinkage would be:

$$\varepsilon_{sh,u} = \left(8.2 \times 10^{-6}\right)\left(1 - 0.06\frac{24 \times 16}{2 \times (24 + 16)}\right)(100 - 25) = 4.38 \times 10^{-6}$$

Comment: This shrinkage value is at the lower bound of the values in the ACI 209 report but also reflects the curing control in a precast plant.

3.3.5 Creep

Creep is the time-dependent deformation of a member under sustained load. Concrete and wood both behave linearly under short duration loading and continue to deform nonlinearly under sustained loading. Figure 3.4 provides schematic deflection versus time behavior of a concrete cylinder subjected to a constant axial load.

Initial elastic strain ε_{in1} occurs when the specimen is first loaded. The solid line represents the total creep strain the member undergoes under continuous sustained load over time. If the load is removed at some intermediate time, a new initial strain ε_{in2} occurs. This rebound strain is followed by a creep recovery, $\varepsilon_{cr,r}$. Reloading the member repeats the process, ε_{in3}.

The creep properties of concrete are apparent in Fig. 3.4. Rebound strain is less than the initial strain due to the higher modulus of elasticity of the cured concrete when the specimen is unloaded. Creep strain rebound is also less because of the

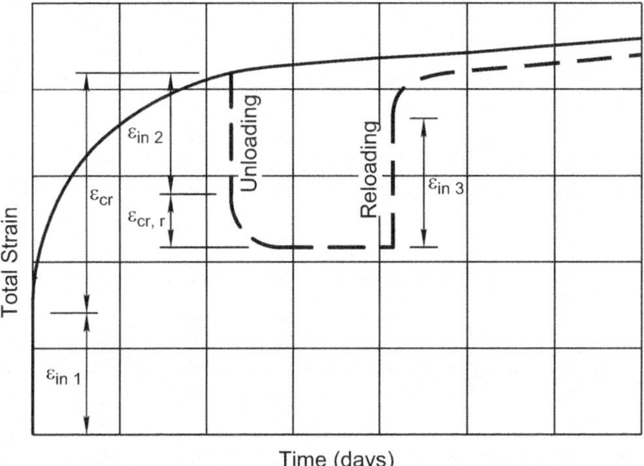

Fig. 3.4 Schematic creep strain versus time (Adapted from Darwin et al., 2015)

Table 3.3 Typical creep parameters

Concrete compressive strength (psi)	Creep coefficient C_c
3000	3.1
4000	2.9
6000	2.4
8000	2.0
10,000	1.6
12,000	1.4

Adapted from Branson (1977)

greater strength and modulus of elasticity, and that some water that was forced out of the member is no longer able to reenter due to the lower porosity of the more fully cured concrete. Figure 3.4 additionally indicates that a member that is loaded and unloaded for extended periods of time has less total creep than if the load is applied continuously.

The magnitude of the total creep is partially a function of the concrete strength. Higher strength concrete has relatively less creep than lower strength concrete (ACI SP-227). **Creep coefficients C_c** are commonly used to calculate the creep effects. Creep coefficient is the ratio of the total deformation ε_{cu} at the end of the specified time period divided by the initial elastic deformation ε_{ci} when subjected to sustained load or

$$C_c = \frac{\varepsilon_{cu}}{\varepsilon_{ci}} \tag{3.5}$$

Typical creep coefficients are given in Table 3.3 are for concrete loaded at 7 days and cured in average humidity.

As seen in Fig. 3.4, creep is nonlinear with time. Branson suggests that the creep coefficient C_{ct} at any time t can be related to the long-term creep by the relationship in Eq. (3.6).

$$C_{ct} = \frac{t^{0.6}}{10 + t^{0.6}} C_c \qquad (3.6)$$

where t is the time in hours after the load is applied.

3.3.6   Temperature Effects

The coefficient of thermal expansion and contraction varies with the aggregate and the mixture design. For temperature ranges of most structures, the coefficient of thermal expansion for concrete varies between 4×10^{-6} and 7×10^{-6} in./in./°F. The coefficient of thermal expansion for steel reinforcement is $6.0\text{–}6.5 \times 10^{-6}$ in./in./°F. The coefficients of thermal expansion are sufficiently close that calculation of differential thermal strains is not commonly required. A value for the coefficient of thermal expansion of $5.5\text{–}6.0 \times 10^{-6}$ in./in./°F for prestressed concrete structures is generally accepted for calculating stresses and deformations caused by temperature change (PCI, 2017).

3.4   Self-Consolidating Concrete

Self-consolidating concrete (**SCC**) contains more fine material than normal concrete. In general, SCC has a slightly lower modulus of elasticity, more creep, and sometimes more shrinkage than normal concrete. *SP-247 Self-Consolidating Concrete for Precast Prestressed Applications* (2007) provides information for variations in material properties. If properties are critical for a project, the project specifications should require that either specified properties are attained or the engineer be provided with the properties for the concrete used in the project prior to initiation of the work.

3.5   Prestressing Steel

Prestressing steel is produced in three different forms: strands, bars, and wires. Strand is the most common form of prestressing reinforcement and typically comes in seven-wire helically wound configurations. The number of wires in the cross section can be increased or decreased as needed for specialty applications such

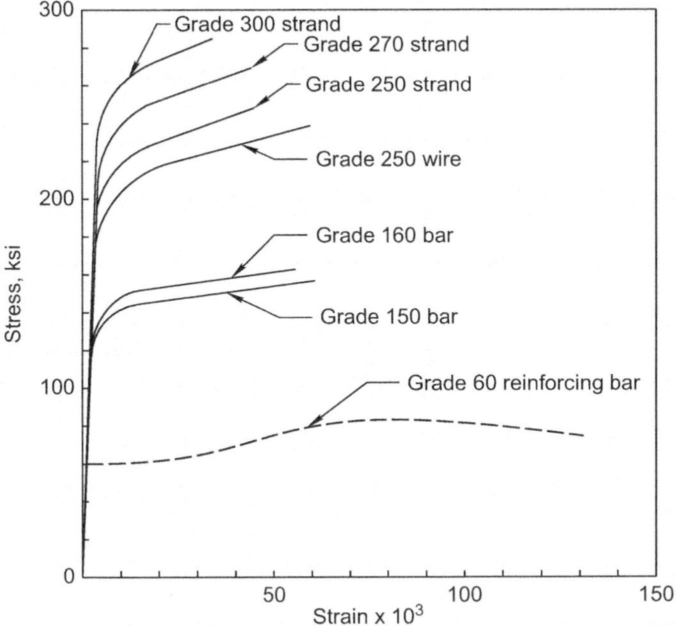

Fig. 3.5 Stress–strain curves for various types of reinforcement

as 21 wire and larger strands that are available for bridge cable stays. High-strength alloy bars are particularly useful for either temporary or permanent post-tensioning in segmental construction because the coupling devices allow segments to be segmentally stressed. High-strength alloy bars are available in smooth or deformed configurations. High-strength wires are used for specialty applications such as tanks and railroad ties.

Figure 3.5 provides the stress–strain relationship for several typical prestressing materials and compares the high-strength materials with deformed ASTM A615 Grade 60 reinforcement. The modulus of elasticity of prestressing reinforcement is not the same for all configurations. Grade 60 reinforcement wires and high-strength alloy bars have a modulus of elasticity of 29,000,000 psi as do other steel elements. Strand has a modulus of elasticity of approximately 28,500,000 psi. The slight reduction in modulus of elasticity is a result of the spiral winding of wires around the center core wire.

3.5.1 Strand and Wire

A seven-wire strand is made of a center wire that is helically wrapped by six smaller diameter wires. A common prestressing strand is a ½ in. nominal diameter seven-wire strand with a tensile strength of 270 ksi and an area of 0.153 in.^{2}. Smaller and larger diameter strand are available and strand strengths can be obtained between 250 and 300 ksi. Tables A.1 through A.5 summarize strand, bar, and wire sizes and strength and Table A.6 contains the properties of nonprestressed reinforcement. Strand comprised of individual wires has cross-sectional areas less than an equivalent solid area based on the nominal diameter.

Strand is available in either regular or low-relaxation treatment styles, discussed under **relaxation** in Sect. 3.5.3. Strand from each supplier has a slightly different stress–strain relationship due to individual treatment methods. The PCI Handbook provides a universal mathematical model for prestressing strand, Eq. (3.7). The stress–strain relationship is provided to assist in design calculations and is used later in this book. If, however, a design is predicated on precise stress–strain relationships, the supplier's data should supersede this universal model for a specific project.

$$f_{ps} = 28,500\,\varepsilon_{ps}\,(\text{ksi}) \quad \text{if} \quad \varepsilon_{ps} \leq 0.0086$$

$$f_{ps} = 270 - \frac{0.04}{\varepsilon_{ps} - 0.007}(\text{ksi}) \quad \text{if} \quad \varepsilon_{ps} > 0.0086 \tag{3.7}$$

Implicit in Eq. (3.7) is that the yield strain of the strand is equal to 0.086 and the modulus of elasticity is 28,500,000 psi.

The modulus of elasticity and the yield strain is specified by the 0.2% offset method, that is, a line parallel to the initial stress–strain deformation. This method retains the modulus of elasticity for 270 ksi strand and clarifies the definition of bars, such as ASTM 1035, that have no sharp yield point.

Individual prestressing wires are used for specialty products. Prestressed concrete railroad ties use wires to take advantage of the shorter development lengths and lower overall prestressing force. Prestressed circular tanks use closely spaced wire and automated wrapping machines to apply the prestressing force. Properties for prestressing wire are given in A-5. Manufacturers' data for the wire stress–strain relationship is preferable for design; however, Fig. 3.6 Grade 270 can be used in lieu of more precise information.

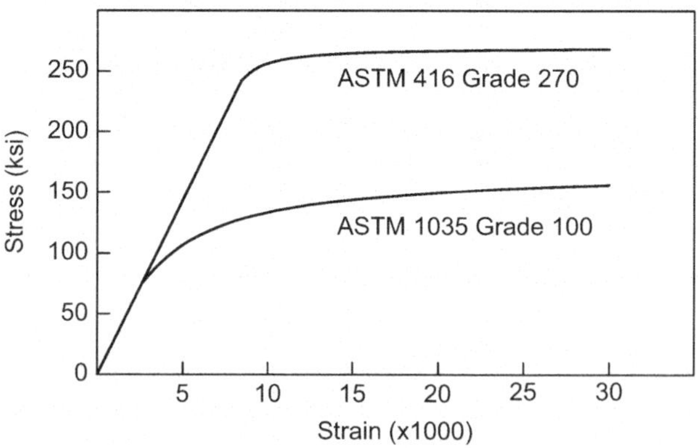

Fig. 3.6 Strand and bar stress–strain curves (based on Eqs. (3.7) and (3.8))

Fig. 3.7 Dywidag bar and
nut

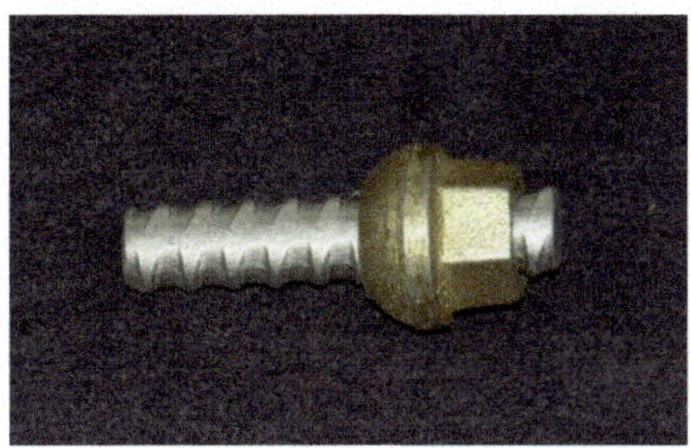

3.5.2 Bars

High-strength alloy bars are used for segmental construction due to their ability to
couple short elements efficiently. They can be either smooth round bars or deformed
bas such as the Freyssibar, Dywidag, or MMFX proprietary systems, Fig. 3.7.
Table A.4 provides data on a 150 ksi deformed bar system, however, exact dimensions of bars vary by manufacturer and should be verified for specific designs.
Equation (3.8) provides a generic stress–strain relationship for a Grade 100 deformed
bars having a tensile strength of 160 ksi (ACI ITG-6R, 2010). The bars do not have a
sharp yield point, and the extended transition zone between yield and tensile strength
is seen in Fig. 3.6.

$$f_{ps} = 29,000\,\varepsilon_{ps}(\text{ksi}) \quad \text{if} \quad \varepsilon_{ps} \leq 0.0024$$

$$f_{ps} = 170 - \frac{0.43}{\varepsilon_{ps} + 0.0019}(\text{ksi}) \quad \text{if} \quad \varepsilon_{ps} > 0.0024 \tag{3.8}$$

3.5.3 Relaxation

Relaxation is the loss of stress in a tendon stressed to a constant strain level.
Relaxation is a time-related phenomenon that results in some string instruments
going out of tune over time. Relaxation reduces the prestressing force and must be
accounted for in design.

Up to the 1970s strand producers processed wire and strand to relieve residual
stresses resulting from the wire drawing process and thus reduce relaxation effects.
The result is a **stress-relieved** strand and wire. Heating and retensioning the strand or
wire for a short period of time further reduced relaxation resulting in **low relaxation**
stress-relieved wire or strand, often referred to as **lo-lax**. ASTM A416, requires that
such steel exhibit relaxation of no more that 2.5% after 1000 h when initially stressed
to 70% of specified tensile strength and not more than 3.5% when loaded to 80% of
tensile strength. For use in prestress loss calculations, relaxation for low relaxation
strand is given in Eq. (3.9).

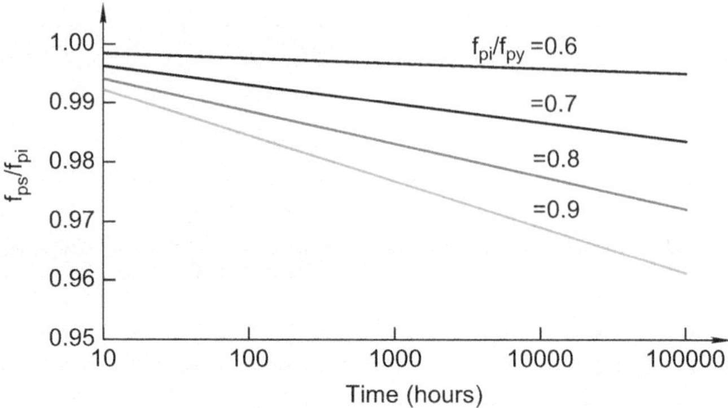

Fig. 3.8 Steel relaxation for low relaxation 270 ksi strand

Table 3.4 Values for C_1

f_{pi}/f_{pu}	C_1
0.70 to 0.75	$1 + 9\left(\frac{f_{pi}}{f_{pu}} - 0.7\right)$
>0.51 to <0.70	$\frac{\left(\frac{f_{pi}}{f_{pu}}\right)}{0.19}\left(\frac{\left(\frac{f_{pi}}{f_{pu}}\right)}{0.85} - 0.55\right)$
≤0.51	$\frac{\left(\frac{f_{pi}}{f_{pu}}\right)}{3.83}$

Adapted from PCI Design Handbook (2017)

$$\frac{f_p}{f_{pi}} = 1 - \frac{\log t}{45}\left(\frac{f_{pi}}{f_{pyi}} - 0.55\right) \tag{3.9}$$

The tests on which Eq. (3.9) is based were carried out on round, stress-relieved wires and are equally applicable to stress-relieved strand. In the absence of other information, results can be applied to alloy steel bars as well. Virtually all current strand production is low relaxation.

The yield stress of a 270 ksi strand occurs at a strain of 0.0086 corresponding to a stress of 245 ksi. Figure 3.8 shows the time-dependent losses for initial jacking stress of 60, 70, 80, and 90% of f_{py}.

The PCI Handbook (2017) addresses relaxation using tabulated coefficients. Unlike Eq. (3.9), the PCI approach recognizes that creep and shrinkage reduce the stress in the strand and this reduction results in a reduced relaxation. The format of the PCI Handbook is given in Eq. (3.10).

$$RE_1 = [K_{re} - J(SH_1 + CR_1 + ES_1)]C_1 \tag{3.10}$$

where RE_1 is the loss due to relaxation, and K_{re} and J are the base loss values given in Table 3.4. The effect of treatment for low relaxation properties is illustrated in the K_{re} and J values where the initial loss is reduced by a factor of four for low relaxation

Table 3.5 Values for K_{re} and J

Type of tendon	K_{re} (psi)	J
Grade 270 low relaxation strand or wire	5000	0.040
Grade 270 stress-relieved strand or wire	20,000	0.15
Grade 145 or 160 stress-relieved bar	6000	0.05

Selected values from PCI Design Handbook (2010)

strand and wire. SH, CR and ES are losses due to shrinkage, creep and elastic shortening and C_l is a correction for the initial stress level given in Table 3.5. For low relaxation strand initially stressed to Code limits, f_{pi}/f_{pu} is in the range of 0.70 to 0.75.

3.5.4 Specialty Prestressing Materials

In addition to the materials mentioned, prestressing reinforcement is provided for specialty applications. These materials include stainless steel, and carbon and glass fiber reinforced tendons. The properties of these materials differ from the predictive models provided and require the engineer to determine their suitability for specific projects. Some examples include the use of stainless steel prestressing strand for highly corrosive or magnetically sensitive applications.

3.6 Anchor Systems

Transfer of force from the prestressing reinforcement to the concrete is critical to effective prestressing. This is accomplished by a bond between the concrete and pretensioned strand or by mechanical anchorage to the concrete for post-tensioned applications. Pretensioned strand is mechanically anchored at the stressing bulkhead and the prestressing force is transferred to the concrete through bond when the strand is detensioned from the bulkhead. Tendons that are mechanically anchored to the concrete can be unbonded, that is they are anchored only at the mechanical anchor, or bonded, where grout is injected into the post-tensioning duct after the tendon is stressed. Typical anchorage systems are discussed below while bonded strand development is addressed in Chap. 6.

3.6.1 Strand Chucks

A **strand chuck** is a mechanical device to grip the strand while stressing and to hold the strand through contact with the bulkhead until the force is transferred to the concrete, Figs. 3.9 and 3.10. Strand chucks are for temporary use and not intended to

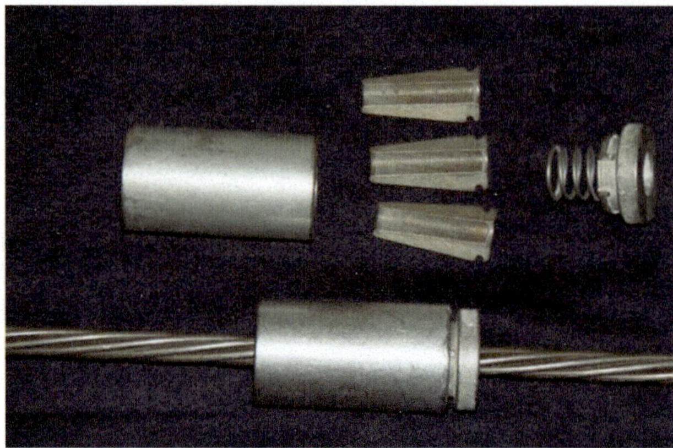

Fig. 3.9 Strand chuck and components

Fig. 3.10 Elongation marks for pretensioning quality control

be long-term anchors. The chuck consists of a barrel, a set of three serrated wedges, a cap and a spring. The spring pushes the wedges forward in the barrel, which keeps the wedges snug against the strand during threading and stressing. When the hydraulic jack is released, the wedges bite into the strand as they are pulled into the barrel. Between 1/8 and 3/8 in. of movement is required to engage or **seat** the wedges. Thus, the strand must either be elongated this additional amount or a portion

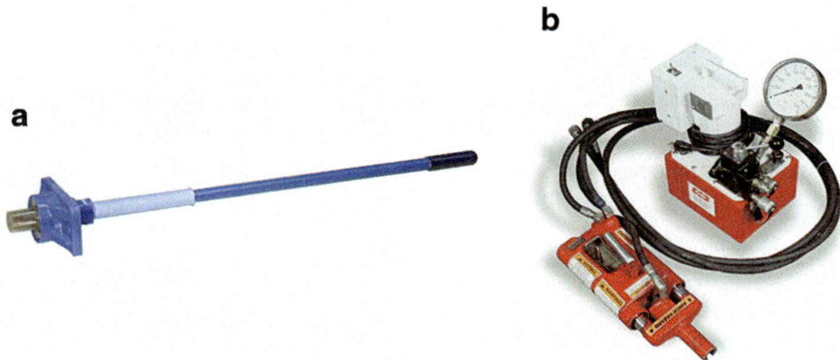

Fig. 3.11 Monostrand anchor and jack. (**a**) Monostrand anchor and wedges (courtesy of VSL). (**b**) Monostrand jack (courtesy of DYWIDAG-Systems International USA Inc.)

of the prestress force is lost during the seating. Strand chucks are typically used in pretensioned construction and are intended to be reuseable.

Individually stressed strands have elongation marks painted on the strand. These marks used in the quality control operation to verify stressing elongation and are visible in Fig. 3.10.

3.6.2 Monostrand Anchors

Most flat-slab building construction use **monostrand,** single-strand unbonded tendons. Monostrand tendons consist of an anchor assembly and strand, which is encapsulated in an extruded plastic sheath filled with a corrosion protectant filler. Encapsulation provides protection against intrusion of moisture, provides corrosion protection for the strand, and reduces friction during stressing. The filler can be seen between the anchor and the strand in Fig. 3.11. The monostrand anchor consists of a ductile cast iron anchor plate and a set of two wedges, Fig. 3.11a. The ductile cast iron has a yield strength of 85 ksi and the wedges are hardened steel. While similar to a strand chuck in appearance, the wedges are specifically designed with deeper teeth to extend through the encapsulation material. The anchor plate is designed to limit bearing stresses during jacking and at initial transfer. Monostrand systems are stressed using specialty jacks as seen in Fig. 3.11b. The jack is mated to a pump, which in turn has a pressure gage to allow monitoring of the jacking force. A full discussion of monostrand post-tensioning systems can be found in Kelley (2003).

3.6.3 Multistrand Anchors

Where large post-tensioning forces are necessary, such as post-tensioned beams and bridge girders, multiple strand tendons are used. Figure 3.12 is a multistrand anchor where all strands are stressed simultaneously using a single hydraulic jack. Once the

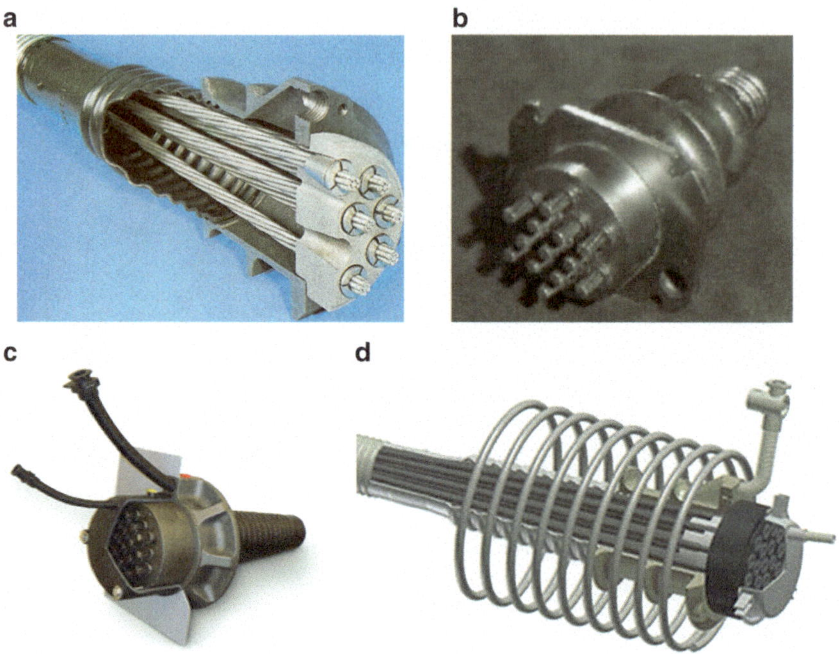

Fig. 3.12 Multistrand anchor systems. (**a**) DYWIDAG-Systems International USA Inc.; (**b**) Freyssinet Inc.; (**c**) Schwager Davis, Inc.; (**d**) VSL

Fig. 3.13 Hydraulic jacks for post-tensioning multistrand tendons (courtesy of Freyssinet Inc. (left) and Schwager Davis, Inc. (right)

strands have been stressed, the tails are cut and the anchor is covered with a grout cap. A post-tensioning filler is then injected into the tendon to provide corrosion protection. If a cementitious grout is used, then the grout also ensures a bonded tendon. If a flexible filler is used, then the tendon is treated as unbonded.

Figure 3.13a illustrates a multistrand stressing operation where all strands are stressed simultaneously. Figure 3.13b is a cut-away section of a grouted multistrand

the anchor with a center plug to anchor six strands. Tendons up to 12 strands are commonly stressed. The stressing force exceeds 360 kips on a 12-strand tendon, implying that protection procedures are essential for safe operation.

3.6.4 Bar Anchors

Prestressing bars systems are typically equipped with threads and nuts to facilitate anchorage, Fig. 3.14. The jack threads onto the bar and reacts against the anchor plate during stressing. Once the target stress level has been reached, the nut is tightened with a wrench system on the jack. This minimizes the seating loss and allows for the use of this system in short tendons (Fig. 3.15).

Fig. 3.14 Multistrand anchor system

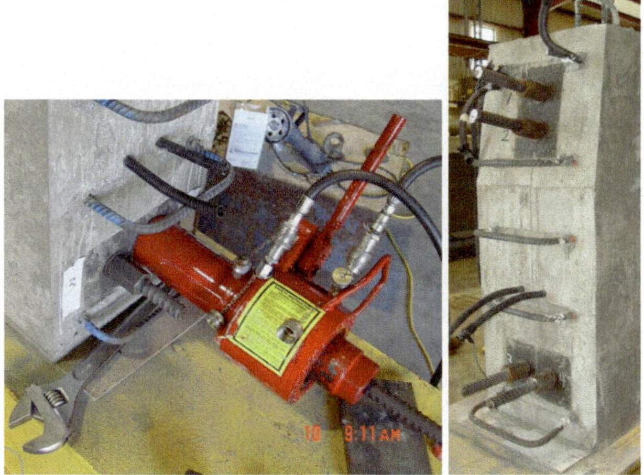

Fig. 3.15 Prestressing bar hydraulic jack. Prestressing bar anchorages after stressing

3.7 Tendon Corrosion Protection

3.7.1 Monostrand Systems

Prestressing tendons are highly stressed and subject to rapid corrosion, especially when in proximity to chloride ions. Early monostrand tendons were coated with grease and wrapped in waterproof paper, so called cigarette wrapped tendons. The grease facilitated reducing friction during stressing and was thought to inhibit corrosion. Years of experience demonstrated that these early systems allowed the grease to escape and moisture to enter the tendon. Corrosion led to loss of prestress (ACI 423.4, 2014). Recognizing these short comings, the industry made strides to improve the performance of monostrand systems and today virtually all monostrand tendons use encapsulation. The details on the jacking end of the tendon permit the strand to pull into the encapsulation coating. The strand extension is cut or burned off leaving a short extension of strand should the strand have to be regripped. A cap filled with corrosion protectant is then placed over the end of the anchor. Proper quality control verifies that the cap is applied immediately after the strand is cut but after a burned strand has cooled sufficiently to prevent the cap from heat distortion or melting. ACI 423.7 (2014) provides detailed specifications for monostrand systems.

3.7.2 Multistrand Systems

Multistrand post-tensioning tendons are installed in ducts cast into the concrete. Upon completion of the post-tensioning, the ducts are grouted to protect the tendon. Early attempts at grouting tendons with a cement paste were not completely successful (Schupack, 1994). Grout segregation and bleed water in the ducts led to locations that were incompletely grouted, especially at tendon high spots (Abu-Yosef et al., 2016). Extensive research into grouting materials resulted in the development of grouts specifically designed for post-tensioning that are less prone to segregation and bleeding (Hamilton et al., 2000; Schokker et al., 2002). ACI 423.4 (2014) addresses corrosion of prestressing tendons and ACI 222.R (2001) addresses protection of metals in concrete.

The inability to replace a grouted tendon continues to be a concern. In the 1990s government agencies considered banning grouted tendons in bridges to be able to replace a corroded tendon (Schupack, 1993). Transportation authorities in the UK experimented with nonmetallic unbonded tendons to address corrosion concerns (Burgoyne, 1993). While unbonded tendons address replacement, they do not address the underlying corrosion issues or the strength gain available with bonded tendons. Wax-based grouting systems are an alternative approach that provides corrosion protection for unbonded tendons (Abdullah et al., 2014).

References

Abdullah, A. B. M., Rice, J. A., & Hamilton, H. R. (2014). Wire breakage detection using relative strain variation in unbonded posttensioning anchors. *Journal of Bridge Engineering, 20*(1), 04014056.

Abu-Yosef, A., Ahern, M., & Poston, R. W. (2016). Concealed threat. *Civil Engineering,* pp. 54–61.

ACI 209. (2008). *Prediction of Creep, Shrinkage, and Temperature Effects in Concrete Structures* (ACI Committee 209R-92, Reapproved 2008, p. 47). Farmington Hills, MI: American Concrete Institute.

ACI 222.R. (2001). *Protection of Metals in Concrete Against Corrosion* (Reapproved 2010, p. 41). Farmington Hills, MI: American Concrete Institute.

ACI 318. (2014, *Building Code Requirements for Structural Concrete* (ACI committee 318-14, p. 519). Farmington Hills, MI: American Concrete Institute.

ACI 423.4. (2014). *Report on Corrosion and Repair of Unbonded Single-Strand Tendons* (p. 28). Farmington Hills, MI: American Concrete Institute.

ACI 423.7-14. (2014). *Specification for Unbonded Single-Strand Tendon Materials* (p. 9). Farmington Hills, MI: American Concrete Institute.

ACI ITG-6R-10. (2010). *Design Guide for the Use of ASTM A1035/1035M Grade 100 (690) Steel Bars for Structural Concrete* (p. 90). Farmington Hills, MI: American Concrete Institute.

ASTM C496/496M. (2017). *Standard Test Method for Splitting Tensile Strength of Cylindrical Concrete Specimens*. West Conshohocken, PA: ASTM.

ASTM C78/C78M. (2015). *Standard Test Method for Flexural Strength of Concrete (Using Simple Beam with Third-Point Loading)*. West Conshohocken, PA: ASTM.

ACI SP-227: Shrinkage and Creep of Concrete. (2005). In N. J. Gardner & J. Weiss, (Eds.), Farmington Hills, MI: American Concrete Institute, p. 319.

ACI 423.10-16. (2016). *Guide to Estimating Prestress Loss* (p. 64). Farmington Hills, MI: American Concrete Institute.

ACI 423.7-14. (2014). *Specification for Unbonded Single-Strand Tendon Materials* (p. 8). Farmington Hills, MI: American Concrete Institute.

Bae, S., & Bayrak, O. (2013). *Examination of Stress Block Parameters for High-Strength Concrete in the Context of the ACI 318 Code* (ACI SP 293-5, pp. 59–77). Farmington Hills, MI: American Concrete Institute.

Branson, D. E. (1977). *Deformation of concrete structures*. New York: McGraw-Hill.

Burgoyne, C. J. (1993). Parafil ropes for prestressing applications. A. Nanni & C. W. Dolan, (Eds.) *Fibre-Reinforced-Plastic (FRP) for Concrete Structures: FRP Reinforcement for Concrete Structures, International Symposium* (ACI SP-138, pp. 333–351). Detroit: ACI.

Darwin, D., Dolan, C. W., & Nilson, A. H. (2015). *Design of concrete structures* (786pp). New York, NY: McGraw Hill Education.

Dolan, C. W., Ballinger, C. A., & LaFraugh, R. W. (1993). High strength prestressed concrete bridge girder performance. *PCI Journal, 38*(3), 88–97.

Hamilton, H. R., Wheat, H., Breen, J., & Frank, K. (2000). Corrosion testing of grout for posttensioning ducts and stay cables. *Journal of Structural Engineering, 126*, 163–170.

Kelley, G. S. (2003). A guide to the components of an unbonded post-tensioning system. *Concrete International, 25*(1), 71–77.

Mindess, S., Young, F., & Darwin, D. (2003). *Concrete* (2nd ed.). New York, NY: Prentice Hall.

Neville, A. (2012). *Properties of concrete* (5th ed.p. 799). Essex, UK: Person Education Limited.

PCI Design Handbook (8th Ed., MNL 120-10). (2017). Chicago, IL: Precast/Prestressed Concrete Institute.

PCI Design Handbook (8th Ed.). (2017). Chicago, IL: Precast/Prestressed Concrete Institute.

Schokker, A. J., Hamilton, H. R., & Schupack, M. (2002). Estimating post-tensioning grout bleed resistance using a pressure-filter test. *PCI Journal, 47*(2), 32–39.

Schupack, M., 1994, "Durability study of 35 year old post-tensioned bridge", Concrete International, 16(2), pg. 54–58.

Schupack, M. (1993). Bonded tendon debate. *Civil Engineering, 63*(8), 64.

SP-247 Self-Consolidating Concrete for Precast Prestressed Applications. (2007). A. K. Schindler, D. Trejo, & R. W. Barnes, (Eds.), *Developed by: ACI Committee 237, Self-Consolidating Concrete, and Joint ACI-ASCE Committee 423, Prestressed Concrete*. Farmington Hills, MI: American Concrete Institute.

Chapter 4
Partial Loss of Prestress

4.1 Introduction

Prestressing a concrete member effectively applies a substantial axial force to the member that is in place for its entire service life. In both pretensioned and post-tensioned methods of prestressing, this prestressing force begins to decline immediately upon its application and continues to decline throughout its service life. This reduction in force is referred to as **partial prestress loss** and is addressed as part of the design of a prestressed member. Partial prestress losses, typically referred to as "prestress losses," are divided into two broad categories: initial and time-dependent effects. Initial losses occur during stressing operation and include anchor seating, elastic shortening, and friction between prestressing steel and post-tensioning ducts or tendon deviators and harped pretensioned strands. Long-term losses occur because of viscoelastic material effects and include concrete shrinkage, creep, and tendon relaxation.

This chapter covers the causes of prestress losses as well as techniques for **estimating** those losses. Numerous methods are available for estimating losses. This is likely due, at least in part, to the difficulty in accurately **predicting losses**. ACI 423-10 (2016) devotes an entire chapter to the variability of loss calculations and the reader is referred there for more detail. In summary, the variability in concrete mechanical properties, curing conditions, and exposure to environment are the primary causes of this difficulty. While there are many different approaches for determining initial and time-dependent effects on prestressed concrete, the most effective method is to construct the element and measure the losses in the field. Corrections are done in the field for friction losses on occasion; other corrections are not common except for research. In many cases, a high level of accuracy is not needed to ensure suitable strength and serviceability. Consequently, the design engineer estimates the losses and bases the design on this estimate.

© Springer Nature Switzerland AG 2019

C. W. Dolan, H. R. Hamilton, *Prestressed Concrete*,
https://doi.org/10.1007/978-3-319-97882-6_4

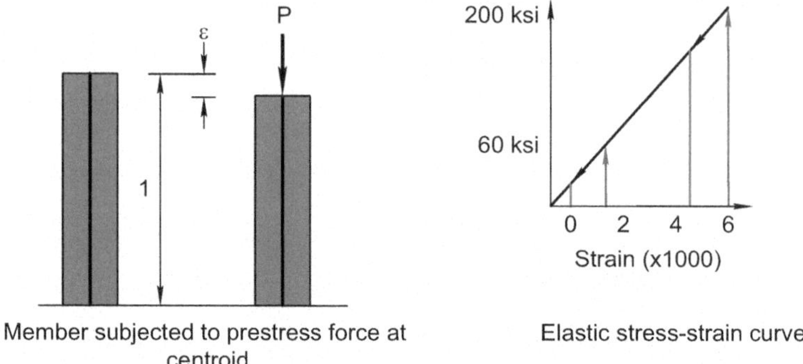

Fig. 4.1 Effect of steel strength on prestress losses. (**a**) Member subjected to prestress force at centroid. (**b**) Elastic stress–strain curve

4.2   Effect of Losses

Early attempts at prestressing were unsuccessful because losses were not fully understood. Eugene Freyssinet, generally recognized as the founder of modern prestressed concrete, was successful because he recognized the value of high-strength prestressing materials and successfully incorporated high-strength materials into his designs. Figure 4.1 illustrates such a need for high-strength reinforcement in prestressed concrete. Figure 4.1a indicates a member one unit long subjected to a prestressing force from a central internal tendon. The axial force produces an initial strain in the concrete and that initial strain increases due to shrinkage and creep ε. For discussion, assume that the total strain due to losses is 0.0015. A Grade 60 reinforcing bar is stressed to 0.002 strain, just below the yield stress of 60 ksi. Figure 4.1b indicates the initial strain and the corresponding loss of stress due to the 0.0015 strain, in this case 0.0015 $E_s = 45$ ksi. The final stress is 15 ksi or a loss of 75% of the initial prestress. Next a high-strength strand is loaded to an initial stress of 200 ksi or a strain of 0.0066. The same loss of 0.0015 is applied. The final stress is approximately 155 ksi corresponding to a loss of 22.5%.

Prestress losses affect the serviceability of prestressed members but have little effect on the strength of a member, unless the tendon is unbonded. The losses directly affect member deflection, camber, cracking, and amount of prestressing reinforcement. Thus, loss calculations are necessary for prudent design. The deformations required to develop the bending strength in beams with bonded tendons generally include sufficient strain to recover losses. The recovery is validated during detailed calculation of member strength or by assumed stress levels if empirical nominal tendon stress equations are used.

4.3   Addressing Losses in Design

All prestressed members are subject to losses resulting from elastic shortening, shrinkage, creep, and relaxation. In addition, post-tensioned members are subject to losses resulting from anchor set and friction (Zia et al. 1979; ACI 423-10 2016). The ACI Building Code (ACI 318-14) requires prestress losses to be considered in the calculation of effective tensile stress in the prestressed reinforcement, f_{se}. The following loss mechanisms have historically been listed by the code for consideration:

(a) Prestressed reinforcement seating at transfer (initial)
(b) Elastic shortening of concrete (initial)
(c) Creep of concrete (long-term)
(d) Shrinkage of concrete (long-term)
(e) Relaxation of prestressed reinforcement (long-term)
(f) Friction loss due to intended or unintended curvature in post-tensioning tendons (initial)

Losses affect the serviceability of the member rather than strength, which is primarily why they are not addressed prescriptively in the ACI Building Code. Although discussed in the commentary, the code provisions do not directly address the effect of losses on serviceability issues such as deflections, camber, and cracking load. The code, however, addresses the effect of shortening on connections.

Losses are calculated at the tendon centroid and include consideration of the stress and strain in the tendon. In the example presented in Fig. 4.1, the tendon is concentric with the member, so no adjustment for tendon position is needed. For beams and other eccentrically loaded members, changes in strain and stress are calculated at the centroid of the tendon.

Depending on the method used, losses can be calculated at a selected section in more simplified methods or along the entire length in more refined methods. For precast pretensioned beams, the losses are usually calculated at the critical sections, which are typically at midspan and the end of the member. For post-tensioned members, the critical sections are at the member end, maximum positive moment locations, typically close to midspan, and maximum negative moment locations, usually over the supports. Applied concentrated loads or abrupt changes in curvature are locations requiring attention and are considered critical locations. In more refined methods, the losses can be estimated along the full length of the tendon.

In all cases, the loss calculations occur at service load levels and the member is assumed to behave linearly elastic. Therefore, reinforcement stress changes may be calculated as the stress in the concrete times the **modular ratio n**, the ratio of modulus of elasticity of the prestressing reinforcement divided by the modulus of elasticity of the concrete. Losses from various sources are cumulative.

The level of effort required to calculate losses varies depending on the experience with a specific design, product, or construction method. For instance, a precast plant that has been fabricating double tees for many years will have a backlog of data on the long-term effects for their products. Likewise, a specialty engineer for a post-tensioning company may use prestress loss data gathered over many years of designing similar post-tensioned slab systems. The most important aspect of loss calculations is to make a best estimate. If the calculated losses are too high, that is, higher than actual losses, then more prestressing reinforcement is used and camber increases. If the calculated losses are too low, the beam sags and cracks under service load. Loss calculations should neither be overly conservative nor ignored.

Methods available for this task vary in complexity and accuracy. In general, the methods are categorized as follows:

1. Lump sum
2. Detailed
3. Time-dependent

For engineers wishing to move forward with their coverage of prestressed concrete design, lump sum losses provide a valid starting point for design and are used in the examples in this book. Use of lump sum losses is presented with the understanding that individual losses should be estimated to complete a design, which can be completed using the remainder of this chapter. For a much deeper background and a detailed coverage of the methods available for calculating losses, the reader is directed to ACI 423.10 (2017). ACI 423.10 includes material such as detailed treatment of creep by alternative approaches, losses affected by composite action, and effects of shrinkage of composite decks.

4.4 Lump Sum Losses

Often detailed calculations can only be performed after the member section has been selected and prestressing levels established. For decades engineers have used **lump sum losses**, the total combined losses in the prestress based on experience or historical data, to select the initial prestress. Later during the design development, the detailed loss calculations are conducted and loss assumptions verified.

Beginning in 1963, the ACI Building Code commentary recognized lump sum losses as an approach to calculating the total prestress loss in a member. The commentary suggested the following values based on an ACI/ASCE Committee 323 report (1958).

$$\text{Pretensioning losses} = 35 \text{ ksi}$$
$$\text{Post-tensioning losses} = 25 \text{ ksi}$$

In keeping with design practice, the lump sum losses noted here are used in most of the examples in this book. Lump sum losses should be used with the understanding that a detailed check of losses is eventually required to complete a design.

Assuming a pretensioned tendon is initially stressed to 200 ksi, a 35 ksi loss represents approximately a 17.5% reduction in prestress. ACI/ASCE Committee report 423.10 (2016) includes measured prestress losses for various members. The report states that losses range between 9 and 39% with a mean value of 19%, or 37.5 ksi, and a coefficient of variation of 32% for normal weight concrete members. The PCI Design Handbook (2017) suggests losses range between 25 and 50 ksi, or 12 and 25%. Both sources are consistent with the experimental data. For sand lightweight concrete, the PCI Handbook suggests losses range between 30 and 55 ksi, or 17 and 27%. The PCI Handbook also suggests that it is sufficiently precise to assume that elastic shortening is 40% of the total and that creep and shrinkage is 50% of the total loss. The AASHTO Bridge Design Specification (2017) specifies slightly higher lump sum losses. The higher losses reflect, in part, the fact that bridge members are often more heavily prestressed than building members.

4.5 Detailed Losses

Understanding how initial prestressing stresses are determined is necessary to calculate prestressing losses. Table 20.3.2.5.1 of ACI 318-14 limits the jacking stress to $0.80 f_{pu}$. Previous versions of the Code further limited the tendon stress to $0.75 f_{pu}$ at the time of transfer. The higher jacking stress allowed prestressing operations to pull the tendon to a slightly higher stress to compensate for anchor seating. The ambiguity in the code resulting from only specifying jacking stress leaves the issue of knowing the stress in the tendon at the time of load transfer to either the engineer or the prestressing contractor.

Post-tensioning tendons are stressed to $0.80 f_{pu}$ to account for seating and friction losses. The ACI Building Code restricts the stress in the post-tensioning tendon to $0.70 f_{pu}$ at the anchor immediately after seating. Pretensioning operations typically are conducted in long-line facilities. The ACI Building Code does not provide a limiting initial tensile stress for pretensioning. Plant operation typically stresses the tendon to about $0.76 f_{pu}$. After the strand chuck seats, the remaining tendon stress is $0.75 f_{pu}$ and no allowance for anchorage seating is calculated for pretensioned members. Loss calculations that follow are based on the industry practice of an initial strand stress of $0.75 f_{pu}$ for pretensioning and $0.70 f_{pu}$ after seating for post-tensioning.

The total amount of loss of prestress is the sum of anchor set f_{anc}, friction f_{fr}, elastic shortening f_{ES}, creep f_{CR}, shrinkage f_{SH}, and relaxation effects f_{RE}. Symbolically, the cumulative loss f_{loss} is:

$$\sum f_{loss} = f_{anc} + f_{fr} + f_{ES} + f_{CR} + f_{SH} + f_{RE} \qquad (4.1)$$

Each of these effects is discussed individually. Simplified methods for calculation of long-term losses follow the detailed discussion.

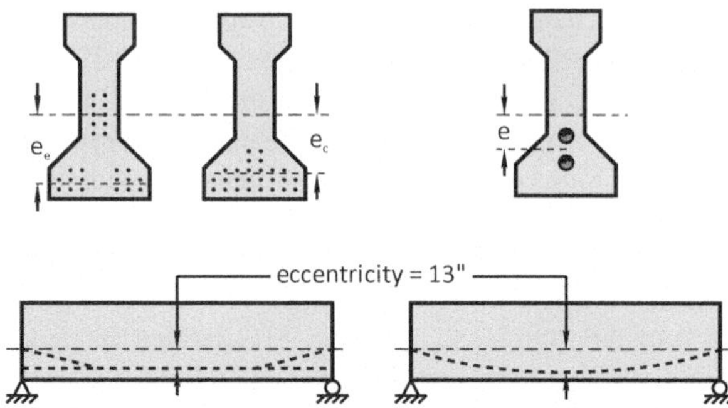

Fig. 4.2 Beam geometry for Example 4.1

4.5.1 Anchor Set

When a prestressing tendon is released from the jack, there is a small amount of movement, or **set**, as the anchor wedges move inward and the teeth on the wedges bite into the steel. The amount of movement ranges between 1/8 in. and 1 in. depending on the anchorage system. Anchor set of ¼ to 3/8 in. is common for single stand systems and the larger values are for some center plug multistrand systems.

In pretensioning operations, anchor set is compensated by overstressing the tendon to account for the seating. In post-tensioning operations, the loss of prestress is calculated by dividing the anchor set by the length of the tendon to arrive at an average strain then multiplying that strain by the strand modulus of elasticity to determine the stress. If there is significant curvature in the tendon, friction affects the distribution of losses.

Example 4.1: Calculate Anchor Seating Losses

In the examples that follow, the loss of prestress is calculated for a 60 ft long AASHTO Type III beam, Fig. 4.2. The beam has a cross-sectional area of 560 in.^{2}, a moment of inertia of 125,390 in.^{4}, a weight of 583 plf, and a surface volume-to-surface ratio of 4.09. Section properties can be found in Appendix C.2. One beam is pretensioned with 16½ in. diameter strands which are deflected to an eccentricity of 13 in. at midspan. A second beam is post-tensioned with two 8-strand tendons placed in a parabolic drape creating an eccentricity of 13 in. at midspan, shown in the figure below. For this, and the examples that follow, the specified transfer strength of the concrete is f'_{ci} of 3500 psi and the 28-day strength is 6000 psi. The beams are cured in a relative humidity of 35%. A ½ in. strand has an area of 0.153 in.^{2}, a tensile strength of 270 ksi, and a modulus of elasticity of $E_p = 28,500$ ksi. The post-tensioned beam uses a corrugated steel duct with a wobble friction of 0.0001 and a curvature friction of 0.20.

Solution pretensioned beam: by stressing the strands to compensate for the anchor set there is no calculated loss. The final stress in the tendon following seating is 0.75 f_{pu} or $f_i = 200$ ksi. The area of a ½ in. strand is 0.153 in.^{2}. For 16 strands the initial prestress force is

$$P_j = 16 \cdot 0.153 \ \text{in.}^2 \cdot 200 \ \text{ksi} = 489.6 \ \text{kips}$$

Solution post-tensioned beam: The post-tensioned tendon is initially stressed to $0.80 \, f_{pu}$ or 216 ksi. The distance to seat the anchor is taken as 3/8 in. The strain resulting from seating is the anchor set divided by the beam length or 0.375 in./$(60 \times 12) = 5.21 \times 10^{-4}$. The corresponding stress loss is the strain times the modulus of elasticity, 28,500 ksi of the strand or

$$f_{anc} = 5.21 \cdot 10^{-4} \cdot 28,500 = 14.8 \ \text{ksi}$$

Thus, the initial stress for the post-tensioned tendon is

$$f_i = 216 - 14.8 \ \text{ksi} = 201.2 \ \text{ksi}$$

corresponding to a force of 242.6 kips per tendon. ACI 318-14 Table 20.3.2.5.1 limits the stress in the tendon at the anchorage device to $0.70 \, f_{pu}$ immediately after transfer. Therefore, the maximum initial prestress force allowed by Code for the post-tensioned beam is $f_i = 270 \cdot 0.70 = 189$ ksi or an initial prestress force of

$$P_i = 16 \cdot 0.153 \ \text{in.}^2 \cdot 189 = 462.7 \ \text{kips}$$

which controls in this case. The jacking stress would be $f_j = 189$ ksi + 14.8 ksi = 203.8 ksi and the initial jacking force of 249.5 kips/tendon is used.

4.5.2   *Losses due to Friction*

A post-tensioned tendon is anchored, or fixed, at one end and jacked from the other end. As the tendon is drawn through the duct the force along the tendon is reduced by friction due to the roughness and unintentional misalignment of the duct and by contact along intentionally curved sections of the duct. Thus, the total loss due to friction is the sum of **wobble friction** due to misalignment and **curvature friction** due to the intentional curvature resulting from the alignment the duct in the member. Wobble friction is a function of the duct material. Wobble friction values k are assumed to be constant along the duct and present even in nominally straight ducts. Total frictional losses due to wobble equal kl. **Curvature friction** is a function of the angle of curvature α intentionally designed into the duct placement and the coefficient of friction μ between the tendon and the duct with a total effect equal to $\mu\alpha$. The effects of wobble and curvature friction are cumulative. Typical friction coefficients are given in Table 4.1.

Calculation of friction losses in post-tensioned members can be complex depending on whether jacking from one end is sufficient or, if the friction is high, secondary jacking is required from the originally fixed end. Full development of

Table 4.1 Friction coefficients

Type of prestressing steel	Corrugated metal duct		Corrugated plastic duct		Smooth steel pipe		Smooth plastic pipe		No duct plastic sheathing	
	μ	k, ft^{-1}	M	k, ft^{-1}	μ	k, ft^{-1}	μ	k, ft^{-1}	μ	k, ft^{-1}
Strand	0.15–0.25	0.00005–0.0003	0.10–0.14	0.00005–0.0003	0.25–0.30	0	0.10–0.14	0		
Strand in precast elements and constant curvature tendons	0.15–0.25	0.00005–0.0003	0.10–0.14	0.00005–0.0003						
External tendons bare dry strand					0.25–0.30	0	0.12–0.15	0		
Lubricated strand	0.12–0.18	0.00005–0.0003			0.20–0.25	0				
Strand coated and extruded[a]	0.01–0.05	0.00005–0.0003	0.01–0.05	0.00005–0.0003	0.01–0.05	0	0.01–0.05	0	0.01–0.07	0.00005–0.0003
Bars, deformed, smooth, and round	0.30	0–0.0002	0.30	0–0.0002						

Adapted from ACI 423.10 (2016)

[a]PT coating in accordance with the performance specification (PTI M10.3-00)

equations for frictional losses is given in Sect. 4.7.2. Derivation of properties for calculating angular change is given in Sect. 4.7.3. Effects of anchorage with higher tendon curvature, members with multiple tendon curvatures, and effects from jacking from both ends are given in Example 4.8. For a simple member with a combined wobble and curvature friction $kl + \mu\alpha$ less than 0.3, the loss of prestress f_f due to friction may be calculated as shown in Eq. (4.2).

$$f_{\mathrm{fr}} = f_{\mathrm{j}}(kl + \mu\alpha) \tag{4.2}$$

The angular change in the tendon is found from Eq. (4.3) where y is the eccentricity or drape of the tendon and l is the tendon length.

$$\alpha = \frac{8y}{l} \tag{4.3}$$

Example 4.2: Calculate Friction Losses
Calculate friction losses using the data from Example 4.1.
Solution pretensioned beam: there is no friction loss to be calculated. Any friction losses due to harping the tendon are corrected in the prestressing plant.
Solution post-tensioned beam: The wobble friction is given as 0.001 and the curvature friction as 0.20, consistent with the values given in Table 4.1. The angle change is

$$\alpha = \frac{8 \cdot 13}{60 \cdot 12} = 0.144 \ \mathrm{rad}$$

and the frictional loss is

$$f_{\mathrm{fr}} = 203.8 \ \mathrm{ksi}(0.001 \cdot 60 + 0.144 \cdot 0.20) = 18.1 \ \mathrm{ksi}$$

Thus, the loss at midspan is 18.1 ksi/2 = 9.1 ksi.

Comment: The combined value of wobble and curvature friction is 0.089, less than the 0.30 limit, thereby validating the use of Eq. (4.2).
Example 4.2 calculates losses assuming the frictional effect is small and extends over the entire length of the member. Section 4.7 develops the equations for frictional effects and presents an example where the frictional losses along a member are higher and jacking is required from both ends of the beam.

4.5.3 Elastic Shortening

Transfer of the prestress force from the tendon to the concrete results in elastic shortening of the member. This shortening reduces the strain in the tendon. Treatment of elastic shortening varies whether the member is pretensioned or post-tensioned.

Pretensioned Members

The entire prestressing force is assumed transferred in a single operation in pretensioned members. The force in the tendons transfers when the stressing bed is detensioned or strands are individually cut or burned. The corresponding stress loss is calculated **at the level of the tendon** as given in Eq. (4.4). The elastic shortening loss f_{ES} is based on the stress in the concrete times the ratio for the modulus of elasticity of the strand and the concrete.

$$f_{ES} = \frac{E_{ps}}{E_{ci}} \left(-\frac{P_i}{A_g} - \frac{P_i \cdot e_p \cdot e_p}{I_g} + \frac{M_g \cdot e_p}{I_g} \right) \tag{4.4}$$

where, E_{ps} is the modulus of elasticity of the tendon, psi; E_{ci} is the modulus of elasticity of the concrete at the time of transfer, psi; P_i is the initial prestress force, lbs; A_g is the gross area of the section, in.^{2}; e_p is the eccentricity of the tendon at the critical section, in.; I_g is the gross moment of inertia of the section, in.^{4}; and M_g is the dead load moment due to girder weight, lb-in.; and the resulting negative value means the prestressing stress is reduced.

The negative signs in Eq. (4.4) indicate a reduction in tendon stress. The prestress eccentricity generates a positive camber; hence, the dead load moment of the girder tends to deflect downward to elongate the tendon and its effect is entered as a positive value.

Equation (4.4) uses the gross section properties, as is common in practice. Some members have ducts that accommodate larger post-tensioning tendons. In situations where there are large longitudinal voids or large amounts of reinforcement present in the member, use of net or transformed section is warranted.

The modulus of elasticity of the concrete is given in Eq. (4.5). This format of the equation comes from the ACI Building Code and is one of many formulations for the modulus of elasticity. A study conducted by members of the ACI Building Code committee concluded that the scatter in the modulus of elasticity did not justify use of an alternative or "more precise" formulation. Thus, Eq. (4.4) is valid for general use. Should a project require more precise determination of the modulus of elasticity, the calculations should be based on project mixture design test data.

$$E_{ci} = 33w_c^{1.5} \sqrt{f'_{ci}} \tag{4.5}$$

where, w_c is the unit weight of the concrete in pcf often taken as 160 pcf for precast concrete and 150 pcf for cast-in-place concrete, and f'_{ci} is the strength of the concrete at the time of transfer in psi.

Post-tensioned Members

Elastic shortening in post-tensioned members is a function of the post-tensioning sequence. Tendons may be stressed to $0.80\ f_{pu}$ but the stress in the tendon immediately after seating must be less than or equal to $0.70\,f_{pu}$. This allows the tendon to be stressed to account for seating losses and possibly elastic shortening. If a single tendon is stressed to include elastic shortening, then the tendon undergoes no axial loss from the initial member shortening. This condition occurs when stressing beams with a single multistrand tendon and all strands are stressed simultaneously. It also occurs for slabs with monostrand tendons spaced such that the stressing overlap is minimized.

Where multiple tendons are stressed individually, each successive tendon stressing affects the previously stressed strand or tendon. The elastic shortening in such cases is calculated based on one-half of the total initial prestress force. This follows because last strand stressed has no effective elastic shortening and the first strand stressed undergoes 100% of the shortening of all subsequent strands. Calculation of elastic shortening then follows the procedure indicated in Eq. (4.4).

Example 4.3: Calculate Elastic Shortening Losses, f_{ES}
The modulus of elasticity of the concrete at the load transfer is

$$E_{ci} = 33 \cdot 160^{1.5}\sqrt{3500} = 3,951,000 \text{ psi}$$

giving a modular ratio of $n = 28{,}500/3951 = 7.2$ and the girder moment is

$$M_g = \frac{wl^2}{8} = \frac{0583 \cdot 60^2}{8} = 262.3 \text{ kip-ft}$$

Solution pretensioned beam: The concrete stress at midspan at the level of the prestress tendon is

$$f_c = \frac{489.6}{560} + \frac{489.6 \cdot 13 \cdot 13}{125,390} - \frac{262.3 \cdot 12 \cdot 13}{125,390} = 1210 \text{ psi}$$

The corresponding elastic shortening loss is the stress in the concrete times the modular ratio

$$f_{ES} = -7.2 \cdot 1210 = 8.7 \text{ ksi}$$

Solution post-tensioned beam: The elastic shortening of the first tendon stressed is the shortening due to stressing the second tendon. Thus, the change in concrete stress at the level of the first tendon stressed due to a jacking force of 462.7 kips/2 is

$$f_c = -\frac{231.3560 - 231.3 \cdot 13 \cdot 13}{125,390} + \frac{262.3 \cdot 12 \cdot 13}{125,390} = -400 \text{ psi}$$

and the loss of prestress is

$$f_{ES} = -400 \cdot \frac{28,500}{3587} = -2.9 \text{ ksi}$$

For ease of calculation, a 1.5 ksi loss will be applied to each tendon resulting in approximately the same total reduction in force.

4.5.4   Creep Losses

Creep is the continued deformation of the concrete under sustained loads. Typically creep deformations are calculated by applying a multiplier from Table 3.2 to the elastic deformation. Creep occurs over an extended time and applying a creep coefficient to the initial elastic losses overestimates the total creep strain. To account for the time effects, two adjustments to the creep calculations are made. First, the elastic shortening is adjusted to use 90% of the initial prestressing force. This adjustment reflects the behavior that the prestressing force is decreasing over the life of the structure. Second, the modular ratio uses the 28-day concrete strength, again reflecting the growth of concrete strength over time.

Example 4.4: Calculate Creep Losses, f_{CR}
The creep coefficient for 6000 psi concrete from Table 3.2 is $C_c = 2.4$. The modulus of elasticity of the concrete at 28 days is

$$E_c = 33 \cdot 160^{1.5}\sqrt{6000} = 5,173,000 \text{ psi}$$

giving a modular ratio of $n = 28{,}500/5173 = 5.5$.

Solution pretensioned beam: The change in concrete stress is

$$f_c = -\frac{0.9 \cdot 489.6}{560} - \frac{0.9 \cdot 489.6 \cdot 13 \cdot 13}{125,390} + \frac{262.3 \cdot 12 \cdot 13}{125,390} = -1054 \text{ psi}$$

$$f_{CR} = nC_c f'_c = 5.5 \cdot 2.4 \cdot -1054 = -13,290 \text{ psi}$$

Solution post-tensioned beam: The tendon stress at midspan is 189 ksi-9.1 ksi frictional loss or 179.9 ksi resulting in a prestressing force of 179.9×16 strands $\times 0.153$ in.^{2} per strand $= 440.4$ kips. This gives a midspan stress in the concrete of

$$f_c = -\frac{0.9 \cdot 440.4}{560} - \frac{0.9 \cdot 440.4 \cdot 13 \cdot 13}{125,390} + \frac{262.3 \cdot 12 \cdot 13}{125,390} = -916 \text{ psi}$$

$$f_{CR} = nC_c f'_c = 5.5 \cdot 2.4 \cdot -916 = -12090 \text{ psi}$$

giving a total loss of prestress of 9.6 ksi.

4.5.5 *Shrinkage Losses*

Shrinkage is the volume reduction of the concrete due to hydration of the cement and loss of water from the concrete as it cures. Linear elements such as beams and columns shorten significantly due to shrinkage, which results in an equal shortening of the tendon and a partial loss of prestress. The loss in stress in the prestressing reinforcement is calculated by multiplying the shrinkage strain occurring after the member is stressed by the modulus of elasticity of the reinforcement. While concrete shrinkage has many causes, it is exceptionally sensitive to the relative surface-to-volume ratio of the cross section and relative humidity of the air surrounding the concrete. Thus, a wide thin member has more shrinkage than a square section and shrinkage in a rain forest is less than that in a desert. The shrinkage loss can be given as:

$$f_{SH} = \varepsilon_{sh,u} E_{ps} (1 - 0.0024 \, V/S)(100 - RH) \tag{4.6}$$

where, f_{sh} = loss due to shrinkage; $\varepsilon_{sh,u}$ = total shrinkage occurring after application of prestress; E_{ps} = modulus of elasticity of the tendon; V/S = volume-to-surface ratio of the member; RH = relative humidity.

For pretensioned members, $\varepsilon_{sh,u}$ is taken as 8.2×10^{-6}. This value of $\varepsilon_{sh,u}$ is valid for post-tensioned members stressed within a few days of casting. If the member is allowed to cure for more than a week, the value of the remaining shrinkage may be reduced based on the shrinkage that has occurred prior to stressing. Similarly, the post-tensioning losses for pretensioned beams integrated into a post-tensioned structure should have the shrinkage losses adjusted accordingly.

Example 4.5: Calculate Shrinkage Losses, f_{sh}
Solution pretensioned beam and post-tensioned beam: Using Eq. (3.4) for the shrinkage strain, a V/S ratio of 4.5 and a 35% relative humidity gives a final shrinkage strain

$$\varepsilon_{sh,u} = -\left(8.2 \cdot 10^{-6}\right)(1 - 0.06 \times 4.05)(100 - 35) = 4.03 \times 10^{-4}$$

And the corresponding loss is

$$f_{SH} = E_{ps}\varepsilon_{sh,u} = 28.5 \cdot 10^{6} \cdot -4.03 \cdot 10^{-4} = -11,490 \text{ psi}$$

Comment: The above calculation assumes the post-tensioned beam is stressed at about the same age as the pretensioned beam. If the post-tensioning is delayed, the shrinkage coefficient may be adjusted to account for shrinkage occurring prior to stressing.

4.5.6   Relaxation of Prestressing Reinforcement

Prestressing reinforcement losses due to relaxation of the steel are calculated using Eq. (3.9), reproduced below for convenience.

$$\frac{f_{\mathrm{p}}}{f_{\mathrm{pi}}} = 1 - \frac{\log t}{45}\left(\frac{f_{\mathrm{pi}}}{f_{\mathrm{pyi}}} - 0.55\right) \tag{4.7}$$

Example 4.6: Calculate Relaxation Losses

The yield stress for a 270 ksi strand occurs at a strain of 0.0086 from Eq. (3.7) giving a yield stress of 245 ksi. Assuming a 50-year service life requires calculating the relaxation loss for 438,000 h.

Solution pretensioned beam:

$$\frac{f_{\mathrm{p}}}{f_{\mathrm{pi}}} = 1 - \frac{\log(438,000)}{45}\left(\frac{200}{245} - 0.55\right) = 0.967$$

Thus, the loss is -0.033×200 ksi $= -6.6$ ksi.

Solution post-tensioned beam:

$$\frac{f_{\mathrm{p}}}{f_{\mathrm{pi}}} = 1 - \frac{\log(438,000)}{45}\left(\frac{189}{245} - 0.55\right) = 0.972$$

Thus, the loss is $f_{\mathrm{RE}} = -0.028 \times 189$ ksi $= -5.3$ ksi.

The losses are cumulative. Each of the above individual calculations is tabulated to determine the best estimate of the total losses in the beam. The total loss in this case is less than the lump sum losses, indicating that calculation of losses is important in the final design.

Example 4.7: Summary of Prestress Losses

Pretensioned beam summary		Post-tensioned beam summary	
Condition	Stress (ksi)	Condition	Stress (ksi)
Initial	200.0	Initial	189.0
Anchor set	−0.0	Anchor set	0.0
Friction	−0.0	Friction	−9.1
Elastic shortening	−8.7	Elastic shortening	−1.5
Creep	−13.3	Creep	−12.1
Shrinkage	−11.5	Shrinkage	−11.5
Relaxation	−6.6	Relaxation	−5.3
Total losses	−35.2	Total losses	−37.0
Final stress	161.4	Final stress	149.5
Percent loss	19.8%	Percent loss	20.9%

Comment: Lump sum losses suggest that post-tensioned members should have less loss than pretensioned members. The underlying assumption in that observations is that the losses are net of friction effects. If friction is treated separately in the above table, then the net post-tensioning losses would be less that then pretensioning losses. The initial and final prestressing force for the pretensioned beam are $P_i = 489.6$ kips and $P_e = 403.4$ kips. The initial and final prestressing force the post-tensioned beam are $P_i = 462.7$ kips and $P_e = 372.1$ kips.

4.6   Time-Step Approach to Losses

Prestressing losses are assumed to have occurred prior to calculation of the final member stresses. In some instances, losses at discrete time-steps are desired. A step-by-step approach to calculating these losses is given in both the PCI Design Handbook and in the ACI 423.10 Guide to Prestressing Losses. Calculation of time-step losses are used for designs where deflections are critical and the losses at specific times affect these deflections. For example, the beams for the Walt Disney World Monorail, Sect. 2.6.6, were designed so the beam deflection was near zero at the time the continuity joints were cast (Dolan and Mast 1972). Similarly, segmental or cable stayed bridges, as seen in Figs. 2.35 and 2.36, require close attention to deflections to assure that the mating ends of the bridge align properly.

Time-step calculations use the initial loss values for anchor set, friction, and elastic shortening. Losses for change in stress levels, creep, and shrinkage are calculated using detailed strength gain, creep, and shrinkage data. Data for strength gain, creep, and shrinkage are developed as part of the concrete mixture design for these fully engineered projects. Losses are then calculated using the differential strains based on the age of the structure being evaluated. For example, in the case of a segmental bridge, the loss and deflection prediction calculations are conducted for each added segment based on the age of the concrete at the time of segment erection. The field deformations of the structure are monitored for compliance and calculations revised if necessary. Time-step calculations are complex and time-consuming and are usually unnecessary. The level of complexity needed when incorporating prestress losses into the design and construction process, however, should be decided on a case-by-case basis.

4.7   Friction Loss Derivation

Wobble friction is the result of the tendon sliding along the duct and is a function of the type of duct and any unintentional misalignment of the duct during construction. It is present even if the duct is theoretically straight. Angular friction results from the tendon sliding across intentional curves place in the duct. While each effect is treated separately, the results are cumulative and are combined when calculating the total frictional loss in the tendon.

4.7.1   Wobble Friction

The incremental force change dP due to wobble friction in a short length dx of tendon can be given as

$$dP = kPdx \tag{4.8}$$

where the prestress force P is a function of the distance along the duct x and k is the wobble coefficient of friction in pounds per foot given in Table 4.1.

4.7.2   Angular Friction

Figure 4.3 shows a curved tendon subjected to a force P_j at the jacking end. A reduced force P_x occurs a distance l from the jacking end. For calculation purposes, the tendon is assumed to be a circular arc.

Figure 4.3 shows the forces in a short length of tendon. The segment is defined by the change in angle $d\alpha$, the change in force by dP resulting from friction along the tendon arc length dx, and P the force at this location. The force vector diagram in Fig. 4.3 indicates that the force normal to the tendon is equal to $Pd\alpha$. Using a coefficient of friction μ between the tendon and the duct, the incremental loss of force due to curvature friction is

$$dP = \mu Pd\alpha \tag{4.9}$$

where values for μ are given in Table 4.1.

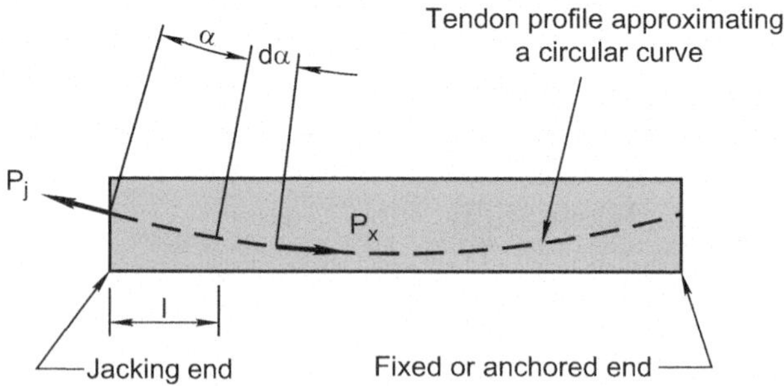

Fig. 4.3 Tendon geometry

Combining Eqs. 4.8 and 4.9 provides the sum of wobble and angular friction forces.

$$dP = kPdx + \mu Pdx \qquad (4.10)$$

The friction loss at a point under consideration is presented as the ratio of dP/P. Integrating between the jacking end and the point of interest gives

$$\int_{P_x}^{P_j} \frac{dP}{P} = \int_0^{l_x} kdx + \int_0^\alpha \mu d\alpha \qquad (4.11)$$

Resulting in

$$\ln \frac{P_j}{P_x} = kl_x + \mu\alpha \qquad (4.12)$$

Taking the exponent of each side, the force at the jacking end of a tendon P_j that is required to produce a force P_x at any point x along the length of the tendon can be found using Eq. (4.13).

$$P_o = P_x e^{kl_x + \mu\alpha} \qquad (4.13)$$

where e is the base of natural logarithms, k is the wobble friction coefficient per ft, μ is the curvature coefficient of friction, and α is the angle change of the tendon. If the sum of wobble and curvature friction is small, that is $kl_x + \mu\alpha$ is less than about 0.30, then the first term of the expansion of the power series for an exponential term can be used in lieu of Eq. (4.13). Thus, the simplified form of frictional losses is

$$P_o = P_x(1 + kl_x + \mu\alpha) \qquad (4.14)$$

The loss of prestress force P_f can be given as

$$P_f = P_j - P_x = P_x(kl_a + \mu\alpha) \qquad (4.15)$$

where l_a is the distance to the point in question. Assuming that for loss calculations that P_j and P_x are close in magnitude the total loss is

$$\Delta P = P_j(kl + \mu\alpha) \qquad (4.16)$$

or in terms of stress loss due to friction as $f_f = f_j(kl + \mu\alpha)$ as was given in Eq. (4.2).

4.7.3 Tendon Geometry

A tendon is composed of straight and curved sections. The frictional losses are calculated cumulatively as the tendon undergoes angular change. This sequential approach allows value of the tendon force to be adjusted for each segment. Equations (4.13) and (4.16) were developed assuming a circular arc. Determination of a circular arc for a tendon is both time consuming and inconsistent with practice since most tendon layout is based on a parabolic curve. For the small angles in most prestressing applications, it is acceptable to use a parabolic shape to approximate the circular curve and to determine the tendon angle change.

Figure 4.4 provides the geometry of a parabola with a drape of amount y. A tangent from the end of the tendon to midspan has an offset m. The angle at the end of the parabola is $\alpha/2$ from which the following can be obtained.

$$\tan\frac{\alpha}{2} = \frac{m}{x/2} = \frac{2m}{x}$$

The geometry of a parabola is such that m is twice y. For small angles, the tangent of the angle is approximately the same as the angle in radians. Therefore

$$\frac{\alpha}{2} = \frac{4y}{x} \tag{4.17}$$

or

$$\alpha = \frac{8y}{x} \tag{4.18}$$

which is the same as given in Eq. (4.3). Similarly, it can be shown by double differentiation of a parabolic line that the minimum radius of a parabola is

$$R = \frac{x^2}{8y} \tag{4.19}$$

The above derivations use parabolic approximations for circular assumptions. Recognizing the mixing of assumptions and the variation in frictions coefficients in Table 4.1, the ACI Building Code requires that the predicted jacking load and tendon

Fig. 4.4 Frictional effect on differential length and force polygon

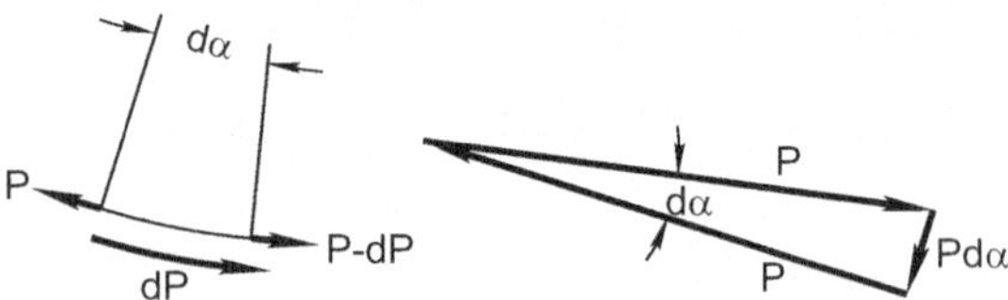

extension agree within 7% or that corrections to either the calculations or the jacking is required.

4.7.4   *Effects of Anchor Set*

In Example 4.1, the anchor seating loss was assumed to occur over the entire length of the beam. If the friction is sufficiently large, the displacement caused by the anchor seating only permeates along part of the length of the member. In such a case, the stress distribution along the tendon resembles Fig. 4.5.

Figure 4.6 assumes a linear loss distribution between the jacking end and the full length of the member l. The anchor seating distance is δ and the total loss in the member is f_t. bcy straight line interpolation

$$f_1 = f_T \frac{x}{l} \tag{4.20}$$

By similar triangles, f_1 is the average loss due to anchor seating which is also due to the stress loss in the tendon due to seating, which is equal to $E_p \delta / l$. Therefore

$$f_1 = f_T \frac{x}{l} = \frac{\delta}{x} E_p \tag{4.21}$$

Solving for x gives

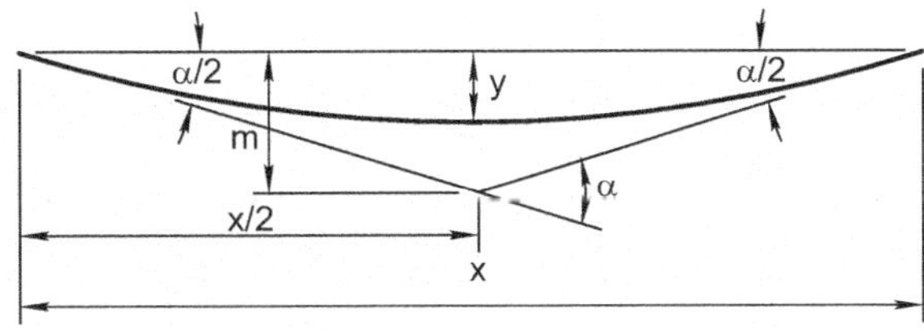

Fig. 4.5 Geometry of a parabola

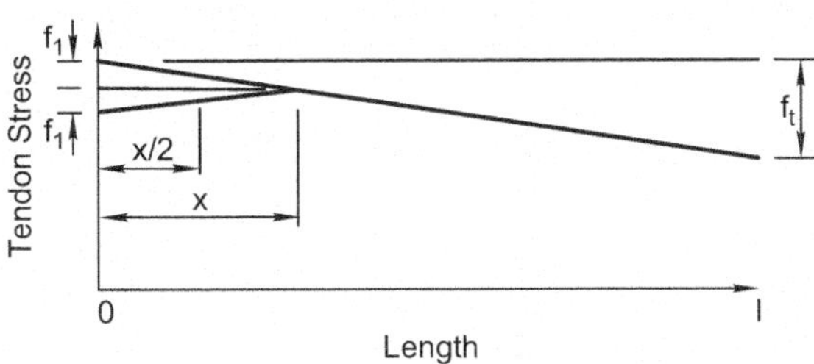

Fig. 4.6 Tendon stress distribution due to anchor seating in length x

$$x = \sqrt{\frac{\delta l E_p}{f_T}} \tag{4.22}$$

where f_T/l is the slope of the stress loss. If the tendon has substantial changes in slope due to multiple curvatures, then the solution may be restrained to the final linear segment or an iterative solution may be required if there are sharp changes in slope between segments. Solving for x provides insight to the tendon profile. If $x > l$ then the total loss remains

$$f_1 = \frac{\delta}{l} E_p \tag{4.23}$$

And the seating loss at the jacking end is $2f_1$ and the loss at the dead end is $f_1/2$.

Example 4.8: Calculate the Loss Due to Anchor Set

A four-span continuous beam is post-tensioned from each end. The beam tendon properties are given in Table 4.2 and the beam geometry is given in Fig. 4.7.
Sample calculations: The first parabolic segment has an eccentricity of 2.0 ft and a length of 40 ft. The angular change is twice the eccentricity divided by the length or

$$\alpha = \frac{2y}{l} = \frac{2 \cdot 2.0}{40} = 0.100$$

The first segment friction is then $\mu\alpha + kl = 0.100 \times 0.20 + 0.0006 \times 40 = 0.044$. The total accumulated friction for the beam exceeds 0.30, therefore the exponential form of losses is required. The summary includes both the linear and the exponential values for comparison, but the exponential values are used for the loss calculations. The remaining losses are calculated segment by segment and summarized in the table below.

Table 4.2 Tendon properties and jacking load

$n =$	28	strands			
$A_{ps} =$	4.284	in.^{2}	$E_{ps} =$	28,500	ksi
$\mu =$	0.2		$f_{pu} =$	270	ksi
$k =$	0.0006	/ft	$f_{pi} =$	202.5	ksi
$\delta_{anc} =$	0.25	in.	$P_j =$	867.51	kip

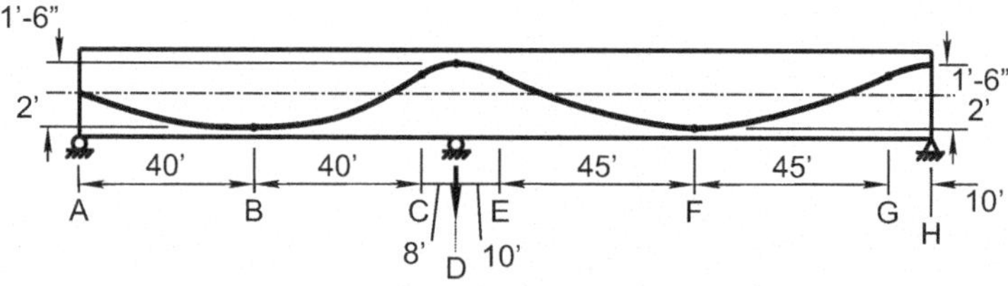

Fig. 4.7 Tendon geometry

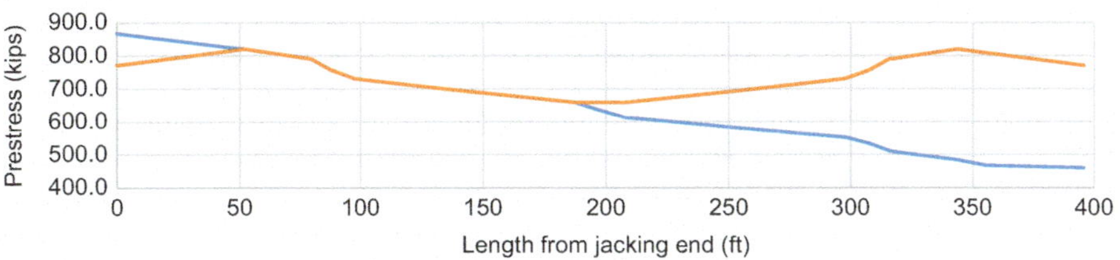

Fig. 4.8 Final tendon stress

Next consider anchor seating. The friction loss in the end section is 202.5 ksi less 193.8 ksi divided by 40 ft or $f_T/l = 0.018$ ksi/in. The length of the anchor set is

$$l_{\text{set}} = \left(\frac{0.25 \cdot 28,200}{0.018}\right)^{0.5} = 626 \text{ in.} = 52.5 \text{ ft}$$

The values for calculation of friction loss and the finals tendon stresses are given in Table 4.2.

The anchor set termination is in the second segment. The change in frictional resistance is small so a linear loss is calculated. The change in force twice the friction loss times the length or

$$\Delta P = 2 \cdot \frac{0.018 \text{ ksi}}{\text{in.}} \cdot 626 \text{ in.} \cdot 4.28 \text{ in.}^2 = 97.5 \text{ kip}$$

giving a force after release of 867.5 kips less 97.5 kips or 770.0 kips. Final tendon forces are illustrated in Fig. 4.8.

The jacking stress is less than $0.8 f_{\text{pu}}$ or 929 kips and the final stress at the anchor after seating is less than $0.7 f_{\text{pu}}$ or 812 kips. Thus, Code requirements are satisfied. The peak interior force is 818 kips or $0.71 f_{\text{pu}}$. The ACI Building Code requires the $0.70 f_{\text{pu}}$ stress limit only at the anchor so this higher stress is allowable.

Comment: The losses over the length of the structure are excessive requiring jacking from the "dead-end" to 867.5 kips then reseating the anchor. The resulting stresses are shown in the figure above. The dead-end stressing calculations terminate when the stress from the live end jacking exceeds the stresses in the dead-end jacking.

Problems

4.1. Prepare an Excel spreadsheet that replicates Table 4.3.
4.2. Confirm the tendon stresses in Table 4.3 for the second end jacking.

Table 4.3 Friction loss data

l (ft)	0	40	40	8	10	45	45	10
Σl (ft)	0	40	80	88	98	143	188	198
$\alpha = 2y/(l/2)$	0	0.100	0.138	0.188	0.150	0.122	0.122	0.150
Δkx	0	0.024	0.024	0.0048	0.006	0.027	0.027	0.006
$\mu\alpha + kl$	0	0.044	0.052	0.042	0.036	0.051	0.051	0.036
$\Sigma(\mu\alpha + kl)$	0	0.044	0.096	0.138	0.174	0.225	0.277	0.313
$1 - \Sigma(\mu\alpha + kl)$	1	0.956	0.905	0.862	0.826	0.775	0.723	0.687
$e^{-\Sigma(\mu\alpha + kl)}$	1	0.957	0.909	0.871	0.840	0.798	0.758	0.731
P_i (kip) $=$	867.5	830.2	788.5	755.8	729.1	692.6	657.8	634.6
f_{pi} (ksi)	202.5	193.8						
$f_t/l =$	0.018 ksi/in.	$l_{set} = 626.37$ in.			$= 52.2$ ft set loss $= 22.7$ ksi		$= 97.5$ kip	
Final P_i (kip) $=$	770.0	830.2	788.5	755.8	729.1	692.6	657.8	634.6
Final f_{pi} (ksi) $=$	179.8	193.8	184.1	176.4	170.2	161.7	153.6	148.1

4.3. If the midspan drape in Example 4.8 was 4 ft instead of 2 ft and all other parameters are the same, what is the final stress in the tendon for one end jacking and what is the percent change in stress from Example 4.8?

4.4. If the midspan drape in Example 4.8 was 4 ft instead of 2 ft and all other parameters are the same, what is the final stress in the tendon after restressing the dead end of the tendon.

References

AASHTO LRFD Bridge Design Specification (8th Ed.). (2017). Washington, DC: American Association of State Highway and Transportation Officials (AASHTO).

ACI 318-14. (2014). *Building Code Requirements for Structural Concrete* (ACI Committee 318-14, p. 519). Farmington Hills, MI: American Concrete Institute.

ACI-ASCE Committee 323. (1958). Tentative recommendations for prestressed concrete. *ACI Journal, 54*(7), 545–578.

ACI-ASCE Committee 423.10R. (2017). *Guide to estimating prestress losses* (p. 110). Farmington Hills, MI: American Concrete Institute.

ACI 423.10-16. (2016). *Guide to Estimating Prestress Loss* (p. 64). Farmington Hills, MI: American Concrete Institute.

Dolan, C. W., & Mast, R. F. (1972). Walt disney world monorail designed for smooth riding. *Civil Engineering (ASCE)*, 4p.

PCI Design Handbook (8th Ed.). (2017). Chicago, IL: Precast and Prestressed Concrete Institute.

PTI M10.3-00: Field Procedures Manual for Unbonded Single Strand Tendons. (2016). Farmington Hills, MI: Post Tensioning Institute.

Zia, P., Preston, H. K., Scott, N. L., & Workman, E. B. (1979). Estimating prestress losses. *Concrete International, 1*(6), 32–38.

Chapter 5
Flexural Basics of Analysis and Design

5.1 Introduction

Analysis and design are two interrelated activities. **Analysis** is the determination of the effect of loads in a member based on the given properties and applied loads. Analysis is typically a deterministic process, that is, for a given set of conditions, a single result is calculated for each external load combination. **Design** is a creative process that establishes the parameters needed to conduct the analysis, including selection of appropriate member sizes, prestress, and connections. Creating a model of a building for computer analysis is an example of a design activity. Running the model is analysis. Examining the computer output and comparing it to the member capacity determines if initial selection of member section properties is adequate for the applied loads or if another iteration is needed is integration of design and analysis. An engineer designs a structure, which can be a building, bridge, or other endeavor. The structure consists of members such as slabs, beams, columns, and foundations. The loads applied to the structure generate internal stresses in these members. Consequently, building codes require that the nominal strength of a member exceed the effects of externally generated loads at all sections. For practical purposes, however, members are checked at selected critical sections to validate overall design adequacy.

Prestressed concrete follows this design and analysis process. During the design phase the initial section, the prestress force, and the tendon location are selected. An analysis evaluates the service stresses and strength to ensure that the section is adequate. Analysis is the starting point because practice with analysis establishes understanding of the prestressing materials, calculations, strength, and mandated code limits. Additionally, analyzing members builds the experience needed to become efficient in selection of properties for design.

Prestressed concrete is required to meet both service level stresses and strength requirements whereas reinforced concrete is mostly predicated on strength design. This makes the analysis and design process slightly more complex. Usually

© Springer Nature Switzerland AG 2019
C. W. Dolan, H. R. Hamilton, *Prestressed Concrete*,
https://doi.org/10.1007/978-3-319-97882-6_5

prestressing is selected for service load conditions then checked for strength. Iteration is sometimes needed to meet both criteria.

5.2   Beam Global Behavior

Structural design involves determining the section geometry, material properties, and prestress magnitude and position to satisfy strength, serviceability, and detailing requirements. Figure 5.1 illustrates the load-deflection behavior of a beam as it is subjected to loads over its full life. For presentation purposes, effective prestress is used. The behavior develops as follows: as soon as the prestress is applied, the beam cambers upward. Simultaneous with the transfer of prestress, the beam girder weight is engaged and the downward deflection due to girder weight δ_0 reduces the camber. The dashed line is the theoretical initial camber, and the curved solid line is the actual behavior. The upward camber is further decreased by the addition of superimposed dead load δ_d. The location of the dead load deflection above or below the balanced

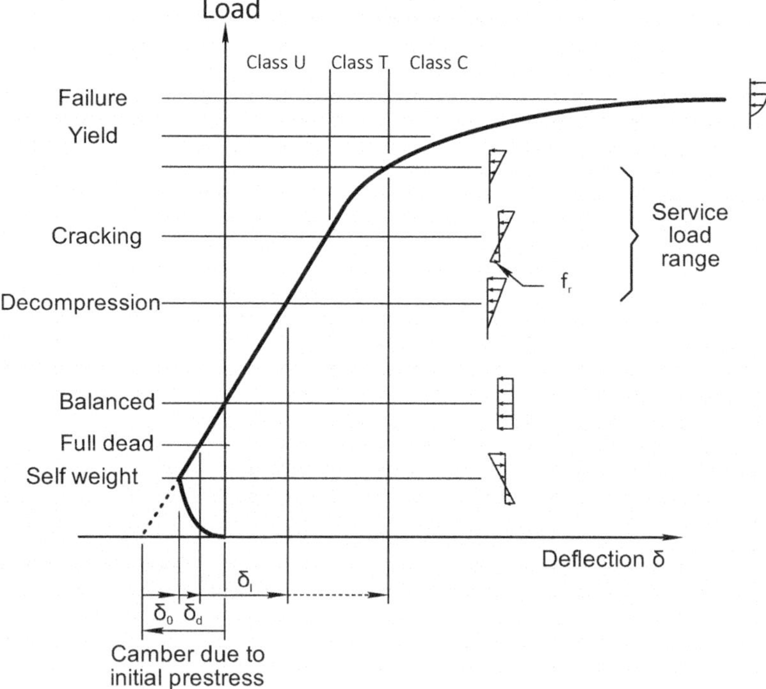

Fig. 5.1 Load-deflection behavior of a beam

state of no camber is a function of the magnitude of the dead load, the live load, and the prestress force and location. A balanced condition occurs when the calculated tensile and compressive stresses are equal over the entire section. Decompression occurs when the combined tensile stress due to prestress and applied live load equals zero. Zero stress at the extreme fiber on the tension face of the beam is the upper bound of service loads for beams with a requirement of no tensile stresses. Beams sustaining tensile stresses continue to deflect and have a service load range above the decompression load. Above this limit, the beam is considered in a transitional condition. The beam is assumed to crack when the stress in the extreme tension fiber of the beam reaches the modulus of rupture f_r of the concrete. Beyond the cracking load, the beam is considered cracked and requires additional calculation effort to determine service level stresses and deflections. Finally, the nominal strength of the beam is reached, usually after yielding of the prestressed reinforcement accompanied by a compression failure of the concrete. The maximum service load is indicated and ranges between a zero-tension criterion and a maximum compression stress in a cracked section. For each level of loading, a schematic of the concrete stress is indicated in Fig. 5.1.

5.3   Service Level Stresses

Service level stresses at a section result from unfactored loads acting on the member. These are the stresses the member is likely to experience during its normal use. Chapter 1 provides a listing of common service loads and the philosophy of load and resistance factor (LRFD) design. A structural analysis determines the moments, shears, and axial forces on a member for both the service conditions and the strength conditions. These loads are used to calculate the stresses at each critical section.

5.3.1   Sign Convention

The sign convention used in this book conforms to common practice in prestressed concrete design. Compression stresses are given as positive and tension stresses as negative.

5.3.2   Calculation of Service Level Stresses

Service level stresses are calculated assuming linear elastic behavior, Fig. 5.2. The stress at any section is given by Eq. (5.1).

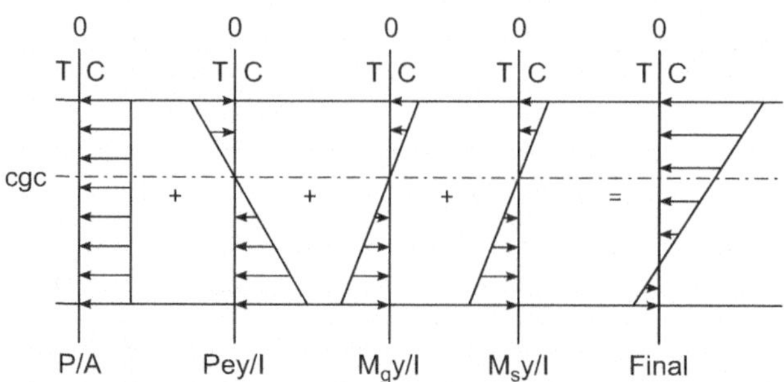

Fig. 5.2 Schematic of service stress distribution at a section

$$f = \frac{P}{A_g} \mp \frac{Pey}{I_g} \pm \frac{M_g y}{I_g} \pm \frac{M_s y}{I_g} \tag{5.1}$$

where f is the concrete stress in a section at a location y from the neutral axis, P is the prestressing force, A is the cross-sectional area, I_g is the gross moment of inertia, e is the tendon eccentricity at the section, M_g is the moment due to the girder self-weight, and M_s is the service load moment. The $\pm$ signs indicate that in most situations, the prestress moment is opposite of the applied moment as indicated in Fig. 5.2 where T signifies tensile stresses and C signifies compressive stresses on the section. The dashed line indicated the center of gravity of the concrete section (cgc).

The sectional area A is usually the gross area of the concrete. Section 4.12.2.4 of ACI 318-14 states "Effect of loss of area due to open ducts shall be considered in computing section properties before grout in post-tensioning ducts has attained design strength." For bonded tendons, the gross area may be used. For unbonded post-tensioned beams, the area of ducts needs to be considered. Typically, the post-tensioned duct is not grouted at the time of initial prestress and a net area may be appropriate; however, the gross section is commonly used. Gross cross-sectional area of concrete is usually used to determine section properties in unbonded monostrand tendons systems.

Going from the generic Eq. (5.1) to a specific section requires several intermediate steps. First, critical sections are identified then stresses are calculated. Stresses must be calculated for conditions when the prestress force is transferred to the section and again after all losses have occurred. Typically, stresses at transfer are calculated for the initial prestress force and girder dead load, Eq. (5.2). This set of stress calculations recognizes that shortly after the transfer of the prestress force to the member, the prestress force decreases due to losses. The stresses are calculated at the top and the bottom of the section, denoted by the subscript t or b in Eq. (5.2). Similar subscripts from the distance from the neutral axis to other locations provide the stresses at intermediate depths.

$$f_t = \frac{P_i}{A_g} - \frac{P_i e y_t}{I_g} + \frac{M_g y_t}{I_g}$$

$$f_b = \frac{P_i}{A_g} + \frac{P_i e y_b}{I_g} - \frac{M_g y_b}{I_g} \tag{5.2}$$

Final stresses are calculated for prestress forces, girder self-weight, and applied service loads Eq. (5.3). The effective prestress force P_e assumes that all losses have occurred. Applied service loads M_s may include long-term superimposed dead loads and live loads, in which case, the service load may be calculated in multiple steps to ensure that the stresses meet the design intent at all stages of loading.

$$f_t = \frac{P_e}{A_g} - \frac{P_e e y_t}{I_g} + \frac{M_g y_t}{I_g} + \frac{M_s y_t}{I_g}$$

$$f_b = \frac{P_e}{A_g} + \frac{P_e e y_b}{I_g} - \frac{M_g y_b}{I_g} - \frac{M_s y_b}{I_g} \tag{5.3}$$

Figure 5.3 summarizes the change in stress on a midspan section through the evolution of the loading. Two items in Fig. 5.3 deserve attention. First, the average stress at the neutral axis decreases between the initial prestress and the final prestress. This is a result of the loss of prestress. Second, the total range of stress change is seen on the top and bottom of the section. The top and bottom stresses can theoretically exceed the allowable stresses f_{ti} and f_{ci} providing the weight of the beam is mobilized during the stress transfer. The concurrent incorporation of the girder moment prevents the concrete from reaching the higher stresses.

Noting that the radius of gyration squared r^2 is equal to I/A, Eq. (5.3) may be rewritten as shown in Eq. (5.4).

Fig. 5.3 Combined stresses at a midspan section

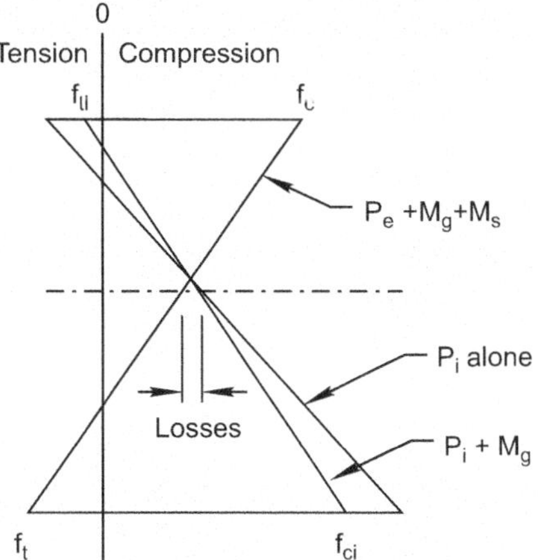

$$f_t = \frac{P_e}{A_g}\left(1 - \frac{ey_t}{r^2}\right) + \frac{M_g y_t}{I_g} + \frac{M_s y_t}{I_g}$$

$$f_b = \frac{P_e}{A_g}\left(1 + \frac{ey_b}{r^2}\right) - \frac{M_g y_b}{I_g} - \frac{M_s y_b}{I_g}$$

(5.4)

The formulation in Eq. (5.4) is slightly more compact than Eq. (5.3), has some advantages for certain calculations and may be preferred for some computer programming alternatives. Equation (5.3) is slightly longer but clearly shows each step in the calculation process and is used in the following examples.

Example 5.1 Calculation of Service Level Stresses

A 60 ft long AASHTO Type III beam has a cross-sectional area of 560 in.[2], a moment of inertia of 125,390 in.[4], a weight of 583 plf, a distance from the neutral axis to the top fiber of 24.73 in., and a distance to the bottom fiber of 20.27 in., Fig. 5.4. The beam is post-tensioned with two 8-strand tendons placed in a parabolic drape so the midspan eccentricity is 13 in. The specified transfer strength of the concrete is f_c' of 3500 psi and the 28-day strength is 6000 psi. A long-term superimposed dead load midspan moment of 1000 in.-kip is applied in addition to a live load moment of 4000 in.-kip. The initial and effective prestress forces and properties are given in Table 5.1.

Solution: Applied loads are summarized and the stresses are calculated in an Excel Spreadsheet based on Eqs. (5.2) and (5.3) and summarized in Table 5.2. In this example, an additional subtotal for the superimposed dead load is used in the final stress calculations to indicate the stress conditions at the intermediate loading. This summary provides insight to the initial and long-term deflection of the beam. In addition to the final stresses, a schematic diagram of the stress distribution in the section at various stages of loading is provided.

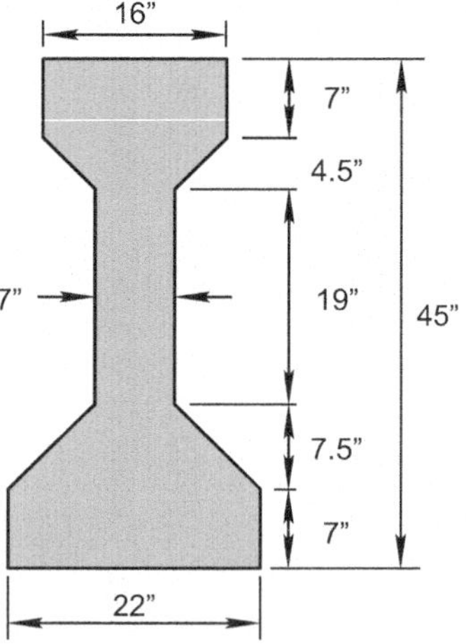

Fig. 5.4 AASHTO Type III beam section geometry

Table 5.1 Properties and loads for Example 5.1

Loads			Properties		
$f_i =$	178.4	ksi	$A_g =$	560	in.^{2}
$f_e =$	159.9	ksi	$I_g =$	125,390	in.^{4}
$P_i =$	436.7	kip	$y_t =$	24.73	in.
$P_e =$	391.4	kip	$y_b =$	20.27	in.
$M_g =$	3148	in.-kip	$e =$	13	in.
$M_{sdl} =$	1000	in.-kip	$L =$	60	ft
$M_l =$	4000	in.-kip	$A_p =$	2.45	in.^{2}
$w_g =$	583	plf	$b =$	16	in.

Table 5.2 Stress summary for Example 5.1

Stresses					
Initial stresses			Final stresses		
	Top (psi)	Bottom (psi)		Top (psi)	Bottom (psi)
P_i/A_g	780	780	P_e/A_g	699	699
$P_i ey/I_g$	−1120	918	$P_e ey/I_g$	−1004	823
$M_g y/I_g$	621	−509	$M_g y/I_g$	621	−509
Sub total	281	1181	Sub total	316	1013
			$M_{sdl} y/I_g$	197	−162
			Subtotal	514	851
			$M_l y/I_g$	789	−647
			Total	1302	204

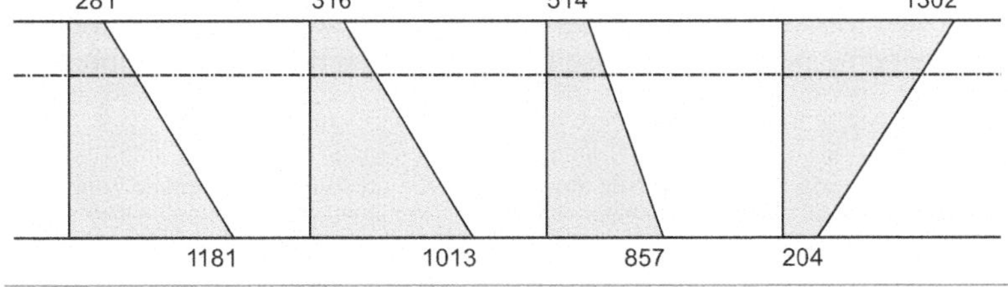

Comment: A schematic diagram of the stress distribution in Example 5.1 provides a useful qualitative assessment of the member performance. The stress gradient is indicative of the curvature of the section because dividing the stress by the modulus of elasticity provides the equivalent strain distribution and corresponding curvature. A constant stress distribution suggests the section and the member is subjected to axial load only. A gradient, as seen in the final stress distribution in Example 5.1 suggests positive curvature and a downward deflection. The stress gradient resulting from the initial prestress is opposite to the long-term gradient, that is, the slope is reversed. This indicates that the beam is subject to an upward deflection or **camber** as soon as it is removed from the form at the time of prestress transfer and may continue camber growth throughout its service life.

The analysis in Example 5.1 provides the calculated service level stresses in the member but does not indicate if these stresses are acceptable. The maximum service stresses allowed at a section are determined by practice and experience and are provided in appropriate building codes. Two codes in wide use are the *ACI Committee 318 Building Code for Structural Concrete* (2014) and the *AASHTO Standard Specification for Bridge Structures* (AASHTO 2017).

5.3.3 ACI 318 Stress Limits

Stress limits are established for both initial and final prestress. Allowable stresses at initial prestress reflect that the concrete is continuing to gain strength and that the prestress force is decreasing with time. ACI 318-14 Chapter 24 provides a classification of prestressed members based on the in-service tensile stress. Class U members are assumed uncracked in service where the maximum tensile stress under full service load is less than the cracking stress of the concrete determined by the modulus of rupture. Class T members have stresses that may crack the member under service load, and Class C members are assumed to be cracked in service. The ACI Building Code stress limits for classification are given in Table 5.3. At transfer the maximum tensile stress is $6\sqrt{f'_c}$ at the member end and $3\sqrt{f'_c}$ elsewhere for all members unless supplemental reinforcement is provided.

Stress limits in compression are dependent on the state of the prestressing and on the location of the section in the member. The compressive stresses allowed by the ACI Building Code are summarized in Table 5.4.

The length of the "End of a simple span beam" is undefined in the Building Code. It is included in the 2014 ACI Building Code to provide some relief at the end of a beam before the bending moment due to self-weight overcomes the precompression stress.

Table 5.3 Beam classification based on tensile stresses f_t in the precompressed tensile zone

Assumed behavior	Class	Limits of f_t
Uncracked	U	$\leq 7.5\sqrt{f'_c}$
Transition between uncracked and cracked	T	$7.5\sqrt{f'_c} < f_t < 12\sqrt{f'_c}$
Cracked	C	$> 12\sqrt{f'_c}$

Table 5.4 ACI allowable compressive stresses at a section

Prestress state	Location	Loading condition	Compressive stress limit
Immediately after transfer	End of simple span beam		$0.70f'_{ci}$
	All other locations		$0.60f'_{ci}$
At Service load after all losses		Prestress plus sustained load	$0.45f'_c$
		Prestress plus full service load	$0.60f'_c$

Example 5.2: Evaluate Service Level Stresses Based on ACI
Evaluate service level stresses in Example 5.1 based on ACI Criteria.

Solution: The calculated midspan stresses from Example 5.1 are compared to the limits established in the ACI Building Code requirements.

Condition and location	Calculated stress (psi)	Allowable stress	Status
Initial prestress			
Top	281	$0.60 \times 3500 = 2100$ psi	Meets ACI Building Code
Bottom	1189		requirements
Final prestress—full load			
Top	1302	$0.6 \times 6000 = 3600$ psi	Meets ACI Building Code
Bottom	204	$-7.5\sqrt{6000} = -581$ psi	requirements and is classified as Class T, transitional

5.3.4   AASHTO Stress Limits

The AASHTO Bridge Design Specification allowable tensile stresses are summarized in Table 5.5 and allowable compressive stresses in Table 5.6. AASHTO stresses are based on units of ksi rather than psi used in the ACI Building Code.

Table 5.5 AASHTO allowable tensile stresses at a section

Prestress state	Conditions	Tensile stress limit
Immediately after transfer	In areas other than Precompressed Tensile Zone and without bonded tendons or reinforcement	$0.0948\sqrt{f'_{ci}} < 0.2$ (ksi)
	In areas with bonded tendons or reinforcement sufficient to resist the tensile force in the concrete computed assuming an uncracked section, where reinforcement is proportioned using a stress of $0.5 f_y$, not to exceed 30 ksi	$0.24\sqrt{f'_{ci}}$ (ksi)
At service load after all losses	In the Precompressed Tensile Zone, assuming uncracked section:	
	• Components with bonded tendons or reinforcement, and/or are not located in areas exposed to chlorides, salts, or sulfates	$0.19\sqrt{f'_{c}}$ (ksi)
	• Components with bonded tendons or reinforcement, and/or are located in areas exposed to chlorides, salts, or sulfates	$0.0948\sqrt{f'_{c}}$ (ksi)
	• Components with unbonded tendons	0
Permanent loads only		0

Table 5.6 AASHTO allowable compressive stresses at a section

Prestress state	Location	Loading condition	Compressive stress limit
Immediately after transfer	All locations		$0.60f'_{ci}$
			$0.60f'_{ci}$
At Service load after all losses		Prestress plus sustained load	$0.45f'_c$
		Prestress plus full service load	$0.60f'_c$

Example 5.3: Evaluate Service Level Stresses Based on AASHTO
Evaluate service level stresses from Example 5.1 based on AASHTO Criteria

Solution: The calculated midspan stresses from Example 5.1 are compared to the limits established in the AASHTO Code requirements.

Condition and location	Calculated stress (psi)	Allowable stress	Status
Initial prestress			
Top Bottom	281 1189	$0.60 \times 3500 = 2100$ psi	Meets AASHTO Code requirements
Final prestress—sustained load	851	$0.45 \times 6000 = 2600$ psi	Meets AASHTO Code requirements
Final prestress—full load			
Top Bottom	1302 204	$0.6 \times 6000 = 3600$ psi $-0.19\sqrt{6} = -0.465$ ksi	Meets AASHTO Code tension requirements

Comment: Comparison of allowable stresses in Example 5.4 illustrates the difference in code requirements whereby the beam meets ACI criteria but does not meet the AASHTO Specification.

Example 5.4: Conduct a Stress Check
Conduct a stress check on a double-T using ACI Building Code working stress calculations. A 30-in. deep pretopped double-T beam is pretensioned with 18-½ in. diameter strands providing an initial prestress of 550.8 kip. Losses are 15% or 82.6 kip. The lightweight concrete transfer stress is 3500 psi, and the specified compressive stress is 5000 psi.

Solution: Section properties and allowable stresses are given in Table 5.7. Calculated stresses, and schematic stress distribution are given in Table 5.8. Calculated stresses use Eqs. (5.2) and (5.3).

Comment: All service level stresses are within the ACI Building Code limits. The sustained compression in the bottom fiber suggests camber growth is likely during the member service life.

Table 5.7 Loads and properties for Example 5.4

Loads			Properties		
$P_i =$	550.8	kip	$A_g =$	928	in.^{2}
$P_e =$	468.2	kip	$I_g =$	59,997	in.^{4}
$M_g =$	2779	in.-kip	$y_t =$	7.06	in.
$M_{sdl} =$	0	in.-kip	$y_b =$	22.94	in.
$M_l =$	4500	in.-kip	$e =$	10	in.
$w_g =$	741	plf	$l =$	50	ft
			$f'_{ci} =$	3500	psi
			$f'_c =$	5000	psi

Table 5.8 Stress summary for example 5.4

Stresses					
Initial stresses			Final stresses		
	Top (psi)	Bottom (psi)		Top (psi)	Bottom (psi)
P_i/A_g	594	594	P_e/A_g	505	505
$P_i e y/I_g$	−648	2106	$P_e e y/I_g$	−551	1790
$M_g y/I_g$	327	−1062	$M_g y/I_g$	327	−1062
Sub total	272	1637	Sub total	281	1232
			$M_{sdl} y/I_g$	0	0
			Subtotal	281	1232
			$M_l y/I_g$	530	−1721
			Total	810	−488

272	281	281		810
1637	1232	1232	−488	

Initial stresses Allowable stresses			Final stresses		
$f_{ci} = 0.6 f'_{ci} =$	2100	psi	$f_c = 0.45\ f'_c =$	3000	psi
$f_{ti} - 3\sqrt{f'_{ci}} =$	−177	psi	$f_t = 7.5\sqrt{f'_c} =$	−530	psi

5.4   Section Flexural Strength

5.4.1   Introduction

The behavior of a section is dependent on the total tensile force and the state of stress in the tendon. Most prestressed beams are **under-reinforced** or **tension controlled**, that is, the tendon yields and a compression failure occurs in the concrete after

Fig. 5.5 Load test of tension controlled prestressed beam

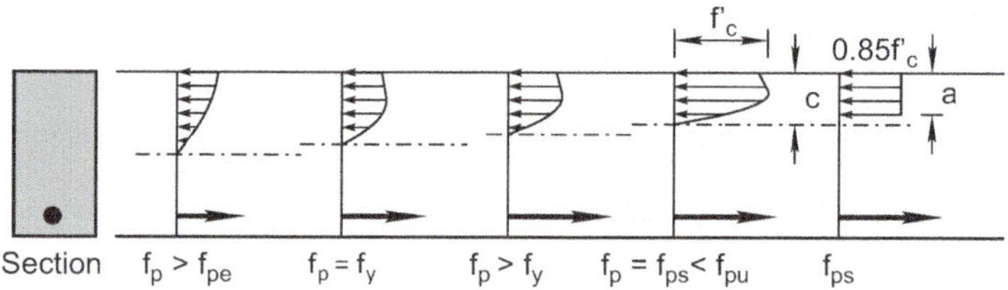

Fig. 5.6 Stress transition in under-reinforced section as load increases

substantial cracking and deformation. The significant deflection and cracking provide warning of impending failure, Fig. 5.5. As the section transitions from service loads to its nominal strength, the strain in the concrete and the prestressing reinforcement increases and the neutral axis migrates upward in an under-reinforced section, Fig. 5.6. The concrete stress–strain behavior remains nonlinear and an equivalent rectangular stress block is used for the concrete compression stress–strain behavior to facilitate the calculation of nominal moment. Integration of the equivalent rectangular stress block over the area in compression provides the same total compression force and force centroid as the integration of the nonlinear stress–strain curve over the same area. The relationship between the rectangular stress block and the nonlinear stress–strain curve requires that the magnitude of the stress be $0.85f'_c$ and that the depth of stress block a be

$$a = \beta_1 c \tag{5.5}$$

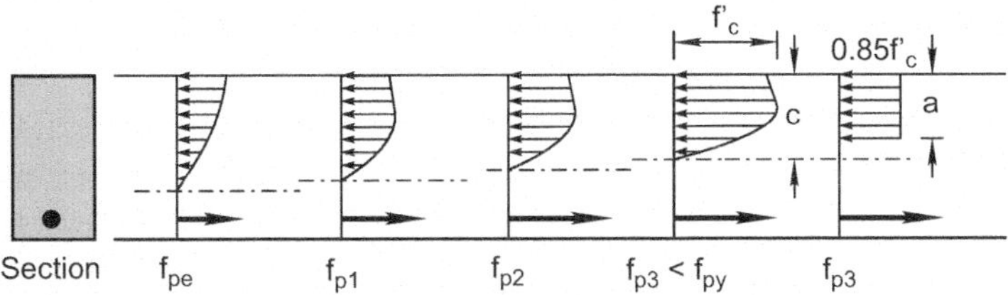

Fig. 5.7 Stress distributions in over-reinforced section as load increases

where the value of β_1 varies between 0.85 and 0.65 for concrete strengths between 4000 psi and 8000 psi and is given by Eq. (5.6).

$$0.85 \geq \beta_1 = 0.85 - \frac{f'_c - 4000}{1000} \geq 0.65 \tag{5.6}$$

When the tensile force is large enough, the section is **over-reinforced** or **compression controlled** and the tendon does not yield. In this case, a sudden compression failure in the concrete occurs with little or no cracking and the beam fails with little warning. In over-reinforced sections, the compression block increases with load and the tendon stress increases from the effective prestress f_{pe} to a final stress f_{ps}, which is less than the yield stress, Fig. 5.7. The stress in the tendon and the location of the neutral axis are both unknown and an iterative solution based on equilibrium of the section is required.

As described in Chapter 1, ACI 318 Building Code adjusts the strength-reduction factor based on the **net tensile strain** in the reinforcement, the strain imparted to the tendon farthest from the compression face due to bending of the member and exclusive of the pretensioning strain. To ensure the adequate ductility and the greatest strength-reduction factor, the net tensile strain must be greater than 0.005. The value of 0.005 is specified to reflect that the prestressed steel is stressed to at least half of its yield strength thus assuring that the steel is in the yield portion of the stress–strain curve. The ACI Building Code requirement that the prestressing reinforcement is stressed to at least half its yield stress assures that the reinforcement strain at nominal strength, including the pretensioning strain, is in the yield region. This provision remains valid for existing strand and for bars with no clearly defined yield point.

Depending on the procedure used to calculate the nominal moment, the value of the net tensile strain may not be readily known. Using the limiting strain of the concrete ε_{cu} and the 0.005 net tensile strain, this condition can be satisfied, providing that the c/d_p ratio is less than 0.375, where c is the distance from the compression face to the neutral axis and d_p is the distance from the compression face to the centroid of the prestress reinforcement, Fig. 5.8.

Calculation of the nominal flexural strength of a member is dependent on the tendon and any supplemental reinforcement. Three conditions are possible. In the first case, the tendons are bonded and supplemental reinforcement is needed to attain

Fig. 5.8 Strain limit for tension controlled section

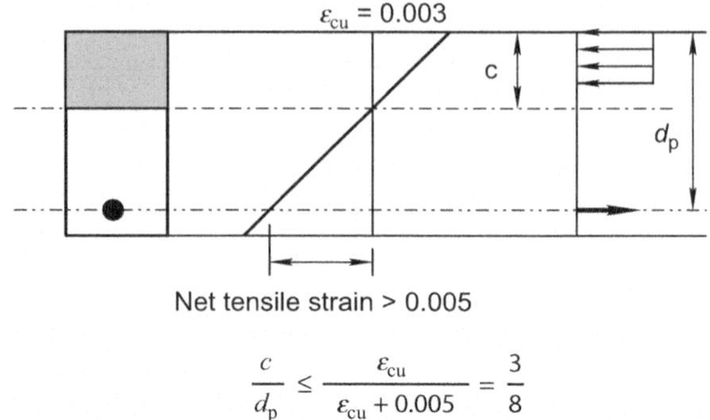

$$\frac{c}{d_p} \leq \frac{\varepsilon_{cu}}{\varepsilon_{cu} + 0.005} = \frac{3}{8}$$

the required strength. For this condition, a strain and equilibrium compatibility solution is needed. Strain compatibility solutions are also required for unusual shapes, when compression blocks move into beam webs, or when needed to ensure that performance meets design intent. In the second case, tendons are bonded and fall within the bounds established in the ACI Building Code. The ACI Building Code then allows an approximate solution for the tensile strength of the tendon. ACI Building Code based solutions are generally conservative. In the third case, tendons are unbonded. The stress in these tendons comes from approximate equations provided in the ACI Building Code because there is no strain compatibility between the tendon and the concrete.

5.4.2   Bonded Tendons: Strain Compatibility Solutions

Sections with unusual cross sections, large or small amounts of prestress force, or if the compression block is in both the flange and web of the member, require a strain compatibility approach to calculate the nominal bending capacity. Strain compatibility solutions examine the cumulative strain in the prestressing reinforcement due to prestressing and flexure.

Strain compatibility solutions tend to be iterative and nonlinear because the section must be in equilibrium for the given strain condition and the stress–strain relationship for the prestressing reinforcement is not linear. Developing a solution requires knowledge of the stress–strain curve for the prestressing reinforcement. The equivalent rectangular stress block is satisfactory for representation of the concrete stress–strain relationship. Equation (3.7) provides a representative stress–strain curve for prestressing strand. Recognizing that Eq. (3.7) is a hybrid of many manufacturers, an engineer may select a specific manufacturer's stress–strain curve should the project require that level of accuracy. A mathematical representation for the prestressing reinforcement stress–strain curve allows a solution to be generated using the solver routines available in EXCEL or MathCAD. Iterative solutions are presented in the following examples to better envision the behavior.

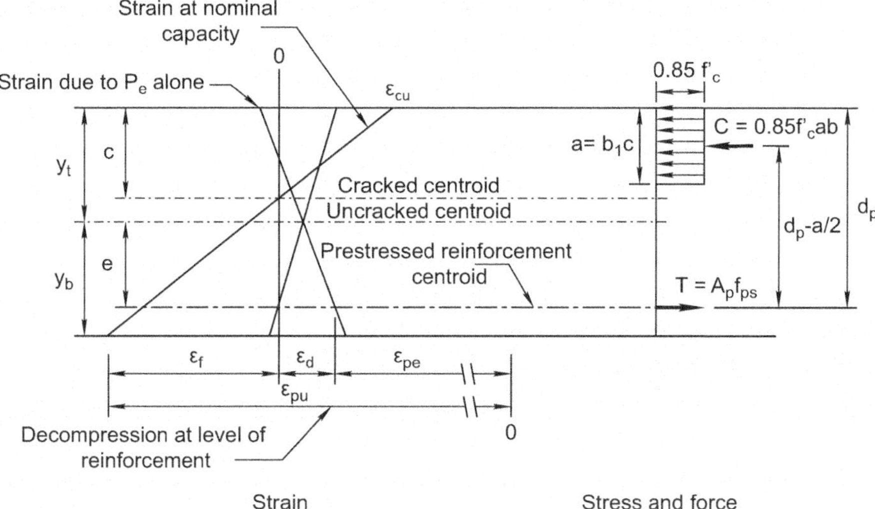

Fig. 5.9 Strain and equivalent stress as a rectangular section is loaded to nominal strength

Figure 5.9 illustrates the evolution of strain and stress at a section as it goes from initial prestress transfer through nominal strength. For development of strain based nominal moment calculations, it is useful to divide the strain in the prestressing steel into three separate conditions. The first condition is the effective strain in the tendon after losses ε_{pe}. This strain is calculated from the effective prestress and is given in Eq. (5.7).

$$\varepsilon_{pe} = \frac{f_{pe}}{E_p} \tag{5.7}$$

The section at the level of the prestress tendon is compressed due to the effective prestress. Therefore, it is necessary to calculate the strain needed to **decompress** the section to a condition of zero strain. The resulting decompression strain ε_d is given in Eq. (5.8).

$$\varepsilon_d = \frac{P_e}{A_g E_c} + \frac{P_e e^2}{I_g E_c} \tag{5.8}$$

The strain due to flexure at the nominal strength ε_f results from strain compatibility. The concrete strain at nominal strength ε_{cu}, typically taken as 0.003, allows for the calculation of the flexural strain as given in Eq. (5.9).

$$\varepsilon_f = \varepsilon_{cu}\left(\frac{d_p - c}{c}\right) \tag{5.9}$$

where c is the distance from the compressive face to the cracked section centroid. The total strain in the tendon is then

$$\varepsilon_{ps} = \varepsilon_{pe} + \varepsilon_d + \varepsilon_f \tag{5.10}$$

Equilibrium on the section requires the tensile force equal the compressive force $T = C$ for a member with a constant width of the compression zone b. Thus,

$$A_p f_{ps} = 0.85 f'_c ba \tag{5.11}$$

where a is the depth of the equivalent rectangular stress block

$$a = \frac{A_p f_{ps}}{0.85 f'_c b} \tag{5.12}$$

and

$$c = a/\beta_1 \tag{5.13}$$

where β_1 is the characteristic value of the equivalent rectangular stress block given in Eq. (5.6). Finally, the nominal moment can be obtained by summing moments about the centroid of the compression force

$$M_n = A_p f_{ps}\left(d_p - \frac{a}{2}\right) \tag{5.14}$$

Equations (5.7) through (5.13) cannot be solved directly because f_{pu} is not known until the strain conditions provide equilibrium on the section. While an equation solver is one approach, the following procedure is used to develop an iterative solution.

1. Calculate the effective prestressing strain ε_{pe}.
2. Calculate the decompression strain ε_d.
3. Assume an initial value for f_{pu}.
4. Calculate the tensile force for the assumed f_{pu}.
5. Calculate the depth to the cracked neutral axis based on horizontal equilibrium using Eqs. (5.11) and (5.12).
6. Calculate the flexural strain ε_f and add it to the effective prestressing strain ε_{pe} and the decompression strain ε_d.
7. Determine the value of f_{pu} for the total strain in step 5.
8. Compare the value of f_{pu} to the assumed value and repeat steps 2 through 6 until satisfactory agreement is attained.
9. With f_{pu} now known, calculate the nominal moment capacity Eq. (5.14).

A comment about iterative solutions and satisfactory agreement is pertinent. Most properties of concrete are known to only two significant figures. Calculating a stress to more than two significant figures is usually not necessary, resulting in an iterative solution requiring only two or three cycles. The above steps are easily programmed in EXCEL as illustrated in Example 5.5. An iterative solution becomes self-checking because convergence is monitored. Equation solvers, while efficient, do not allow the same level of behavioral understanding and checking.

Example 5.5: Calculate a Nominal Moment Capacity Using Strain Compatibility

Determine the nominal strength of the beam in Example 5.1 using strain compatibility and the prestressing reinforcement stress–strain relationship from Eq. (3.7).

Solution: The factored applied moment is:

$$M_u = 1.2 M_D + 1.6 M_L = 1.2 \cdot (3148 + 1000) + 1.6 \cdot 4000 = 11,378 \,\text{in.-kip}$$

The structural depth is the distance from the top of the beam plus the eccentricity so

$$d_p = y_t + e_p = 24.73 + 13 = 37.73 \,\text{in.}$$

The results on the EXCEL spreadsheet are given in Table 5.9 and the convergence is noted between the two bold values for f_{pu} and f_{ps}.

Comments: (1) The decompression strain is an order of magnitude less that the pretensioning and flexure strains. This is common, and some engineers ignore the decompression strain. (2) The convergence of the trial stress and final stress increase difference is less than 1%, certainly within accuracy for the known material properties. (3) The value of c/d_p is less than 0.375 indicating the beam is acting in a ductile fashion and a ϕ factor of 0.90 is valid. (4) Adding steel reinforcement would be reflected in the calculation of T assuming the steel yields. Validation of the steel strain through strain compatibility ensures that the assumption is correct. (5) the depth of the stress block is less than 1 in. below the 7 in. available but results in less than 1% difference in M_n.

5.4.3   Bonded Tendons: ACI Approach

The ACI Building Code offers an approximate method for calculation of the stress in prestressing reinforcement at nominal flexural strength in bonded prestress applications, Eq. (5.15).

$$f_{ps} = f_{pu}\left[1 - \frac{\gamma_p}{\beta_1}\left(\rho_p \frac{f_{pu}}{f'_c} + \frac{d f_y}{d_p f'_c}(\rho - \rho')\right)\right] \qquad (5.15)$$

Table 5.9 Nominal strength worksheet for Example 5.9

	Properties			Calculations		
$A_g =$	560	in.^{2}	$\varepsilon_{pe} =$	$f_{pe}/E_p =$	0.0056	
$I_g =$	125,390	in.^{4}	$\varepsilon_d =$	$(P_e/A_g + P_e e^2/I_g)/E_p =$	4.30×10^{-5}	
$e =$	13	in.				
$d_p =$	37.73	in.	**Try:**	$f_{pu} =$	**265**	ksi
$b =$	16	in.	$T =$	$A_p f_{pu} =$	649	kips
$A_p =$	2.45	in.^{2}	$a =$	$T/(0.85f'_c b) =$	7.95	in.
			$c =$	a/β_1	10.6	in.
$P_e =$	391	kips	$\varepsilon_f =$	$\varepsilon_u (d_p - c)/c =$	0.011356	
$f'_c =$	6000	psi	$\varepsilon_{PS} =$	$\varepsilon_{pe} + \varepsilon_d + \varepsilon_f =$	0.01701	
$\beta_1 =$	0.75		$f_{ps} =$	$270 - 0.04/(\varepsilon_{ps} - 0.007)$	**266.7**	ksi
$f_{pe} =$	159.9	ksi		$c/d_p =$	0.320	<0.375 OK
$\varepsilon_{cu} =$	0.003					
$E_p =$	28,500	ksi	$\boldsymbol{M_n =}$	$\boldsymbol{T(d_p - a/2)}$	**21,898**	**in.-kips**

where γ_p is a factor used to define the type of prestressing reinforcement used. Values for γ_p are $\gamma_p = 0.28$ for f_{py}/f_{pu} not less than 0.90, which includes low relaxation wire and strands $\gamma_p = 0.40$ for f_{py}/f_{pu} not less than 0.85, which includes ordinary stress relieved wire and strands, and $\gamma_p = 0.55$ for f_{py}/f_{pu} not less than 0.85, which includes prestressing bars and rods.

Equation (5.15) is valid for members where the initial prestress is at least 50% of the yield stress and can be considered in two parts. The term beginning with ρ_p relates to the prestressing contribution to the stress and the term beginning with d/d_p adds the contribution of non-prestressed reinforcement. The non-prestressed reinforcement may be either tensile reinforcement or compressive reinforcement. The latter term may be ignored if there is no supplemental reinforcement present. Equation (5.15) was developed by examining the parameters affecting the nominal strength of many prestressed members in available test reports. Most of these tests were for members prestressed with strand and the strand stressed to at least 50% of the yield stress. The equation remains valid for members prestressed with all strand sizes and Grades. While there is no restriction in the ACI Building Code on the use of Eq. (5.15), a strain compatibility approach is prudent for members prestressed with high-strength bars.

Example 5.6: Calculate the Nominal Strength of a Bonded Prestressed Tendon Beam

Determine the nominal strength of the girder in Example 5.1 using the ACI approach assuming the tendon is bonded.

Solution: The beam uses 270 ksi low relaxation strand, giving $\gamma_p = 0.28$ and for $f'_c = 6000\,\text{psi}$, $\beta_1 = 0.80$. From Eq. (5.15) and noting that no nonprestressed reinforcement is present

$$f_{ps} = f_{pu}\left[1 - \frac{\gamma_p}{\beta_1}\left(\rho_p\frac{f_{pu}}{f_c'}\right)\right] = 270\left[1 - \frac{0.28}{0.75}\left(0.0044\frac{270}{6}\right)\right] = 250.0\,\text{ksi}$$

giving

$$a = \frac{A_p f_{ps}}{0.85 f_c' b} = \frac{2.45 \cdot 250.0}{0.85 \cdot 6 \cdot 16} = 7.51\ \text{in.}$$

which is less than the depth of the top flange. The nominal moment is then

$$M_n = A_p f_{ps}\left(d_p - \frac{a}{2}\right) = 2.45 \cdot 250.0\left(37.73 - \frac{7.51}{2}\right) = 20825\ \text{in.-kip}$$

Comment: The nominal moment is within 5% of the strain compatibility approach, as expected for a beam with "normal" prestressing placement and stress levels.

5.4.4   Unbonded Tendons

A fundamental assumption in reinforced concrete behavior is that there is no slip between the concrete and the reinforcement. This assumption allows the calculation of the strain in the reinforcement based on the strain in the concrete. An **unbonded tendon** is a tendon in which prestressed reinforcement is prevented from bonding to the concrete. The prestressing force is permanently transferred to the concrete at the tendon ends by the anchor assembly only. The assumption of the same strain in the concrete and the tendon is not valid for unbonded tendons. As a beam deforms in bending, the total tendon elongation is a function of the elongation of the tendon between the beam ends. For a simple span beam with a straight tendon, the total elongation is approximately the change in length due to the rotation at the end of the beam, Fig. 5.10.

A tendon in a beam with substantial structural depth will have a greater elongation that a tendon in a thin slab. Recognizing these effects, Section 20.3.2.4.1 of ACI 318-14 provides a series of limits for the stress gain in unbonded tendons that is dependent on the span-to-depth ratio of the member. Historically, the span to depth ratio was used to indicate increases in stress in unbonded tendon; however, equations based on span to depth ratios were found to overestimate the increase in tendon stress. The prestressing reinforcement ratio ρ_p is a better indicator of the final stress when using the ACI approach. Strength limits in Table 5.10 are based on data from experimental evaluations (Mojtahedi and Gamble 1978). The equations further require that the effective prestress f_{se} be at least 50% of the reinforcement tensile strength or $f_{se} \geq 0.5\,f_{pu}$.

Fig. 5.10 Beam end indicating tendon elongation

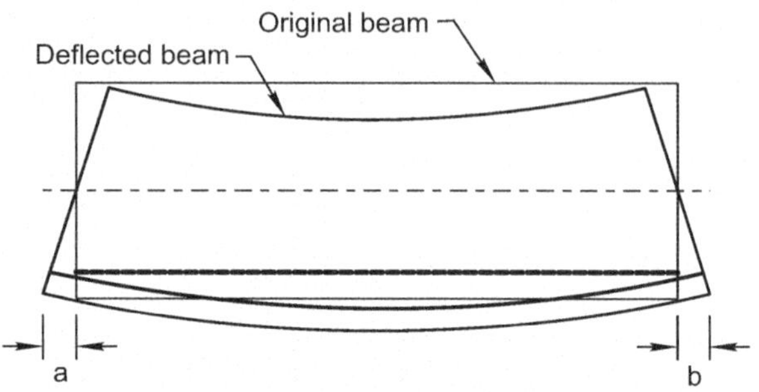

Table 5.10 Approximate values of f_{ps} at nominal flexural strength for unbonded tendons

ℓ_n/h	f_{ps} (psi)	
≤ 35	The least of:	$f_{pe} + 10,000 + f'_c/(100\rho_p)$
		$f_{pe} + 60,000$
		f_{py}
>35	The least of:	$f_{pe} + 10,000 + f'_c/(300\rho_p)$
		$f_{pe} + 30,000$
		f_{py}

Adapted from Section 20.3.2 of ACI 318-14

Example 5.7: Calculate the Nominal Strength of a Beam with an Unbonded Prestressed Tendon

The beam from Example 5.1 has a cross-sectional area of 560 in.^{2}, a moment of inertia of 125,390 in.^{4}, a weight of 583 plf, a distance from the neutral axis to the top fiber of 24.73 in., a distance to the bottom fiber of 20.27 in. and a width of the top flange of 16 in. The beam is post-tensioned with an unbonded 16-strand tendon placed in a parabolic drape so the eccentricity is 13 in. at midspan. Service level moments are $M_g = 3148$ in.-kip, $M_{sdl} = 1000$ in.-kip, and $M_l = 4000$ in.-kip. The specified concrete strength is 6000 psi and the effective prestress is 159.9 ksi.

Solution: From Examples 5.1 and 5.2, service level stresses are within ACI specified limits. The factored applied moment is:

$$M_u = 1.2 M_D + 1.6 M_L = 1.2 \cdot (3148 + 1000) + 1.6 \cdot 4000 = 11378 \,\text{in.-kip}$$

The structural depth is the distance from the top of the beam plus the eccentricity so

$$d_p = y_t + e_p = 24.73 + 13 = 37.73 \,\text{in.}$$

and $\rho_p = 2.45/(20 \cdot 37.73) = 0.0044$. The span to depth ratio is 60 ft $\times$ 12 in./ft/34. 73 in. $= 20.7$, therefore, the first equation from Table 5.10 is applicable and the tendon stress is:

$$f_{ps} = f_{se} + 10,000 + \frac{f'_c}{100\rho_p} = 159.9 + 10 + \frac{6}{100 \cdot 0.0037} = 183.5\,\text{ksi}$$

which is less than both $f_{se} + 60$ ksi and f_{py}. Use $f_{ps} = 183.5$ ksi to calculate the nominal moment.

$$a = \frac{f_{ps} \cdot A_{ps}}{0.85 f'_c \cdot b} = \frac{183.5 \cdot 2.45}{0.85 \cdot 6 \cdot 16} = 5.51\,\text{in.}$$

The compression block is within the compression flange so the nominal moment is

$$M_n = f_{ps} A_{ps}\left(d_p - \frac{a}{2}\right) = 183.5 \cdot 2.45\left(37.73 - \frac{5.51}{2}\right) = 15,723\ \text{in} \cdot \text{kip}$$

Applying the strength reduction factor of $\phi = 0.90$, $\phi M_n = 0.9 \times 15{,}723 = 14{,}150$ in.-kip, which is greater than M_u, and therefore the design is satisfactory.

Comment: The stress in the bonded tendon from Example 5.6 is substantially higher than the stress in the unbonded tendon and the moment capacity is up to 39% higher than the beam with the unbonded tendon. Grouting tendons ensures strain compatibility, which will provide a greater flexural strength than the same size unbonded tendon. Grouting requires careful inspection to ensure the long-term serviceability of the tendon (Hamilton et al. 2000; Abdullah et al. 2014).

5.4.5   Flanged Sections

Example 5.6 was solved using a rectangular cross section as the compression block remained in the top flange. If the depth of the compression block exceeds the depth of the top flange, then the strain compatibility procedure needs to be modified to calculate the depth of the compression block in steps, similar to the calculation of compression blocks in reinforced T-beams. Figure 5.11 shows a schematic beam section with a compression block extending below the top flange and into the web. The assumed rectangular stress block allows the tapered parts of the section to use average widths rather than dividing the section of the prism into smaller components.

For the case indicated in Fig. 5.11, the compression resultant would be calculated in three sections. Section 1 is the width of the flange times its average depth with a centroid at the center of the rectangle. Section 2 is a rectangular section the depth of

Fig. 5.11 Compression
areas of flanged section

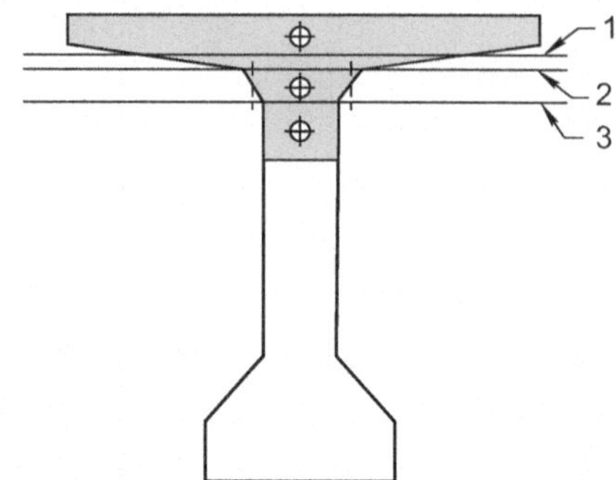

the sloped area times the average width. Section 3 is a rectangular section in the web. Calculation of the nominal moment can be done by summing moments about the tension centroid to each of the compression elements rather than determining the composite centroid of the compression zone.

5.5   Stresses in Class T and C Beams (Partial Prestress)

Chapter 1 noted the initial development of prestressed concrete did not allow tension under service loads. Members with tension stresses were historically considered **partially prestressed**, whether or not they cracked. This led to considerable debate within the design profession regarding design procedures for partial prestressing (ACI 423.5-99). The debate effectively ended with the classification of prestressed members into Class U, T, and C for uncracked, transitional, and cracked members. This classification unifies the treatment of prestressed members and effectively eliminated the term partial prestress. Class U members are considered uncracked; stresses and deflections are calculated on the gross sectional area. The engineer has the option of using the transformed gross area if there is a considerable area of prestressing reinforcement present. Class C members are considered to be cracked and Class T members may be cracked. Section 24.4 of the ACI Building Code requires deflection calculations for Class T and C members to be based on the cracked section properties. Section 24.5 of the ACI Building Code allows calculation of stresses in Class U and Class T members based on uncracked section properties while Class C members stress calculations are based on cracked section properties.

5.5.1   Cracked Section Properties

When a prestressed concrete member cracks, the tension force carried by the portion of the concrete in tension is released, the neutral axis moves upward until a new

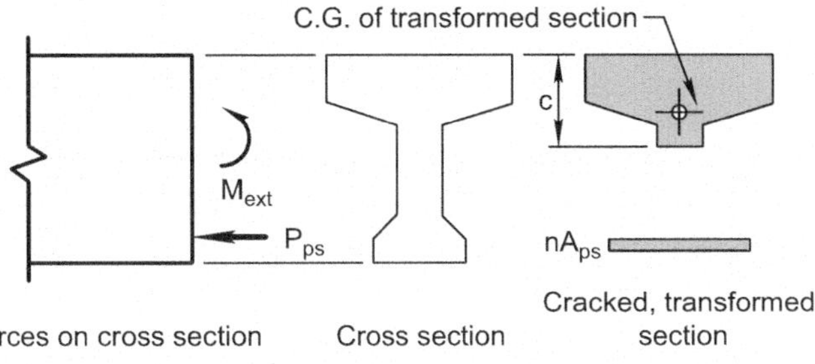

Fig. 5.12 Forces on a cracked prestressed section

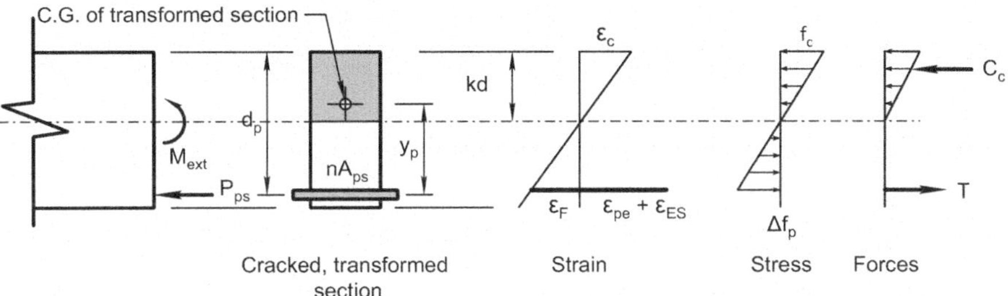

Fig. 5.13 Stress strain and equilibrium on cracked section

equilibrium condition is reached. In the process of reaching equilibrium, the strain in the prestress reinforcement and the strain in the remaining concrete in compression increases, Fig. 5.12. Thus, calculation of cracked section properties of prestressed members is based on equilibrium on the section and strain compatibility. The calculation is an iterative procedure similar to finding the nominal strength of the section. The primary difference is that the concrete stress remains elastic.

The iterative procedure to find section properties and stresses begins with a stress check on the gross section to determine if the section cracks, that is the tensile stress exceeds $f_t > 7.5 \sqrt{f'_c}$. Two procedures are presented. First, a solution for a rectangular section with a single layer of prestress is developed based on first principles and then extended to a general solution. Next, a solution method from the *PCI Design Handbook* (2017) is presented.

When the section cracks, the remaining section must be in equilibrium, Fig. 5.13. The strain in the prestress reinforcement is equal to the effective strain after losses ε_{pe} plus the strain due to elastic shortening ε_{ES}, which brings the section to a neutral position. Lastly, the flexure of the section increased the strain ε_f. The stress levels remain low allowing linear elastic behavior for the development of the equations.

The flexural strain in the prestressed reinforcement is found by strain compatibility and substitution of the stress divided by the modulus of elasticity for the strain, where the depth from the compression face to the neutral axis is noted as *kd*. For a trial value of *kd*:

$$\varepsilon_{\mathrm{f}} = \frac{\varepsilon_{\mathrm{c}}}{kd}\left(d_{\mathrm{p}} - kd\right) = \frac{f_{\mathrm{c}}}{E_{\mathrm{c}}}\left(\frac{1-k}{k}\right) \tag{5.16}$$

In Eq. (5.16), the depth to the prestress face d_{p} is assumed to be the same as the average structural depth. If there are multiple levels of prestress, Eq. (5.16) can be modified to account for the differences in the prestressed reinforcement. The resulting prestress force is then

$$f_{\mathrm{ps}} = f_{\mathrm{pe}} + f_{\mathrm{ES}} + E_{\mathrm{p}}\varepsilon_{\mathrm{f}} = f_{\mathrm{pe}} + f_{\mathrm{ES}} + nf_{\mathrm{c}}\left(\frac{1-k}{k}\right) \tag{5.17}$$

Equilibrium on the section requires that the compression force C_{c} equal the tension force T given in Eqs. (5.18) and (5.20).

$$C_{\mathrm{c}} = \frac{1}{2}f_{\mathrm{c}}bkd \tag{5.18}$$

and

$$T = f_{\mathrm{ps}}A_{\mathrm{p}} = \left(f_{\mathrm{pe}} + f_{\mathrm{ES}} + nf_{\mathrm{c}}\left(\frac{1-k}{k}\right)\right)A_{\mathrm{p}} \tag{5.19}$$

Setting the forces equal and solving for f_{c} gives

$$f_{\mathrm{c}} = \frac{2\left(f_{\mathrm{pe}} + f_{\mathrm{ES}}\right)}{bkd - n\frac{1-k}{k}} \tag{5.20}$$

finally, the internal moment for the section is

$$M_{\mathrm{int}} = f_{\mathrm{ps}}A_{\mathrm{p}}\left(d_{\mathrm{t}} - \frac{kd}{3}\right) \tag{5.21}$$

The internal moment is compared to the applied moment. If the internal moment is less than the external moment, kd is adjusted to be smaller. If the internal moment is larger than the external moment, kd is increased and the calculation repeated until convergence is achieved.

When the moments converge, the centroid of the cracked section is determined by finding the distance from the top of the beam.

$$y_{\mathrm{t}} = \frac{\frac{b(kd)^2}{2} + nA_{\mathrm{p}}d_{\mathrm{t}}}{bkd + nA_{\mathrm{p}}} \tag{5.22}$$

Finally, the cracked moment of inertia is found

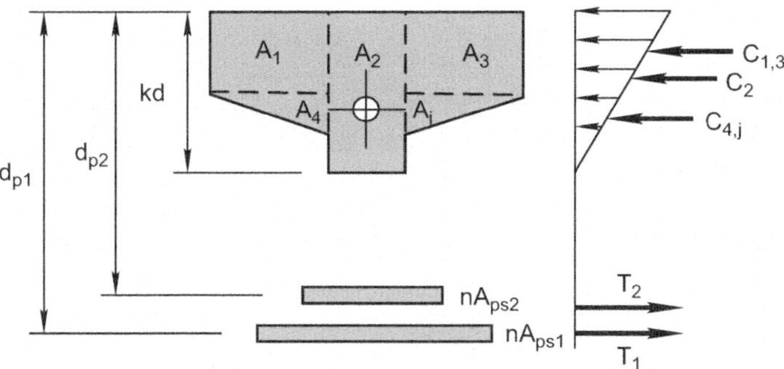

Fig. 5.14 Sectional elements and forces

$$I_{cr} = \frac{b(kd)^3}{3} + bkd\left(y_t + \frac{kd}{2}\right) + nA_p\left(d_p - y_t\right)^2 \tag{5.23}$$

Equations (5.18) through (5.23) are for a rectangular section. For sections of varying cross section, the approach is still valid; however, the compression zone must be divided into individual segments, A_1 through A_j, Fig. 5.14. Similarly, if there are multiple levels of prestress or non-prestressed reinforcement present, each element must be tabulated separately. If the beam has composite topping, the process is modified by including the stresses and forces in the composite topping after the topping has been modified to an equivalent transformed area corresponding to the base beam properties.

Example 5.8: Calculate the Cracked Moment of Inertia
A 12-in. wide by 32 in. deep rectangular beam 40 ft long is post-tensioned with a 12-strand grouted tendon. For the properties, loads, and prestress level given, calculate the cracked moment of inertia of the section. Convergence occurs when a value for kd returns a zero difference between M_{int} and M_{ext}, indicated in bold font. **Solution**: The first trial for the neutral axis is $kd = 20$ in. The final value is shown in Table 5.11.

Comment: In this example the uncracked moment of inertia of the gross section is $bh^3/12$ or 32,768 in.[4] The cracked section has an effective moment of inertia 34% of the uncracked section and the compressive stress increases by a factor of nearly 3.

5.5.2   PCI Design Handbook Approach

The *PCI Design Handbook* (2017) has a slightly different approach based on a paper by Mast (1998). This approach transfers the compression force to the uncracked

Table 5.11 Cracked moment of inertia worksheet

	Properties			Calculations		
			Check cracking			
$A_g =$	348	in.^{2}	$M_d =$	$w_d L^2/8$	3360	kip-in.
$I_{ut} =$	32,768	in.^{4}	$M_L =$	$w_d L^2/8$	3000	kip-in.
$y_t =$	16	in.	$M_{ext} =$	$M_d + M_L$	6360	kip-in.
$y_b =$	16	in.	$f_b =$	$M_{ext}\, y_b/I_g$	−798	psi
$b =$	12	in.	$f_r =$	$7.5\sqrt{f'_c}$	581	psi
$d_p =$	26	in.		$f_b =$	10.3	$\sqrt{f'_c} > f_r$
$e =$	10	in.		***Conclusion***		
				Class T member and section is cracked		
$A_p =$	1.836	in.^{2}				
$f_{se} =$	162	ksi	**Find neutral axis**			
$P_e =$	297	kips	**Try:**	$kd =$	20.405	in.
$f_{ES} =$	16	ksi	$k =$	kd/d_p	0.785	
			$f_c =$	$2(f_{se} + f_{ES})/(kdb - n(1 - k)/k)$	1.46	ksi
$f'_c =$	6000	psi	$f_{PS} =$	$f_{se} + f_{ES} + n f_c ((1 - k)/k)$	180.4	ksi
$E_p =$	28,500	ksi	$M_{int} =$	$f_{PS} A_p (d_p - kd/3)$	6360	kip-in
$E_c =$	4696	ksi		$M_{int} - M_{ext}$	**0.0**	kip-in
$n =$	6.07					
			Moment of inertia			
$L =$	40	ft	$A_{ct} =$	$kd\, b + n A_p$	252.8	in.^{2}
$w_g =$	400	plf	$y_t =$	$(b(kd)^2/2 + nA_p\, d_p)/A_{ct}$	11.0	in.^{2}
$w_{sdl} =$	1000	plf	$I_{ct} =$	$b(kd)^3/12 + b\, kd(y_t - kd/2)^2 +$		
$w_L =$	1250	plf		$nA_p(d_p - y_t)^2$	11,035	in.^{4}

portion of the section. The same notation as the PCI Design Handbook Example 4.2.2.5 is used in this section and the terms are given in Fig. 5.15.

Once cracking of the section is determined, estimate the decompression force P_{ps} by calculating the decompression stress $f_{dc} = f_{se} +$ *elastic shortening loss* and noting $P_{ps} = f_{dc} A_{ps}$. If there is non-prestressed reinforcement in the section, estimate its decompression force resulting from creep and shrinkage times the area of the reinforcement. This decompression force is opposite in sign to the prestressing force. Calculate the modular ratio for the concrete and the reinforcement then calculate transformed areas of the reinforcement. Begin the iteration by estimating the location of the neutral axis of the cracked section and calculating the cracked transformed A, I, and the centroid of the transformed section. Calculate the product of P_{ps} times its eccentricity from the neutral axis y_p. Combine this moment with the external moment to obtain the internal moment M_{int}. Finally, calculate the stress at the neutral axis. Adjust the estimate of c until the neutral axis stress is effectively zero. Once convergence is achieved, these transformed section properties may be used to calculate stresses in Class C beams and deflections in Class T and C beams.

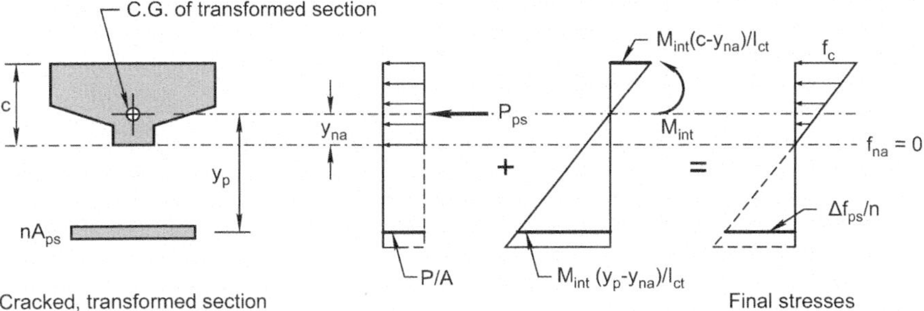

Fig. 5.15 Notation used in PCI Design Handbook for cracked prestressed concrete sections

The PCI Design Handbook example and Example 5.7 are within 3% of each other, which is acceptable considering the assumptions and iterations involved.

5.5.3   Unbonded Tendons

The procedure for bonded tendons is also suitable for cracked sections with unbonded tendons. Unbonded tendons do not have strain compatibility with the concrete; therefore, two revisions to the above procedure are necessary. First, the strains due to elastic shortening ε_{ES} and flexure ε_f may conservatively be assumed to be zero. This follows because bending in the member does not result in appreciable strain changes in the tendon at service level loads. Second, calculation of the cracked section moment of inertia is not included in the transformed area calculation. The result is more cracking of the section and a smaller cracked moment of inertia, Example 5.7.

Example 5.9: Calculate the Cracked Section Properties with Unbonded Tendons
The section in Example 5.8 is repeated assuming that the tendon is unbonded. Revised Eqs. (5.17)–(5.21) are indicated in the spreadsheet in Table 5.12. The same effective prestress is used in both examples. Convergence occurs when a value for kd returns a zero difference between M_{int} and M_{ext}, indicated in bold font.

Comments: The lack of strain compatibility results in a lower prestress force, a higher concrete compressive stress, a higher location for the neutral axis, and a smaller cracked moment of inertia.

Table 5.12 Cracked properties worksheet for unbonded tendons

	Properties			Calculations		
			Check cracking			
$A_g =$	348	in.^{2}	$M_d =$	$w_d L^2/8$	3360	kip-in.
$I_{ut} =$	32,768	in.^{4}	$M_L =$	$w_d L^2/8$	3000	kip-in.
$y_t =$	16	in.	$M_{ext} =$	$M_d + M_L$	6360	kip-in.
$y_b =$	16	in.	$f_b =$	$M_{ext} y_b/I_g$	−798	psi
$b =$	12	in.	$f_r =$	$7.5 \sqrt{f'_c}$	581	psi
$d_p =$	26	in.		$f_b =$	10.3	$\sqrt{f'_c} > f_r$
$e =$	10	in.		***Conclusion***		
				Class T member and section is cracked		
$A_p =$	1.836	in.^{2}				
$f_{se} =$	162	ksi	**Find neutral axis**			
$P_e =$	297	kips	**Try:**	$kd =$	13.85	in.
$f_{ES} =$	16	ksi	$k =$	kd/d_p	0.533	in.
			$f_c =$	$2(f_{se})/(kdb - n(1 - k)/k)$	2.01	ksi
$f'_c =$	6000	psi	$f_{PS} =$	f_{se}	162.0	ksi
$E_p =$	28,500	ksi	$M_{int} =$	$f_{PS} A_p (d_p - kd/3)$	6360	kip-in
$E_c =$	4696	ksi		$M_{int} - M_{ext}$	**0.1**	kip-in
$n =$	6.07					
			Moment of inertia			
$L =$	40	ft	$A_{ct} =$	$kd\,b$	174.1	in.^{2}
$w_g =$	400	plf	$y_t =$	$kd/2$	6.0	in.^{2}
$w_{sdl} =$	1000	plf	$I_{ct} =$	$b(kd)^3/12 + b\,kd(y_t - kd/2)^2$	2692	in.^{4}
$w_L =$	1250	plf				

Problems

5.1. A 40 ft beam has a 6-½ in. diameter low relaxation strands with a constant eccentricity of 5-½ in. The initial prestress is 178.4 ksi and the final effective prestress is 151.6 ksi. The beam carries a superimposed dead load moment of 500 in.-kip in addition to its self-weight and a live load moment of 820 in.-kip. Section properties are $A_g = 176$ in.^{2}, $I_g = 12,000$ in.^{4}, $y_t = y_b = 12$ in. and the width of the compression face is 10 in. Determine the service level stresses in the beam.

5.2. A 10 ft by 24 in. (Appendix B.2) 50-foot-long double-T beam has 10-½ in. diameter low relaxation strands with a constant eccentricity of 9 in. The initial prestress is 306 kip and the final effective prestress is 260 kip. The beam carries its self-weight plus a midspan live load service moment of 1500 in.-kip. Determine the service level stresses in the beam and comment about the stresses at the end of the beam.

5.3. What is the required end eccentricity of the strands in Problem 5.2 to bring the stresses within allowable limits if the transfer strength is 3500 psi and the design strength is 5000 psi?

5.4. Determine the nominal moment strength of the beam in Problem 5.1 using the ACI approximation for the tendon strength (f_{ps}) given $f'_c = 5000\,\text{psi}$ and determine if the beam has adequate design moment strength for the factored moments.

References

AASHTO LRFD Bridge Design Specification (8rd Ed.). (2017). Washington, DC: American Association of State Highway and Transportation Officials (AASHTO).

Abdullah, A. B. M., Rice, J. A., & Hamilton, H. R. (2014). Wire breakage detection using relative strain variation in unbonded posttensioning anchors. *Journal of Bridge Engineering, 20*(1).

ACI 318-14. (2014). *Building Code Requirements for Structural Concrete* (ACI Committee 318-14, p. 519). Farmington Hills, MI: American Concrete Institute.

Hamilton, H. R., Wheat, H., Breen, J., & Frank, K. (2000). Corrosion testing of grout for posttensioning ducts and stay cables. *Journal of Structural Engineering, 126*(2), 163–170.

Mojtahedi, S., & Gamble, W. L. (1978). Ultimate steel stresses in unbonded prestressed concrete. *Proceedings, ASCE, 104*(ST7), 1159–1165.

Mast, R. F. (1998) Analysis of Cracked Prestressed Concrete Sections: A Practical Approach. *PCI Journal 43(4)*:80–91

PCI Design Handbook (8th Ed.). (2017). Chicago, IL: Precast and Prestressed Concrete Institute.

Chapter 6
Flexure: Design

6.1 Practical Flexural Design Approach

Structural design involves determining the section and material properties followed by prestress details to satisfy serviceability, strength, and detailing. The process typically begins by validating that service level stresses are adequate for the selected section and materials. The service level stress check is followed by design of the nominal flexural strength. If the nominal flexural strength is inadequate with only prestressing reinforcement, additional strength is obtained by increasing the amount of prestressing steel or adding nonprestressed reinforcement, or both. Once the flexural demands are satisfied, the design effort continues to address shear capacity, bond and development, and serviceability requirements of deflection, camber, and vibration. The order of resolving these issues is dependent on the member being designed.

In Chap. 2 structures or members are classified as standardized precast prestressed elements, fixed cross section elements, and fully engineered structures. For the first two structure types, the selection of cross section geometry is limited so that determination of the magnitude, quantity, and placement of the prestressing force reinforcement is often all that is required. The industry practice of determining the prestressing is selected varies throughout the country and among design firms. Selection of a cross section and the corresponding prestress force to meet the structural requirements is standard practice in many firms. In some structural engineering firms, the selection of the prestressing details is delegated to the precast plant. If delegated, the principal design firm validates the selection of prestress force and location when checking shop drawings. Projects having loadings other than the ordinary uniformly distributed loads discussed in Chap. 1 require a detailed description and selection of prestressing forces.

Bridges are typically designed by state departments of transportation or their consultants. In either case, the bridge design requires the engineer to prepare

© Springer Nature Switzerland AG 2019

C. W. Dolan, H. R. Hamilton, *Prestressed Concrete*,

https://doi.org/10.1007/978-3-319-97882-6_6

Table 6.1 Span-to-depth ratios for building prestressed members

Member	L/h
I-beams and single-T beams	24–36
Double-T beams	30–40
One-way solid slabs	35–50
One-way hollowcore slabs	40–50
Two-way solid flat plates	40–50

Table 6.2 AASHTO girder span ranges

AASHTO standard girder	Span range (ft)	Approximate L/h
I	30–45	13–20
II	40–50	13–17
III	55–80	15–21
IV	70–100	15–22
V	80–120	15–23
VI	110–140	18–23

comprehensive prestressing details. These details include the magnitude of initial prestress, reinforcement placement, anchorage details, and stressing specifications.

6.1.1 Selection of Section

Selection of a section is dependent on the application. Resources such as the PCI *Design Handbook* (2017) provide span load tables for many common standard shapes. These include solid and hollowcore planks, double-T beams, and inverted T-beams. Table 6.1 provides ranges of span-to-depth ratios for many common concrete sections. These span-to-depth ratios are for usual loadings such as those listed in Chap. 1. Heavier or unusual loads require adjustment of span or member depth to ensure strength and serviceability limits are satisfied. For example, a post-tensioned two-way floor slab thickness is established by span-to-depth ratios in the range of 40–50 from Table 6.1 and then checked for deflection. Thus, a 30-ft post-tensioned slab would have a trial thickness of 8 in. The approximate span range of AASHTO standard sections is provided in Table 6.2. The cast-in-place bridge deck and heavier bridge loadings are reflected in the lower span-to-depth ratios.

Specialty designs often impose restrictions on selection of a section. For example, the Walt Disney World monorail beam required a 26-in. width and minimum 48-in. overall height to conform to the vehicle wheel and support geometry. The minimum depth in a segmental bridge is often set by the span-to-depth ratio to control deflections and vibrations. The section over the pier is established for overall strength. Water storage tank wall thickness is set by the size of the post-tensioning duct and cover requirements for long-term durability.

Lin and Burns (1981) and Nilson (1987) have used the available service level stress range indicated in Fig. 6.1 to derive equations to establish required section

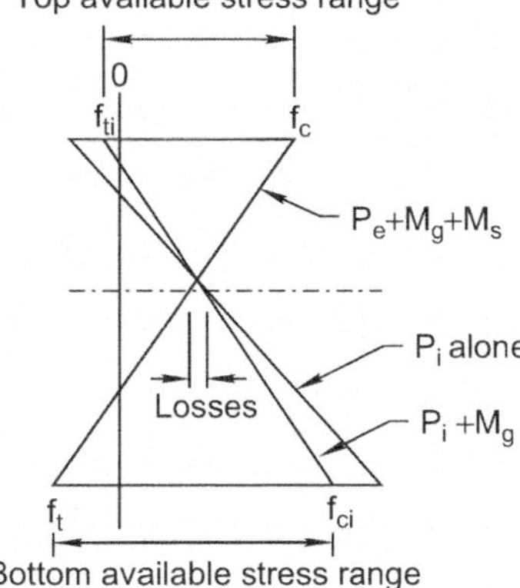

Fig. 6.1 Combined sectional stresses due to loads indicated

moduli for given loads and concrete stresses. Their proposed equations include a factor to account for the losses. The equations then provide an "exact value" for required section modulus and tendon location to optimize a section. In ordinary design cases, the actual member section has at least one property that exceeds these optimal solutions leading to over or under stress conditions at other locations. The losses used in the equations can only be accurately calculated once the section and prestress is known. Lastly, these equations only address the allowable stresses but not strength or serviceability conditions, thus making the equations a first approximation of sectional requirements. These equations from Lin and Burns or Nilson are not included in this text in favor of a rational selection process.

A logical design approach is to follow a three-step process: (1) select a section based on project requirements or span-to-depth ratio, (2) select the initial prestress force (a) using the maximum available tendon eccentricity, (b) providing an average prestress force at the member centroid of approximately one half of the allowable prestress at the extreme fibers, or (c) using on load balancing techniques, and (3) complete the detailed check of the design revising it as needed.

6.1.2   Selecting a Prestress Force and Tendon Location

The prestress force is selected based on the member use, loading, and cost. A beam designed to maximize one of the section properties, e.g., tension in the concrete or compression in the concrete, uses as much of the section capacity as possible without exceeding stress limits. Determining the maximum available tendon eccentricity then calculating the prestress force often provides the lowest required prestress force. The maximum possible eccentricity is the distance to the bottom fiber less the cover and placement requirements for the tendon, an approach used in Example 6.1. This approach is beneficial when standard elements are selected and the full stress range is not used. For example, a bridge girder whose span is at the lower end of the span-to-depth ratios in Table 6.2 would not be stressed to the maximum compressive capacity under service load.

Using the largest eccentricity may lead to excessive tensile stresses at the time of prestress transfer. The **kern point k** is the location of the maximum eccentricity that results in a net stress of zero at the extreme fiber of the precompressed tensile zone. In a rectangular section, this is referred to as the middle third rule. Any axial load placed within the middle third of a rectangular section generates no tension at the extreme fiber. The kern points for non-rectangular sections are calculated using Eq. (6.1). The equation is derived by setting the top and bottom stresses to zero, and ignoring the applied loads, thus:

$$0 = \frac{P_e}{A_g} - \frac{P_e k_b y_t}{I_g}$$

$$0 = \frac{P_e}{A_g} + \frac{P_e k_t y_b}{I_g}$$

Noting that the prestressing force cancels out and solving for the top and bottom kern points k_t and k_b respectively gives

$$k_t = -\frac{I_g}{A_g y_b} = -\frac{r^2}{y_b}$$

$$k_b = \frac{I_g}{A_g y_t} = \frac{r^2}{y_t}$$

$$(6.1)$$

Any tendon eccentricity below the neutral axis greater than k_b results in tension in the top fiber of the member and any eccentricity above the neutral axis greater than k_t creates tension in the bottom of the member. The minus sign in Eq. (6.1) indicates that the tendon eccentricity is above the neutral axis. Using the radius of gyration, $r^2 = I_g/A_g$, gives a slightly more compact form of Eq. (6.1).

The zero tension criteria can be maintained with a greater eccentricity if the girder self-weight is applied simultaneously with the prestressing force. In a simple span member, where the prestress provides upward camber, the eccentricity may be increased by the amount M_g/P_i so that the stresses generated by the girder dead load are compensated by the increase in tendon eccentricity. Sufficient space must be available to lower the tendon below the kern point, or below the point generating maximum tension or compression. Under these conditions the self-weight of the beam is carried "free," that, is no additional prestress force is required.

An alternative starting point for a beam to be fully stressed is to select an initial prestressing force that produces approximately half of the allowable compression stress at transfer. Examining Fig. 6.1, either the top or the bottom of the section is used to determine the average prestress. Half the allowable service stress, that is $(f_{ci} - f_{ti})/2$, allows the use of the maximum stress range between maximum compression and maximum tension stresses but can result in more prestress than is needed. The amount of prestress is adjusted if the section is highly asymmetric, as the asymmetry results in either the top or bottom fibers reaching maximum stress limits before the other limit is reached. This approach can result in greater amounts of prestressing reinforcement and is most beneficial when the full section capacity is required for the applied loading.

Floor systems are often designed using load balancing. Based on the floor thickness, a maximum drape results when the tendon is at the limits of cover requirements. The load to be balanced is then selected. Post-tensioned floor systems typically use unbonded monostrand tendons. Therefore, given a single strand tendon and an equivalent load, the tendon spacing is calculated. In these designs, the ACI Building Code requires an average prestressing stress of 125 psi on all sections, including drop panels and shear panels. The 125-psi minimum effective prestress is a convenient trial prestress force. The prestress force required to maintain a 125-psi stress level placed at their maximum eccentricity results in the maximum tendon spacing. This approach is used in Example 6.2.

The prestress force in structures such as water tanks is selected to provide a minimum effective prestress, often in the range of 300 psi, under full hydrostatic load (ACI 372R-13 2013). Such approaches ensure the structure remains watertight in service.

6.1.3 Perform Detailed Check of Design

Once the section geometry, tendon force and its location are established, all stress checks are completed, and then nominal strength requirements are checked. When the flexural design is complete, the shear and serviceability design is undertaken.

Example 6.1: Design the Prestress for a Class U Double-T Beam
A 10-ft-wide 40-ft-long double-T beam is to be designed to carry a live load of 32 psf and a superimposed dead load of 20 psf. Provide the flexural design of the beam. The concrete transfer strength is 3500 psi and the specified design strength is 5000 psi. The tendon is ½ in. diameter 270 ksi low relaxation strand.

Solution: The *PCI Design Handbook* suggests that a 10DT24 or 10DT32 would be satisfactory. The preliminary design tables in the Handbook indicated that the 32-in. deep beam has excessive capacity for this application. Therefore, a 10DT24 that is 10-ft wide and 24-in. deep is selected. The span-to-depth ratio of the beam is 40 ft/ 2 ft = 20, which is in the range to provide reasonable deflection performance in later checks.

The load tables in the *PCI Design Handbook* indicate that the section has more capacity than needed for this application. Therefore, an eccentricity and the initial prestress are selected. Using the *PCI Design Handbook* as a guide, an eccentricity of 8 in. and an initial prestress of is selected based on ½ the initial stress range. Without the PCI Handbook, selection of the eccentricity would be based on the strand location. The distance from the neutral axis to the bottom of the double-T is

Table 6.3 Properties, loads and initial trial

Properties			Loads			
$A_g =$	449	in.^{2}	$q_{sdl} =$	20	psf	
$I_g =$	22,469	in.^{4}	$q_{ll} =$	32	psf	
$y_t =$	6.23	in.	$M_g =$	1123	in.-kip	
$y_b =$	17.77	in.	$M_{sdl} =$	480	in.-kip	
$k_b =$	8.0		$M_l =$	768	in.-kip	
$b_w =$	6	in.				
$b_f =$	120	in.	$M_u =$	3153	in.-kip	
$l =$	40	ft				
$w_g =$	468	plf	Losses	28.6	ksi	Assumed
$f'_{ci} =$	3500	psi				
$f'_c =$	5000	psi	$\beta_1 =$	0.80		
			Trial			
$f_{ci} = 0.60 f'_{ci} =$	2100	psi	$e =$	**8**	**in.**	
$f_t = 3\sqrt{f'_{ci}} =$	177	psi	$P_i =$	**431.6**	**kip**	
				14.1	**strands**	
$f_c = 0.60 f'_c =$	3000	psi				
$f_t = 7.5\sqrt{f'_c} =$	530	psi	**Try**	**8**	**strands**	
			$P_i =$	**244.8**	**kip**	
			$P_e =$	**209.8**	**kip**	
			$\rho =$	**0.0072**		

17.77 in. Assuming 2-1/2 in. for cover and 7 strands 2 in. on center, the eccentricity would be 17.77 in. less 2 in. cover less 7 in. to the centroid of the strands giving an eccentricity of 8.27 in. Using one half the initial stress range, a trial initial prestress force of 432 kips and 14-½ in. diameter strands are needed to provide this force. This initial prestress force produces excess compression at transfer so the trial is adjusted down to 8 strands for satisfactory initial stress levels.

Section properties and loads are summarized in an Excel sheet in Table 6.3. The trial portion of the solution is indicated in bold type. Two preliminary trials preceded this solution. A final stress check is given in Table 6.4, and the final strength check is given in Table 6.5.

Comment: The selection of 50% of the average compression resulted in an initial estimate of 14 strands, which provides more prestress than required. This selection of 50% of the stress range is made to demonstrate that a simple Excel table allows rapid evaluation of trial designs and how a solution may be derived without external design aids. By trial, 8 strands are finally selected. In this example, an even number of strands are chosen to ensure symmetric prestressing because half of the strands are placed in each leg of the double-T. A further trial indicates that 6 strands are insufficient. The PCI Design Handbook suggests 8 strands for the load range. The

Table 6.4 Service level stress checks

Service level stress check						
	Initial midspan stresses		Final midspan stresses			
	Top	Bottom		Top	Bottom	
	(psi)	(psi)		(psi)	(psi)	
P_i/A_g	545	545	P_e/A_g	467	467	
$P_i ey/I_g$	−543	1549	$P_e ey/I_g$	−465	1327	
$M_g y/I_g$	311	−888	$M_g y/I_g$	311	−888	
Sub total	314	1206	Sub total	313	906	
			$M_{sdl}y/I_g$	133	−380	
			Subtotal	446	527	
			$M_l y/I_g$	213	−607	
	Initial end stressed		Total	659	−81	
	Top	Bottom				
	(psi)	(psi)				
P_i/A_g	545	545				
$P_i ey/I_g$	−543	1549				
$M_g y/I_g$	0	0				
Sub total	2	2094				

Midspan stresses

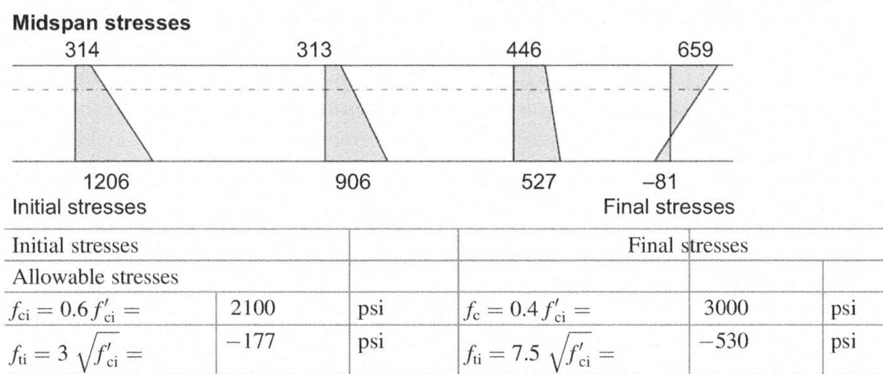

Initial stresses			Final stresses		
Allowable stresses					
$f_{ci} = 0.6 f'_{ci} =$	2100	psi	$f_c = 0.4 f'_{ci} =$	3000	psi
$f_{ti} = 3 \sqrt{f'_{ci}} =$	−177	psi	$f_{ti} = 7.5 \sqrt{f'_{ci}} =$	−530	psi

Table 6.5 Nominal strength check

Nominal strength using ACI equation and ductility check					
$f_{ps} =$	266	ksi	Eq. (5.7)		
$a =$	0.64	in.			
$\phi M_n =$	4081	in.-kip	$>M_u =$	3153	OK
$c =$	0.80	in.			
$c/d_p =$	0.06	<0.375 OK			

design effort could have been shortened by using the PCI Handbook recommendation as a first trial.

A check of c/d_p validates the use of $\phi = 0.90$. Lastly, this example includes a check of the stresses at the end of the beam for the initial prestress. At the beam end

Table 6.6 Cover requirements for cast-in-place prestressed concrete members

Concrete exposure	Member	Reinforcement	Specified cover (in.)
Cast against and permanently in contact with ground	All	All	3
Exposed to weather or in contact with the ground	Slabs joists and walls	All	1
	All other	All	1-1/2
Not exposed to weather or in contact with the ground	Slabs, joists and walls	All	3/4
	Beams, columns, and tension ties	Primary reinforcement	1-1/2
		Stirrups, ties, spirals, and hoops	1

Adapted from Section 20.6 of ACI 318-14

the moment due to self-weight of the beam is not available to balance prestress forces and this location proved to be critical for the selection of the eccentricity.

6.2   Cover and Spacing Requirements

The ACI Building Code specifies minimum cover and spacing for reinforcement and prestress tendons. The cover requirements meet multiple objectives. First, the cover provides protection against corrosion. Second, cover provides a basic level of fire protection. Lastly, cover and clearance provide sufficient concrete around the prestress steel to ensure bonded stand development.

6.2.1   Cover

The ACI Building Code provides specified cover requirements for prestressed concrete for cast-in-place members, Table 6.6, and for plant fabricated members, Table 6.7. ***Specified cover*** is the value provided by the engineer and does not include tolerances. The typical industry tolerances associated with specified cover are -0, $+1/4$ in. implying that increased cover is acceptable but decreased cover is not (ACI 117-10 2015). Values in these tables account for both exposure and basic fire resistance as specified in ACI 216 (ACI 216.1-14 2014). Bundling tendons is not allowed at the end of a member but is allowed in the middle portion of the span. In such cases the minimum cover for the bundle is at least the equivalent diameter of the bundle or 2 in. and for concrete cast against the ground 3 in. Class T and C prestressed concrete members exposed to corrosive environments or other severe exposure categories such as sea water, soils with sulfates, or severe freeze-thaw conditions require that the specified cover be increased to one and a half times the values given in Tables 6.6 and 6.7.

Table 6.7 Cover requirements for precast nonprestressed or prestressed concrete members manufactured under plant conditions

Concrete exposure	Member	Reinforcement	Specified cover (in.)
Exposed to weather or in contact with the ground	Walls	No. 14 and No. 18 bars; tendons larger than 1-1/2 in. diameter	1-1/2
		No. 11 bars and smaller; W31 and D31 wire and smaller; tendons and strands 1-1/2 in. diameter and smaller	3/4
	All other	No. 14 and No. 18 bars; tendons larger than 1-1/2 in. diameter	2
		No. 6 through No. 11 bars; tendons and strands larger than 5/8 in. diameter through 1-1/2 in. diameter	1-1/2
		No. 5 bar, W31 or D31 wire, and smaller; tendons and strands 5/8 in. diameter and smaller	1-1/4
Not exposed to weather or in contact with the ground	Slabs, joists and walls	No. 14 and No. 18 bars; tendons larger than 1-1/2 in. diameter	1-1/4
		Tendons and strands 1-1/2 in. diameter and smaller	3/4
		No. 11 bar, W31 or D31 wire, and smaller	5/8
	Beams, columns, pedestals, and tension ties	Primary reinforcement	Greater of d_b and 5/8 and need not exceed 1-1/2
		Stirrups, ties, spirals, and hoops	3/8

Adapted from Section 20.6 of ACI 318-14

Table 6.8 Minimum center-to-center spacing of pretensioned strands at ends of members

f'_{ci} (psi)	Nominal strand diameter (in.)	Minimum spacing (in.)
<4000	All	$4d_b$
$\geq$4000	<0.5 in.	$4d_b$
	0.5 in.	1-3/4
	0.6 in.	2

Adapted from Section 25.6 of ACI 318-14

6.2.2   Minimum Spacing Requirements

Spacing requirements for strands and tendons is defined as a center-to-center distance of the pretensioned strand and is dependent on the strength of the concrete at the time of prestress transfer, Table 6.8. Spacing of post-tensioning ducts at the end of a member is established by the anchor spacing and the anchor zone reinforcement. Spacing of bundled of post-tensioning ducts is based on performance requirements. Section 25.6.2 of ACI 318-14 states:

> Bundling of post-tensioning ducts shall be permitted if shown that concrete can be satisfactorily placed and if provision is made to prevent the prestressing steel, when tensioned, from breaking through the duct.

This leaves the engineer with the task of determining an appropriate distance between ducts. Equivalent bearing loads on draped tendon ducts can be reduced to local bearing stresses by dividing the equivalent load by the tendon diameter. The stress calculated with this approach provides an indication of stress concentrations and loads on the concrete between tendons.

6.2.3 *Maximum Spacing Requirements and Crack Control*

Crack widths in structures are highly variable. The ACI Building Code provisions for spacing are intended to limit surface cracks to a width that is generally acceptable in practice but may vary widely in a structure. The role of cracks in the corrosion of reinforcement remains controversial. Research shows that corrosion is not clearly correlated with surface crack widths in the range normally found with reinforcement stresses at service load levels (Darwin et al. 1985; Oesterle 1997). Consequently, the ACI Building Code does not differentiate between interior and exterior exposures. Maximum tendon spacing requirements are specified to reduce cracking (Beeby 1979; Frosch 1999; ACI Committee 318 1999). Control of crack width has long been an issue for both reinforced and prestressed concrete.

Early attempts to predict and limit crack width were only partially successful due to the variation in crack width (Gergely and Lutz 1968). Beginning in 2011, the ACI Building Code replaced crack width provisions with maximum spacing requirements. These limits apply to both nonprestressed and prestressed members. The maximum spacing requirements for nonprestressed reinforcement in nonprestressed and Class C prestressed members are given in Table 6.9. Class U and Class T beams are assumed uncracked so provisions of Table 6.9 do not apply to members meeting these criteria.

Only tension reinforcement nearest the tension face need be considered in selecting the value of c_c used in calculating spacing requirements. The stress in the deformed reinforcement f_s is for the reinforcement closest to the tension face at service loads and is calculated based on the unfactored moments. Alternatively, f_s is taken as $2/3\, f_y$. The value for Δf_{ps} is calculated based on strain compatibility of the cracked section analysis minus the decompression stress f_{dc} as seen in Example 5.5. Section 24.3.2.2 of ACI 318-14 allows the value of f_{dc} to be the effective stress in the prestressed reinforcement; however, the value of Δf_{ps} is not allowed to be greater than 36,000 psi. Based on historical performance of members designed using working stress methods, if the value of Δf_{ps} is less than 20,000 psi, the spacing limits in Table 6.9 need not be satisfied. In cases where there is only one bonded strand or tendon nearest the tension face, the maximum width of the extreme tension face is limited to the value of s in Table 6.9.

Table 6.9 Maximum spacing of bonded reinforcement in nonprestressed and Class C prestressed one-way slabs and beams

Reinforcement type	Maximum spacing, s (in.)	
Deformed bars or wires	Lesser of:	$15\left(\dfrac{40,000}{f_s}\right) - 2.5\,c_c$
		$12\left(\dfrac{40,000}{f_s}\right)$
Bonded prestressed reinforcement	Lesser of:	$\left(\dfrac{2}{3}\right)\left[15\left(\dfrac{40,000}{\Delta f_{ps}}\right)\right] - 2.5\,c_c$
		$\left(\dfrac{2}{3}\right)\left[12\left(\dfrac{40,000}{\Delta f_{ps}}\right)\right]$
Combined deformed bars or wires and bonded prestressed reinforcement	Lesser of:	$\left(\dfrac{5}{6}\right)\left[15\left(\dfrac{40,000}{\Delta f_{ps}}\right)\right] - 2.5\,c_c$
		$\left(\dfrac{5}{6}\right)\left[12\left(\dfrac{40,000}{\Delta f_{ps}}\right)\right]$

Adapted from Section 24.3.2 of ACI 318-14

f_s is the stress in the reinforcement, Δf_{ps} is the change in stress in the prestressing tendon, and c_c is the clear cover to the reinforcement

If the flange of a T-beam is in tension, two factors affect the design. First, if the reinforcement is distributed across the entire effective width of the flange, excessively wide cracks may occur over the web. Conversely, if the bonded reinforcement is grouped over the web, excessive cracking may occur at the outer edges of the flange. To address this, the ACI Building Code requires that bonded reinforcement be placed over a width corresponding to one tenth the span length $l_n/10$. If $l_n/10$ is less than the effective width or actual width of the T-beam, additional reinforcement is placed in the outer portion of the flange using the maximum spacing limits given in Table 6.9.

6.3 Effective Flange Width

As a flanged beam is loaded, a finite distance is required for the stresses in the stem to reach the outer limits of the flange. If the flange is sufficiently wide and thin, the beam may not be able to mobilize the full width of the flange. The ACI Building Code places limits on **effective flange width**, the width effective at full stress levels to resisting bending, to address this condition. Section 6.3.2 of ACI 318-14 provides guidance for calculating the effective flange width. The effective flange width consists of the width of the stem plus the overhang width beyond the stem. The limitations for overhang dimensions are provided in Table 6.10. In Table 6.10, h is the thickness of the flange, s_w is the clear distance to adjacent stems, and l_n is the clear span length.

The values in Table 6.10 are empirical and were developed for nonprestressed beams. ACI 318-14 commentary R6.3.2.3 reads in part:

Table 6.10 Dimensional limits for effective flange width for T-beams

Flange location	Effective overhanging flange width beyond the face of the web	
Each side of web	Least of:	$8h$
		$s_w/2$
		$l_n/8$
One side of web	Least of:	$6h$
		$s_w/2$
		$l_n/12$

Adapted from Section 6.3.2 of ACI 318-14

Fig. 6.2 Reinforcement in a prestressed beam

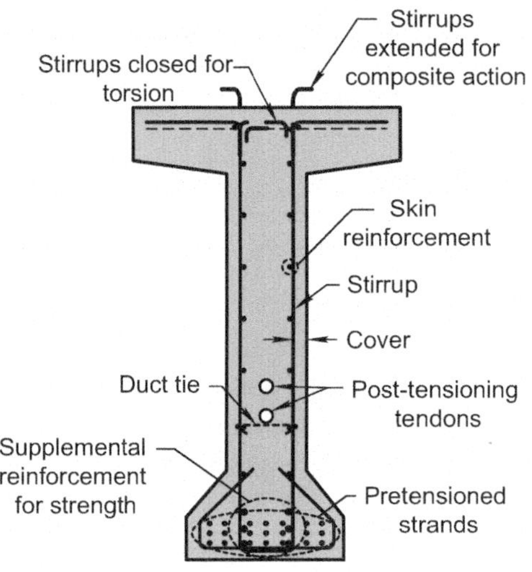

"The flange widths in 6.3.2.1 and 6.3.2.2 should be used unless experience has proven that variations are safe and satisfactory. Although many standard prestressed products in use today do not satisfy the effective flange width requirements of 6.3.2.1 and 6.3.2.2, they demonstrate satisfactory performance. Therefore, determination of an effective flange width for prestressed T-beams is left to the experience and judgment of the licensed design professional. It is not always considered conservative in elastic analysis and design considerations to use the maximum flange width as permitted in 6.3.2.1."

Virtually all commercial precast double-T and single-T products use the full flange width based on industry tests and successful implementation.

6.4 Contributions of Nonprestressed Reinforcement

A prestressed concrete beam contains prestressed and nonprestressed reinforcement. Figure 6.2 is a composite sketch of the reinforcement that may be present including both pretensioned and post-tensioned tendons. Many beams, double-T beams for example, have only pretensioned tendons while other beam types contain only post-tensioned tendons. Example 6.2 includes the contribution of No. 4 bars to attain the necessary design strength.

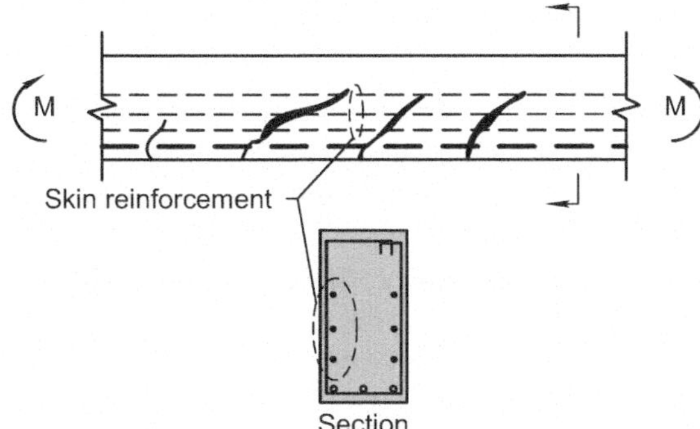

Fig. 6.3 Crack width growth at mid-height without skin reinforcement

6.4.1 Longitudinal Reinforcement

Longitudinal reinforcement consists of supplemental reinforcement for strength and skin reinforcement. Reinforcement for strength is placed as far from the compression face as possible. In Fig. 6.2, the supplemental reinforcement is intermixed with the prestressing strands. Exact placement is dependent on space availability and clearance requirements. If there is insufficient side clearance, supplemental reinforcement would be placed above or below the prestressing strands.

The strength contribution of supplemental reinforcement placed as indicated is calculated assuming the reinforcement has yielded. As the depth of the reinforcement decreases, its efficiency decreases and the assumption of yielding must be checked using a strain compatibility approach described in Section 5.3.4 of ACI 318-14.

Skin reinforcement is required by the ACI Building Code and is distributed over the portion of the beam in tension; it is required to control cracking and the growth of crack width at mid-height of the beam, Fig. 6.3 (Frantz and Breen 1980; Frosch 2002). For Class C beams with depth greater than 36 in., the skin reinforcement is placed in the tension zone of the beam at a spacing not to exceed the values given in Table 6.9. Skin reinforcement is not required for Class U members and may not be required for Class T members if cracking is restricted to the lower portion of the member.

6.4.2 Stirrups

Stirrups provide shear reinforcement. Cover is measured to the stirrup from the face of the concrete. Cover requirements for strands and tendons are greater than the cover requirements for stirrups; however, because these elements are typically inside the stirrups, the additional cover created by the thickness of the stirrup provides the additional distance needed to meet these requirements.

6.4.3   Minimum Reinforcement

Prudent design practice dictates that the flexural strength of a member exceed the cracking strength of the member. This philosophy assures that, should a member crack, the member would have sufficient ductility to permit large deflections and cracking prior to collapse. The requirement was initially developed for deep architectural members where strength was not a governing criterion. Minimum reinforcement is required to prevent sudden failure of a member immediately following initial cracking. Reinforced concrete minimum reinforcement requirements are given as minimum reinforcement ratios. Prestressed concrete members have higher cracking moments because of the prestressing. For beams and slabs with bonded tendons, the total amount of prestressed and nonprestressed reinforcement must develop a factored load at least 1.2 times the moment that causes cracking M_{cr} based on Eq. (6.2).

$$M_{cr} = \frac{(f_{pe} - f_r)I_g}{y_t} \tag{6.2}$$

where the modulus of rupture f_r is defined in Chap. 3, f_{pe} is the compressive stress in the loaded concrete section at the extreme tension fiber after all losses, and y_t is the distance from the neutral axis to the extreme tension face. Thus, for beams and slabs with bonded tendons, $M_n \geq 1.2\ M_{cr}$ satisfies this minimum reinforcement requirement.

Members with unbonded tendons have not experienced sudden failure at the onset of cracking. The unbonded tendon allows distribution of strain along the length of the tendon thereby relieving the stress concentrations present in bonded reinforcement. One consequence of this behavior is that beams with unbonded tendons form a few large cracks. To provide narrower crack widths, the ACI 318 Building Code requires minimum bonded reinforcement in beams $A_{s,min}$ equal to

$$A_{s,min} = 0.004 A_{ct} \tag{6.3}$$

where A_{ct} is the area of that part of the beam between the neutral axis of the gross section and the flexural tension face. ACI committee 318 is considering similar provisons for One-way slabs. Two-way slabs have separate provisions discussed in Chap. 11.

Example 6.2: Design of a Two-Span One-Way Slab
Design a two-span continuous 6 in. thick one-way floor slab spanning 20 ft and carrying a superimposed dead load of 10 psf and a live load of 40 psf. The concrete is initially stressed when it reaches 2500 psi and the specified 28-day strength is 4000 psi. Table 6.11, Use unbonded monostrand tendons. The slab is shored during construction so all the dead and live load is carried on the continuous structure.

Solution: The maximum negative moment is $wl^2/8$ at the interior support due to both spans loaded. The positive moment is $9wl^2/128$ with both spans loaded. A maximum

Table 6.11 Properties and loads

Properties			Loads			
	1 ft wide strip			1 ft wide strip		
$A_g =$	72	in.^{2}	$w_{slab} =$	75	plf	
$I_g =$	216	in.^{4}	$w_{sdl} =$	10	plf	
$y_t =$	3	in.	$w_{ll} =$	40	plf	
$y_b =$	3	in.	$w_{eq} =$	56.8	plf	
$t =$	6.0	in.	**Positive moments**			
$b =$	12	in.	$M_{initial} =$	3406	in.-lb/ft	
$s =$	18	in.	$M_{final} =$	26,613	in.-lb/ft	
$L =$	20	ft				
$q_g =$	112.5	psf	$M_u =$	42,610	in.-lb/ft	1.2 D + 1.6 L
$e =$	3.5	in.	**Negative moments**			+1.0 PS
$f'_{ci} =$	2500	psi	$M_{initial} =$	6055	in.-lb/ft	
$f'_c =$	4000	psi	$M_{final} =$	40,916	in.-lb/ft	
$l/d =$	48					
$\rho_p =$	0.002125		$M_u =$	65,516	in.-lb/ft	1.2 D + 1.6 L
						+1.0 PS
$P_i =$	**30.6**	**kip**				
$P_e =$	**26.8**	**kip**				
$s =$	**2.75**	**ft**	**Spacing initially set at 3 ft to maintain**			
			maintain 125 psi effective prestress			

positive moment due to one span loaded is a combination of the maximum moment due to dead load plus one span carrying live load.

$$M_+ = \frac{1}{2}\left[\frac{7}{16}w_l + \frac{3}{8}(w_d + w_{sdl})\right]lx'$$

where

$$x' = \frac{\frac{7}{16}w_l + \frac{3}{8}(w_d + w_{sdl})}{w_d + w_{sdl} + w_l}l$$

The design is based on a 1-ft wide strip. For a 6-in. thick slab with ¾ in. cover, plus ¼ in. to the center of the tendon, the tendon eccentricity is 2 in. at the critical positive moment and 3 in. at the support before the reverse curve is implemented. Assuming a zero eccentricity at the end, the tendon drape is 2 in. + ½ × 3 in. = 3.5 in., Fig. 6.4 In this case, the drape is assumed to continue to the top of the slab, and the reverse curve of the cap tendon is ignored. A complete discussion of tendon geometry and continuity is presented in Chap. 9. For strength design, the actual tendon location is used.

The slab self-weight is 150 pcf × 6 in./12 in. = 75 psf. A ½ in. diameter 270 ksi strand is initially stressed to 200 ksi and after losses has an effective stress of 175 ksi,

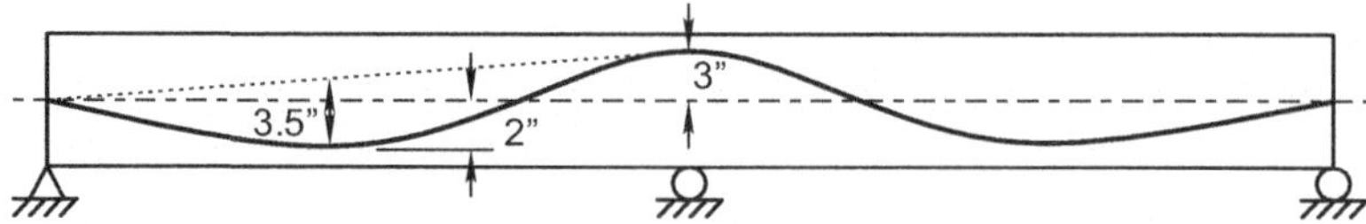

Fig. 6.4 Tendon geometry

Table 6.12 Stress check for positive moments

Stress check positive moments					
	Initial midspan stresses		Final midspan stresses		
	Top (psi)	Bottom (psi)		Top (psi)	Bottom (psi)
P_i/A	155	155	P_e/A	135	135
$-M_{\text{initial}}y/I$	47	−47	$-M_{\text{final}}y/I$	370	−370
Sub total	202	107	Subtotal	505	−234

resulting in an effective prestress force of 175 ksi $\times$ 0.153 in.^{2} = 26.78 kips. The ACI Building Code requires 125 psi average compression stress for two-way slabs and as low as 100 psi for one-way slabs. A spacing of 2.75 ft on centers gives an average compression stress of 135 psi at midspan after losses and is used for this example. The corresponding equivalent load for the effective prestress on a one foot wide strip is

$$w_{\text{eq}} = \frac{8P_e e}{l^2} = \frac{8 \times 26.78/2.75 \times 3.5/12}{20^2} = 58.6 \text{ plf}$$

The maximum positive moment occurs 7.9 ft in from the end and has a maximum service moment of 24,593 in.-lb/ft and a factored moment of 40,590 in.-lb/ft, which is about 26% higher than both spans loaded. The factored moment is calculated based on a load combination of 1.2 D + 1.6 L + 1.0 PS.

A stress check is performed for the difference between the equivalent load and the applied load. Initially, the applied load is the weight of the slab. Final stresses are checked for the slab weight, the superimposed dead load and the live load, Tables 6.12 and 6.13. The first check is for initial conditions before any superimposed dead load is applied. In this example, intermediate stresses are not checked as they fall between the limits of the initial and final stresses.

The initial check of the stresses using a 3-ft tendon spacing resulted in the tensile stresses exceeding the code allowable limits. The spacing of the tendons was reduced to 2.75 ft and the stresses recalculated as indicated and found to be acceptable.

The initial strength check indicated that the prestress alone has insufficient negative moment capacity but that the positive moment capacity is adequate. The design is modified to add No. 4 Grade 60 reinforcement at 24 in. on center to provide supplemental negative moment strength. The addition of the nonprestressed reinforcement is sufficient to meet the overall strength requirement. Table 6.14 shows a value of A_s = 0.10 in.^{2} for positive moment and 0.10 in.^{2} for negative moment.

Table 6.13 Service level tress check for negative moments

Stress check negative moments					
	Initial support stresses		Final support stresses		
	Top (psi)	Bottom (psi)		Top (psi)	Bottom (psi)
P_i/A	155	155	P_e/A	135	135
My/I	−84	84	$−M_1y/I$	−568	568
Sub total	70	239	Sub total	−433	704

Positive stresses   **Negative support stresses**

202   505   70   −433

155   −234   239   704

Allowable stresses					
$f_{ci} = 0.6\,f'_{ci} =$	1500	psi	$f_c = 0.4\,f'_{ci} =$	2400	psi
$f_{ti} = 3\sqrt{f'_{ci}} =$	−150	psi	$f_{ti} = 7.5\sqrt{f'_{ci}} =$	−474	psi

Table 6.14 Nominal strength calculations

Nominal strength using ACI equation and ductility check					
Positive moment					
$f_{ps} =$	191,275	psi	Table 5.5	for $l/d > 35$	
$A_s =$	0.00	in.^{2}			
$a =$	0.24	in.			
$\phi M_n =$	42,848	in.-lb/ft	$>M_u =$	42,610	in.-lb/ft
$c =$	0.30	in.			
$c/d_p =$	0.06	<0.375 OK			
Negative moment					
$f_{ps} =$	191,275	psi	Table 5.5	for $l/d > 35$	
$A_s =$	0.10	in.^{2}	No. 4 at 24 in.		
$a =$	0.39	in.			
$\phi M_n =$	69,202	in.-lb/ft	$>M_u =$	65,516	in.-lb/ft
$c =$	0.48	in.			
$c/d_p =$	0.10	<0.375 OK			

Comments: The slab is lightly reinforced and the strength check indicated additional reinforcement is needed for nominal strength. In this case No. 4 Grade 60 bars at 2 ft on center are needed for negative moment reinforcement. The load balancing method needs only the difference in load between the applied load and the balanced load when adjusted for losses. Secondary moments due to the prestress are incorporated directly in the load balancing method. Per the ACI Building Code, a load factor of 1.0 is used on the balance load.

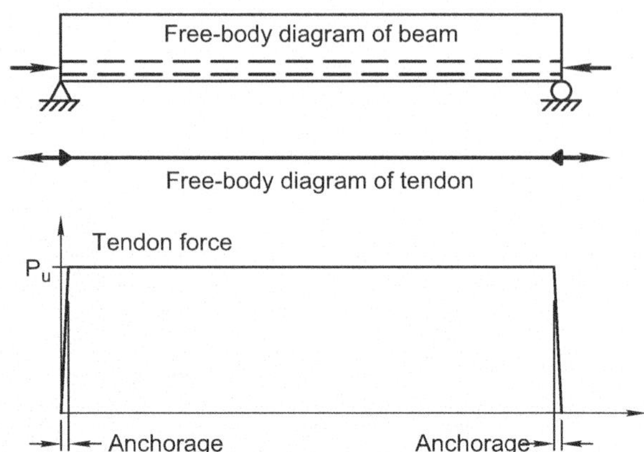

Fig. 6.5 Development of effective prestress force at end of post-tensioned members

6.5 Transfer of Prestress

In prestressed concrete, **transfer** of the effective prestressing force from the tendon to the concrete section is critical to the performance of the member. In post-tensioned concrete, tendons are connected to a steel anchorage that is embedded in the concrete. In pretensioned concrete, the force is delivered directly from the prestressing strands to the concrete through bond.

6.5.1 Post-tensioning Anchorage

Post-tensioned tendons are stressed against the concrete member itself, Fig. 6.5. Steel anchors, whether unbonded or unbonded then grouted, are used to transfer the force directly between the prestressing strands and anchor. The anchor is typically embedded in the concrete when the member is cast. This configuration results in a stress concentration at the anchor, which causes transverse splitting stresses as the concentrated force is distributed into the full section. The concrete surrounding post-tensioning anchors must be reinforced to resist these stresses.

As with pretensioned members, the force in the tendon increases abruptly to the effective prestress force, which is transferred to the concrete. In straight tendons, however, the effective prestress force varies due to friction from tendon wobble. Tendon deviation or drape results in greater variation of the effective prestress. These topics are covered in Chap. 4.

For design purposes, the concrete surrounding the post-tensioning anchor of the concrete is subdivided into the general zone and local zone, Fig. 6.6. The general zone is the portion of the member in which the prestressing force is distributed from a concentrated force to a linearly varying distribution of stresses as described by St. Venant's principle. ACI 318 defines the general zone as the "...portion of the member portion of the member through which the concentrated prestressing force is

Fig. 6.6 Illustration of
general and local zones

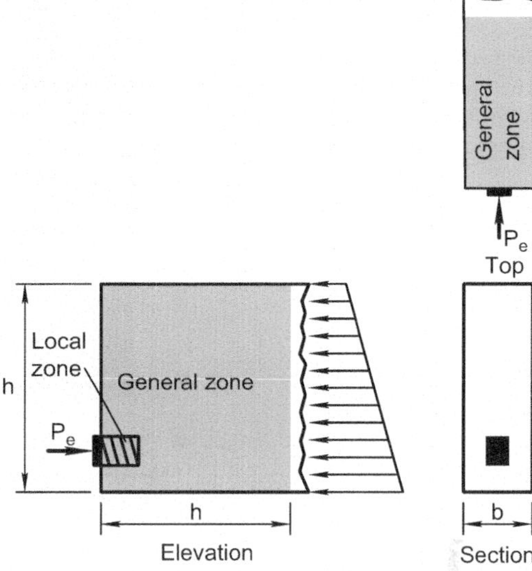

transferred to the concrete and distributed more uniformly across the section." This
definition does not provide the engineer with guidance on the actual proportions of
the general zone, but the commentary does suggest that the size of the zone can be
estimated by using the largest dimension of the cross section.

Located within the general zone in close proximity to the anchorage is the local
zone, which is defined as "...a rectangular prism (or equivalent rectangular prism for
circular or oval anchorages) of concrete immediately surrounding the anchorage
device and any confining reinforcement." This region of the general zone carries a
particularly high bearing stress directly from the anchorage, which may be well
above the unconfined concrete compressive strength. If not confined properly, this
area is susceptible to severe cracking and damage during stressing. Specific details of
the anchorage device and confinement reinforcement has significant effect on the
behavior and strength of the local zone. Post-tensioning anchorage devices do not
have a standard configuration, but rather are custom designed by the manufacturer.
Frequently the engineer does not know which post-tensioning manufacturer will be
used. Furthermore, because of the complexities of the local zone behavior, it is
sometimes more economical and expedient to conduct testing on the particular
anchorage device and ancillary confinement reinforcement. An example detailing
for post-tensioning anchors is given in Chap. 14.

Fig. 6.7 Pretensioned tendon in bottom flange of bridge girder

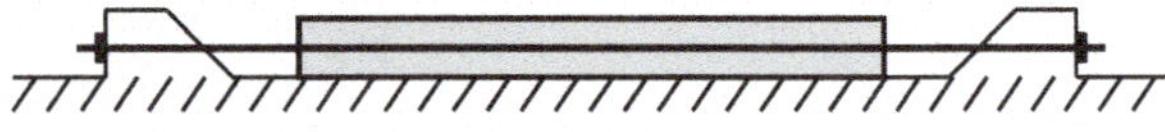

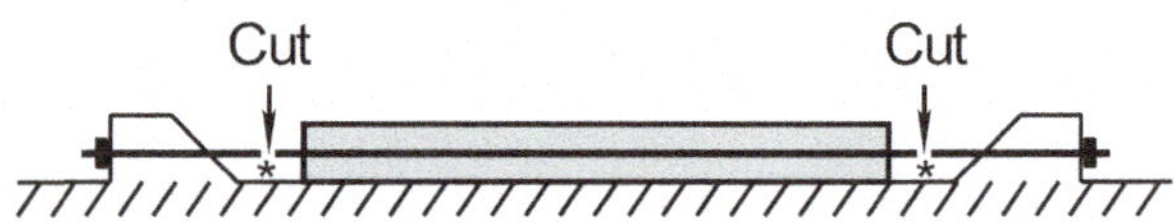

Fig. 6.8 Prestress transfer in precast, pretensioned member

6.5.2 Pretensioning Bond, Transfer Length, and Development Length

Pretensioned tendons are typically used in precast plants and are composed of groups of 7-wire prestressing strands that are anchored in a grid pattern at abutments during fabrication, Figs. 6.7, 6.8, and 6.9. Once the concrete has reached f'_{ci}, then the strands are released from the abutment and the prestressing force is **transferred** to the concrete. The prestressing strands are released using one of two methods. One is to apply heat to the individual strands using a torch. The torch is waved back and forth over the strand until the strand is heated sufficiently to rupture. In long-line applications where multiple members are cast in a single bed, multiple construction personnel are placed with torches at strategic locations between the members. Synchronized by a flag drop or by radio, each torch operator cuts the same strand (within the pattern) at the same time, providing an incremental transfer of prestressing. A less abrupt method of transfer is to hydraulically release the strands from the abutment either individually or simultaneously. The abrupt application of prestress force that occurs when the strands are torch-cut can lead to a longer transfer length and end-cracking (Kannel et al. 1997).

Bond in pretensioned reinforcement is considered in two parts. First, the **transfer length** is the length of the reinforcement required to convey the initial prestressing force in the reinforcement to the concrete, shown schematically in Fig. 6.10. Second,

Fig. 6.9 Heating of prestressing strand with torch to induce rupture. Prestress is transferred to the concrete when the strand ruptures

Fig. 6.10 Development of effective prestress force at end of pretensioned members

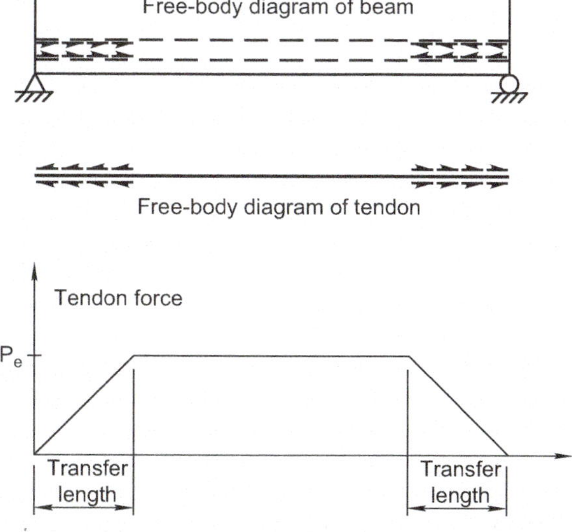

the **development length** l_d is the length of reinforcement required to mobilize its nominal tensile strength. Development length is necessary where the member is subjected to a bending moment approaching the nominal strength of the section and the stress in the reinforcement exceeds the effective prestress stress, Fig. 6.11. The transfer length is less than the development length because it does not have to carry the full tensile capacity of the reinforcement.

It is instructive at this point to examine the behavior of a seven-wire strand. The development length of the strand is due to friction and chemical bonding between the strand and the concrete. Strand, with its helical construction, develops friction from two sources. First, the strand twists under tension as it is pulled longitudinally. This

Fig. 6.11 Development of f_{ps} in pretensioned beam

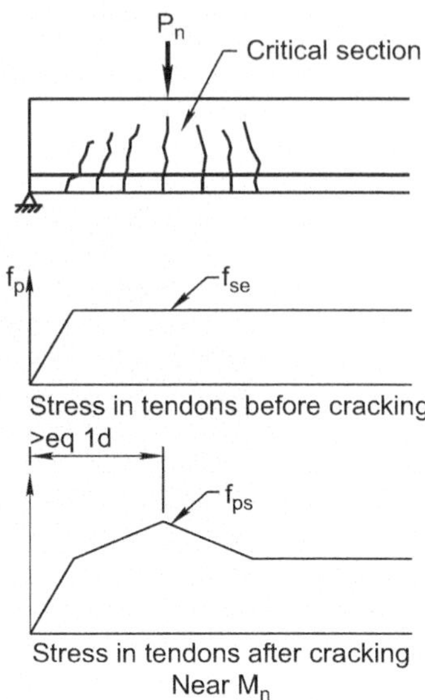

Fig. 6.12 Hoyer effect on strand bond, transfer, and development

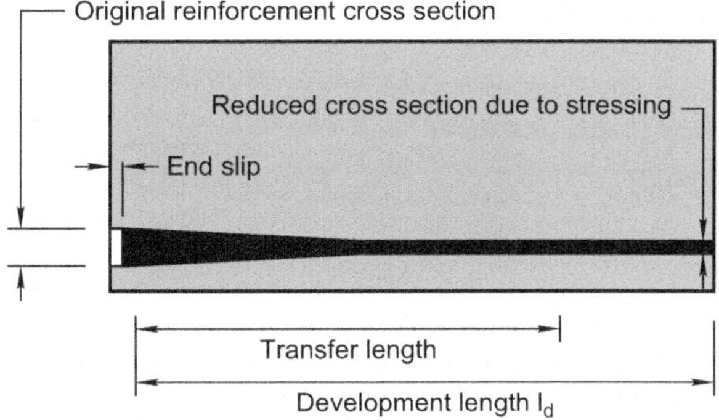

creates a frictional force along the length of the strand. Second, as the prestress decreases, the cross-sectional area of the strand increases establishing an additional normal force with the concrete. The cross-sectional increase and corresponding locking of the strand against the concrete is called the Hoyer effect, and this effect reflects the recovery of the reduced area due to the Poisson effect when strand is initially stressed, Fig. 6.12. Bars and wires have a Hoyer effect but not the benefits of the spiral winding. Hence, strand develops in a shorter length than bars or wires of comparable diameter.

To develop bond with the concrete the strand must move relative to the concrete at the member end. A dial gage placed on the strand before it is released shows this slip. The effect of strand bond development is visible in precast prestressed members, such as hollowcore, that are sawn to length at the plant. The strand at the saw cut slips into the concrete as the force transfer occurs.

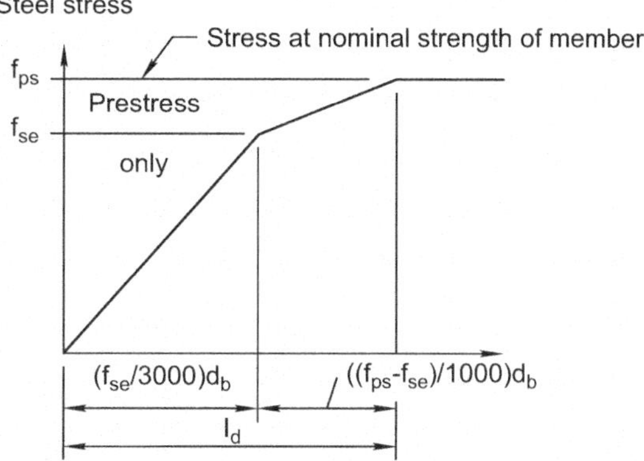

Fig. 6.13 Idealized strand transfer and development length

This behavior of the strand results in a two-stage prediction of transfer and development length given in Eq. (6.4) and shown in Fig. 6.13.

$$l_d = \left(\frac{f_{se}}{3000}\right)d_b + \left(\frac{f_{ps}-f_{se}}{1000}\right)d_b \qquad (6.4)$$

The steep initial curve for the transfer length reflects the Hoyer effect and initial stress transfer. The remaining length is the distance required to develop the nominal strength of the strand. The development length is measured from the end of the member.

The equations for transfer length and for the additional bonded length necessary to develop an increase in stress of $(f_{ps} - f_{pe})$ are based on tests of members prestressed with clean, 1/4, 3/8, and 1/2 in. diameter strands for which the maximum value of f_{ps} was 275,000 psi and the concrete strength was 3000 psi (Kaar and Magura 1965; Hanson and Kaar 1959; Kaar et al. 1963). The tests were conducted with 2 in. of cover in well consolidated normalweight concrete containing no admixtures. These tests may not fully represent the behavior of strand in no-slump or self-consolidating concrete. Consequently, concrete placement operations are monitored to ensure consolidation of concrete around the strand to obtain contact between the steel and concrete.

In the 1990s a series of incidents of excessive strand slip brought into question the development length equations (Buckner 1994; Martí-Vargas et al. 2006). Two concurrent activities impacted bond. First, strand lubricants used in drawing the wires changed resulting in lower friction. Second, self-consolidating concrete was coming into commercial use. The strand coating issue is largely resolved and researchers are developing a strand-bond protocol to validate strand behavior. Research into self-consolidating concrete is continuing and ACI Building Code committee has retained Eq. (6.4) pending the outcome of this research.

Development requirements for pretensioned strand are intended to provide bond integrity needed to attain the nominal strength of the member. Structural integrity requires the full development length extend l_d beyond any critical section. The ACI

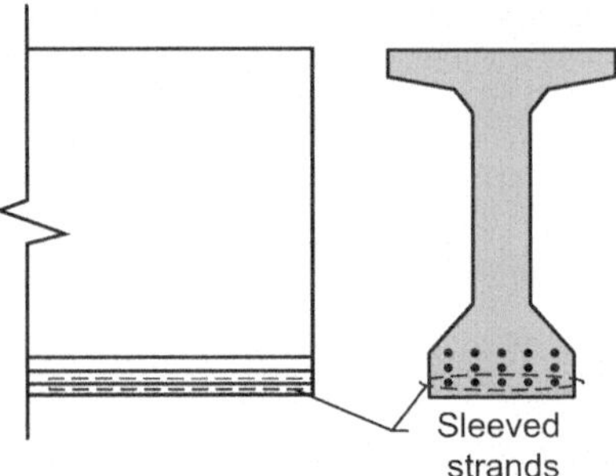

Fig. 6.14 Illustration of sleeved strands

Building Code allows less than the full development length, providing that the stress in the strand does not exceed the values in Fig. 6.9. As a practical issue, providing less than the full development length leaves the structure vulnerable to sudden failure, should an unanticipated overload occur that requires the nominal stand capacity.

6.6 Control of Stresses at Pretensioned Beam Ends

Because the weight of the beam is engaged when the pretensioning force is applied to pretensioned beams, it is possible to meet the service stress limits at midspan but not at the beam end. Two techniques are available to reduce service stresses at the beam end. First, the strands are draped or harped at the beam end to ensure that the eccentricity results in service stresses within the allowable ranges. Second, strands are **sleeved**, that is, a cover is applied over the strand so the prestress transfer initiation is moved inward along the beam, Fig. 6.14. Pretensioned beams use harped strands or strand sleeves to control end stresses. Post-tensioned beams typically use draped tendons.

Strand sleeves are tight fitting plastic wrapped around the strand for the prescribed length. Sleeved tendons reduce the stresses at the end of the beam and move the centroid of the reinforcement upward. The ACI Building Code recognizes the benefit of sleeves by permitting a higher initial compressive stress "at the end of beams." The ACI Building Code also recognizes that there is an internal stress concentration at the end of the sleeve. If the strand does not begin at the end of the member, that is, it is sleeved or otherwise debonded, the transfer and development lengths given in Eq. (6.4) are doubled.

Harping stands and installing sleeves have advantages and liabilities. Harping strands is a complex operation. Depressed strands are initially tensioned then jacked down to their final eccentricity, Fig. 6.15. The initial stress in the strand is adjusted to allow for the increase in stress as the tendon is depressed. Alternatively, strands are threaded through a **hold down** set at the proper eccentricity. The strands are then stressed and the friction of pulling the strand through the hold down accounted for in

Fig. 6.15 Harping
hardware

the stressing operation. Both operations are part of the initial stressing, require additional calculation effort, and create potential safety hazards in the prestressing plant.

Using a sleeve on the strands allows the strands to run straight and avoids the hazards of harping. This is done at the expense of reduced longitudinal reinforcement at the end of the member. Sleeved strands are also accounted for in shear calculations. Reduced shear and moment capacity at the beam end results from the reduction of prestressing. This reduction is important if there are concentrated loads or other abnormalities near the beam end.

An alternative to tendon sleeves adds reinforcement at the end of the beam to control cracking and provide resistance to tensile stresses. The critical section occurs at the end of the transfer length where the strand fully transfers the prestress to the beam. Design of the reinforcement calculates the total tension force in the tension zone, then provides reinforcement to compensate for that force. Because the stresses are service level, an allowable stress in the reinforcement is set at $2/3\,f_y$ or 40,000 psi for Grade 60 reinforcement. Use of a higher stress may be considered because losses immediately reduce the initial prestress force.

Example 6.3: Design of End Reinforcement

An AASHTO I beam has a 10-½ in. diameter low relaxation strands with a constant eccentricity of 8 in. The initial prestress is 178.4 ksi. Determine the service level stresses at the beam end and design the reinforcement to control the tension stress. Transfer stress is 3500 psi and $f_y = 60{,}000$ psi.

Solution: For the Appendix, $A_g = 276$ in.^{2}, $I_g = 22{,}750$ in.^{4}, $y_t = 15.41$ in., $y_b = 12.59$ in. and the width of the compression face is 12 in. and the depth of the rectangular portion is 4 in. The initial prestress is $10 \cdot 0.153\ \text{in}^2 \cdot 178.4\ \text{ksi} = 273$ kip. The stress at the top of the beam is

$$\frac{P}{A_g} - \frac{Pey}{I_g} = \frac{273}{276} - \frac{273 \cdot 8 \cdot 15.42}{22,750} = 490 \ \text{psi}$$

The allowable stress at transfer is 444 psi, so the allowable stress is exceeded and supplemental reinforcement is required. The stress at the bottom of the section is

$$\frac{P}{A_g} + \frac{Pey}{I_g} = \frac{273}{276} + \frac{273 \cdot 8 \cdot 15.42}{22,750} = 2200 \ \text{psi}$$

The neutral axis is then $\text{NA} = h \cdot f_t/(f_t + f_b) = 28 \cdot 490/(490 + 2200) = 5.1$ in.
This is slightly below the 4-in. thick top flange. For simplicity, a rectangular section of 12-in. by 5.1-in. is used to calculate the tension force, or

$$T = \tfrac{1}{2} \cdot 12 \cdot 5.1 \cdot 490 \ \text{psi} = 15 \ \text{kip}.$$

The corresponding required area of reinforcement is

$$A_s = T/f_y = 15,000/60,000 = 0.25 \ \text{in.}^2$$

Select 3 No. 3 bars, for which $A_s = 0.33$ in.^{2}.
Lastly, check to confirm the bars are adequately developed. The transfer length is taken as $f_{pi} \cdot d_{ps}/3000 = 178,400 \cdot 0.153/3000 = 9.1$ in. The development length of a No. 3 bar is 8 in., therefore no hooks or other anchorage is required.

6.7   Handling and Erection

Design of hardware for handling and erection of precast elements is usually the responsibility of the contractor's specialty engineer. Awareness of member stiffness and stability during handling and erection assists the design by reducing the chance that inadvertent cracking or damage occurs.

Beams are removed from the form when and the concrete strength is low. Lifting points are located inward from the location of the final supports. This reduces the positive moment and may generate negative moment in addition to the prestressing camber, thereby requiring supplemental reinforcement in what eventually is the positive moment region, Fig. 6.16a.

Figure 6.16b shows the lifting arrangement for a beam that may later be post-tensioned. The lifting points are selected to minimize the positive and negative moments. Supplemental reinforcement is provided to control cracking.

Dynamic inertial forces during transport create negative moment on the beam. Figure 6.16c indicates supplemental reinforcement to preclude cracking during transport.

Fig. 6.16 Handling and erection considerations

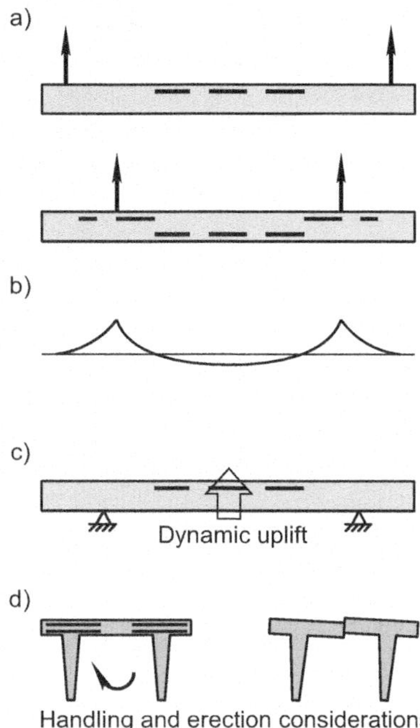

Figure 6.16d shows a thin web and deck member subjected to torsion during shipping. The twist can fracture the deck, and supplemental reinforcement is provided in the deck as a precaution.

Problems

6.1. Redesign the beam in Example 6.1 to span 50 ft and remain as Class U, harping the strands if needed.

6.2. Design a 60-ft long 10DT 34 Class U double-T beam for the loadings given in Example 6.1. Loads, section, and material properties are in Table 6.15:

6.3. For the girder shown in Fig. 6.17:

(a) Calculate the service stresses at midspan with the full live load in place.
(b) Calculate the change in stress in the prestressing steel at midspan when the full live load is applied. Comment
    $P1 = 32$ kip, $a = 16$ ft.

6.4. For a one-way slab and girder system shown in Fig. 6.18, determine the post-tensioning force required to balance 95% of the structure's self-weight. Girders are spaced at 16 ft center-to-center. Compute the section properties necessary to solve the problem. $b_w = 16$ in., $h_f = 5.5$ in., $h = 36$ in., $L = 58$ ft.

6.5. For the double tee shown in Fig. 6.19, determine the classification in accordance with ACI 318.

Table 6.15 Properties and loads for Problem 6.2

	Properties		10DT34	Loads		
$A =$	855	in.^{2}	$q_{sdl} =$	25	psf	
$I =$	80780	in.^{4}	$q_{ll} =$	40	psf	
$y_t =$	8.93	in.	$M_g =$	4811	in.-kip	
$y_b =$	25.07	in.	$M_{sdl} =$	1080	in.-kip	
$k_b =$	10.6		$M_l =$	1728	in.-kip	
$b_w =$	6	in.				
$b_f =$	120	in.	$M_u =$	9834	in.-kip	
$L =$	60	ft				
$w_g =$	891	plf	Losses	25	ksi	Assumed
$f'_{ci} =$	3500	psi				
$f'_c =$	5000	psi	$\beta_1 =$	0.8		

The strand may be harped to control service level stresses

Fig. 6.17 Information for Problem 6.3

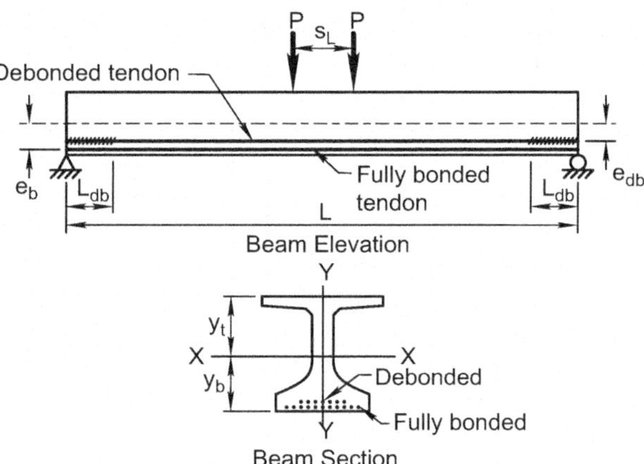

Fig. 6.18 Problem 6.4

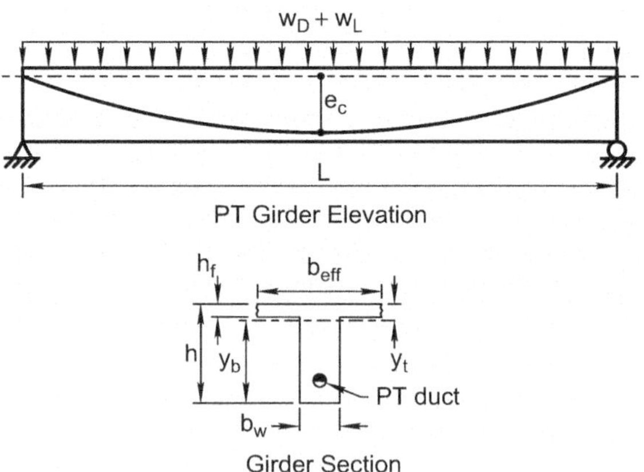

Fig. 6.19 Problem 6.5

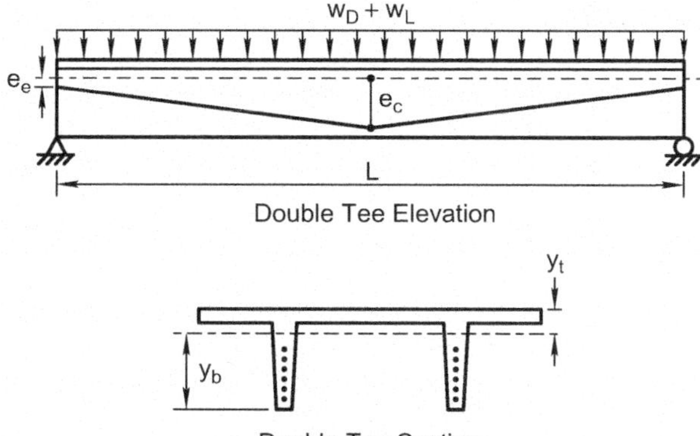

Fig. 6.20 Problem 6.6

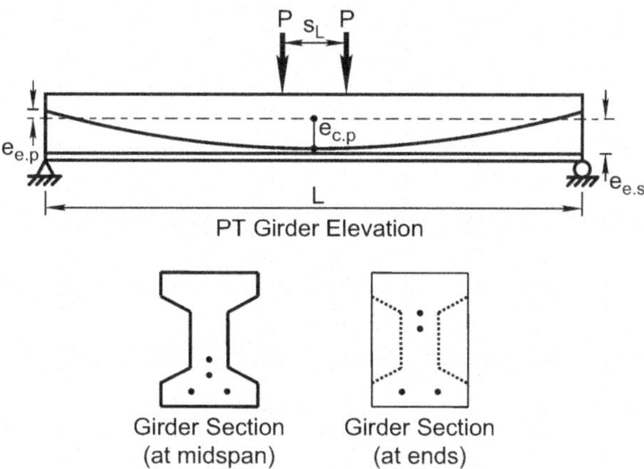

Fig. 6.21 Problem 6.7

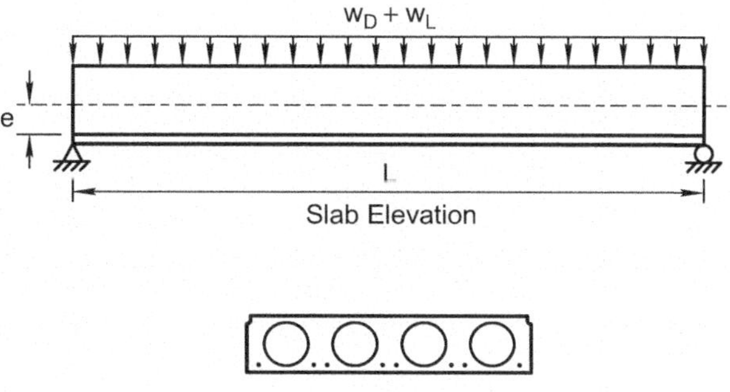

6.6. For the post-tensioned bridge girder shown in Fig. 6.20, determine the classification in accordance with ACI 318.

6.7. For the hollow core section shown in Fig. 6.21, determine the classification in accordance with ACI 318.

References

ACI 117-10. (2015). *Specification for Tolerances for Concrete Construction and Materials (ACI 117-10) and Commentary-Reapproved 2015* (Reported by ACI Committee 117, 76p). Farmington Hills, MI: ACI.

ACI 216.1-14. (2014). *Code Requirements for Determining Fire Resistance of Concrete and Masonry Construction Assemblies* (Reported by ACI Committee 216, 28p). Farmington Hills, MI: ACI.

ACI 318-99. (1999). *Building Code Requirements for Structural Concrete* (ACI Committee 318, 391p). Farmington Hills, MI: American Concrete Institute.

ACI 318-14. (2014). *Building Code Requirements for Structural Concrete* (ACI Committee 318-14, 519p). Farmington Hills, MI: American Concrete Institute.

ACI 372R-13. (2013). *Guide to design and construction of circular wire-and-strand-wrapped prestressed concrete structures* (Reported by ACI committee 372). Farmington Hills, MI: ACI.

ACI 423.5. (1999). *Report on Partially Prestressed Concrete* (Reported by ACI committee 216, 37p). Farmington Hills, MI: ACI. (ACI withdrew the report withdrawn but it is available for informational purposes.)

Beeby, A. W. (1979). The prediction of crack widths in hardened concrete. *The Structural Engineer, 57A*(1), 9–17.

Buckner, C. D. (1994). *An analysis of transfer and development lengths for pretensioned concrete structures*. Federal Highway Administration.

Darwin, D., Manning, D. G., & Hognestad, E. (1985). Debate: Crack width, cover, and corrosion. *Concrete International, 7*(5), 20–35.

Frantz, G. C., & Breen, J. E. (1980). Cracking on the side faces of large reinforced concrete beams. *ACI Journal Proceedings, 77*(5), 307–313.

Frosch, R. J. (1999). Another look at cracking and crack control in reinforced concrete. *ACI Structural Journal, 96*(3), 437–442.

Frosch, R. J. (2002). Modeling and control of side face beam cracking. *ACI Structural Journal, 99*(3), 376–385.

Gergely, P., & Lutz, L. A. (1968). *Maximum crack width in reinforced concrete flexural members* (Causes, Mechanism, and Control of Cracking in Concrete, SP-20, pp. 87–117). Farmington Hills, MI: American Concrete Institute.

Hanson, N. W., & Kaar, P. H. (1959). Flexural bond tests pretensioned beams. *ACI Journal Proceedings, 55*(7), 783–802.

Kaar, P. H., LaFraugh, R. W., & Mass, M. A. (1963). Influence of concrete strength on strand transfer length. *PCI Journal, 8*(5), 47–67.

Kaar, P., & Magura, D. (1965). Effect of strand blanketing on performance of pretensioned girders. *PCI Journal, 10*(6), 20–34.

Kannel, J., French, C. E., & Stolarski, H. K. (1997). Release methodology of strands to reduce end cracking in pretensioned concrete girders. *PCI Journal, 42*(1).

Lin, T. Y., & Burns, N. H. (1981). *Design of prestressed concrete structures*. New York: Wiley.

Martí-Vargas, J. R., Serna-Ros, P., Fernández-Prada, M. A., Miguel-Sosa, P. F., & Arbeláez, C. A. (2006). Test method for determination of the transmission and anchorage lengths in prestressed reinforcement. *Magazine of Concrete Research, 58*(1), 21–29.

Mast, R. F. (1998). Analysis of cracked prestressed sections: A practical approach. *PCI Journal, 22*(1).

Nilson, A. H. (1987). *Design of prestressed concrete* (2nd ed.). New York: Wiley.

Oesterle, R. G. (1997). *The role of concrete cover in crack control criteria and corrosion protection* (RD Serial No. 2054). Skokie, IL: Portland Cement Association.

PCI Design Handbook (8th Ed.). (2017). Chicago, IL: Precast and Prestressed Concrete Institute.

Chapter 7
Shear and Torsion

7.1 Introduction

The behavior of beams and slabs in flexure is well understood and is generally predicted with satisfactory margin of error as reflected by the strength reduction factor. Flexure is considered on a section basis. The application of principles learned in mechanics of materials with some modification for nonlinear material behavior is more than suitable to obtain a reasonable solution. The modeling of shear behavior in reinforced and prestressed concrete, however, does not lend itself to consideration at the sectional level. Before cracking, as we shall explore in this chapter, the element is analyzed using first principles. After cracking, though, flexure is still analyzed on a sectional basis, but not shear. In fact, shear behavior of cracked concrete beams has historically been idealized as a truss by visualizing a series of concrete diagonals and steel vertical members that transfer stresses through the element. The concrete diagonal members have traditionally been assumed to be at an angle of 45° from the beam axis. The contribution of the concrete tensile strength was customarily ignored, so the method underestimated the shear capacity.

Current design practice in both the ACI Building Code and AASHTO LRFD Design Specification utilizes a plastic truss analogy for both reinforced and prestressed concrete. These codes include a concrete contribution to supplement the strength derived from the plastic truss analogy. In this chapter, the concrete and steel contribution is discussed along with the effect of prestressing on the shear strength and design procedures. Incorporating torsion into shear design is covered, including a discussion of compatibility and equilibrium torsion as is covered by the ACI Building Code.

© Springer Nature Switzerland AG 2019
C. W. Dolan, H. R. Hamilton, *Prestressed Concrete*,
https://doi.org/10.1007/978-3-319-97882-6_7

Fig. 7.1 Example shear
stress calculation (**a**) shear
and moment diagram and
(**b**) stresses calculated at the
selected location

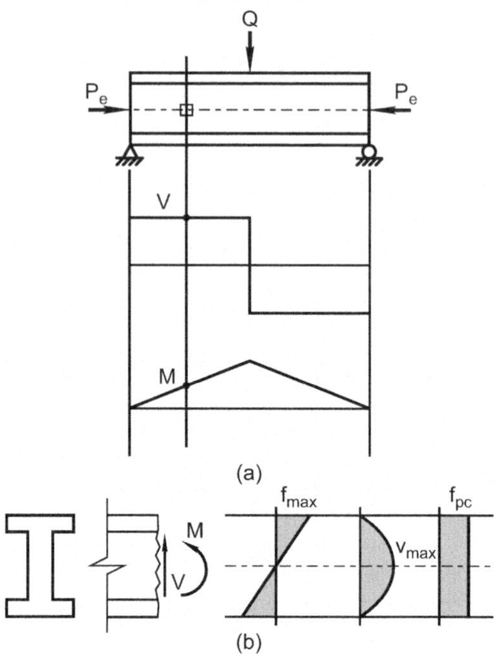

7.2   Effect of Shear and Torsion Before Cracking

Prior to cracking, the concrete behaves in a reasonably linear and elastic manner.
Shear stress is then analyzed using first principles. Although knowledge of the shear
stresses is important, the beam strength is dependent on the combination of concrete
and transverse reinforcement for shear integrity, which means the concrete must
crack to engage the reinforcement. Similar to non-prestressed concrete, prestressed
concrete sections rely on the formation of a plastic truss when the concrete cracks
transferring the shear stresses through a series of struts and ties following concrete
cracking. It is instructive, however, to consider the nature of the stresses that leads up
to the formation of diagonal cracking. Indeed, some form of these equations make up
the concrete contribution that are used to calculate the overall shear strength.

Figure 7.1 shows a simply supported beam carrying a concentrated load at
midspan and a prestress force at the centroid of the section. To illustrate calculation
of stresses, consider the flexure and shear stress at an arbitrary point along the length
of the beam. The shear, moment, and axial force at the point of interest are V, M, and
P. Assuming that the concrete remains linear and elastic, then the stresses caused by
each of these actions are calculated as follows.

Flexural stress is

$$f_{max} = \frac{Mc}{I_g} \qquad (7.1)$$

where M is the bending moment, c is the distance to the point of interest on the section, and I is the second moment of inertia of the section.

Shear stress is

$$v_{max} = \frac{VQ}{I_g t} \qquad (7.2)$$

where V is the shear force, Q is the first moment of inertia of the section above the point of interest, I is the second moment of inertia of the entire section, and t is the thickness of the section at the location of interest.

Prestress is

$$f_{pc} = \frac{P_e}{A} \qquad (7.3)$$

where P_e is the effective prestress force, and A is the cross-sectional area.

Figure 7.2 shows the Mohr's circle analysis of the stresses at the neutral axis of the section. The stresses computed from the actions can be transformed into their principal orientation, which is 45° from the beam axis. This results in a diagonal principal tension stress that causes a diagonal crack that matches the crack that would form in the web under this stress state.

The principal tensile stress is f_1 and is a function of the applied shear and moment. As shown here it is also affected by the prestressing force. Rearrange this equation such that the shear stress is the dependent variable and the prestress is the independent variable. In addition, normalize by f_1.

$$v_{max}/f_1 = \sqrt{1 + f_{pc}/f_1} \qquad (7.4)$$

Substitute the tensile strength of concrete f_t for f_1 and plot the function for typical ranges of shear and prestress, Fig. 7.3. This plot illustrates the effect of prestress on the web-shear cracking capacity. When the prestress is zero ($f_{pc} = 0$), then principal tensile stress is equal to the applied shear stress ($f_t = v_{cr}$). When the prestress is three times the principal tensile stress, then the principal tensile stress is equal to half of the shear stress. Consider the case when the applied prestress is zero. Based on the transformation of stresses for this case, the principal tensile stress (f_1) is equal to the applied shear stress (v_{max}). When a prestress force is applied, however, the principal angle and stresses change. As precompression is added, the amount of shear required to reach the principal tensile stress increases proportionally. For example, if the applied prestress is three times the concrete tensile strength, then that same principal

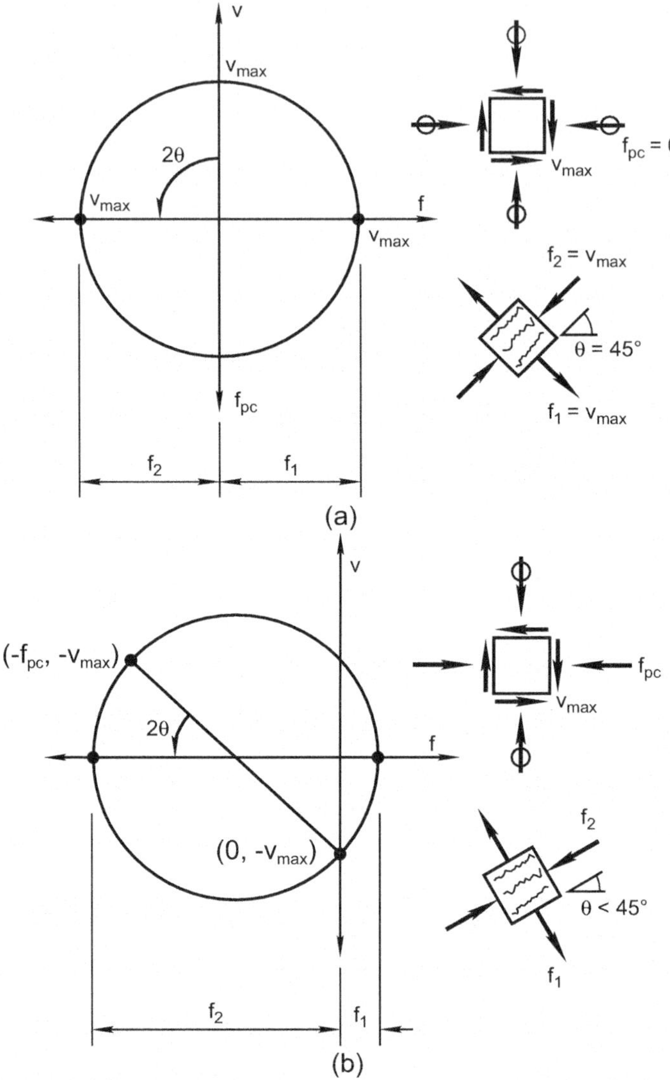

Fig. 7.2 Principal stresses at neutral axis in beam with (**a**) no prestressing (**b**) prestressing

tensile stress is twice the applied shear stress. This effectively increases the cracking stress of the concrete by a factor of two. Consequently, prestressing the concrete has a beneficial effect on the web-shear cracking strength. This effect is addressed by making use of these equations to formulate the concrete contribution to the overall shear strength.

The stress transformation can be extended to examine the effect of prestress on the principal angle and provide an estimate of the effect that prestressing might have on the shear crack angle, Fig. 7.4. Using the equations from the stress transformation

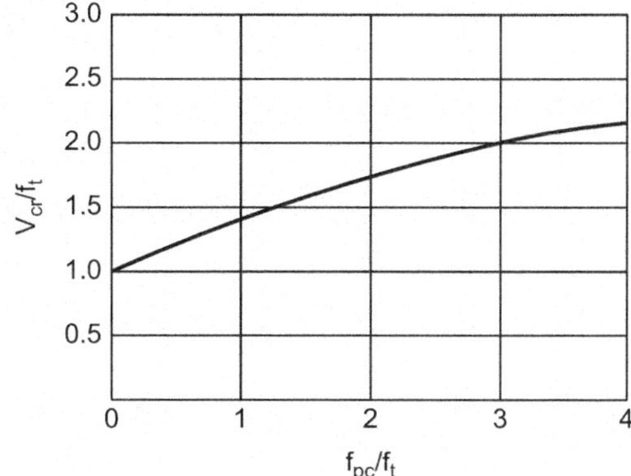

Fig. 7.3 Increase in applied shear stress required to cause cracking

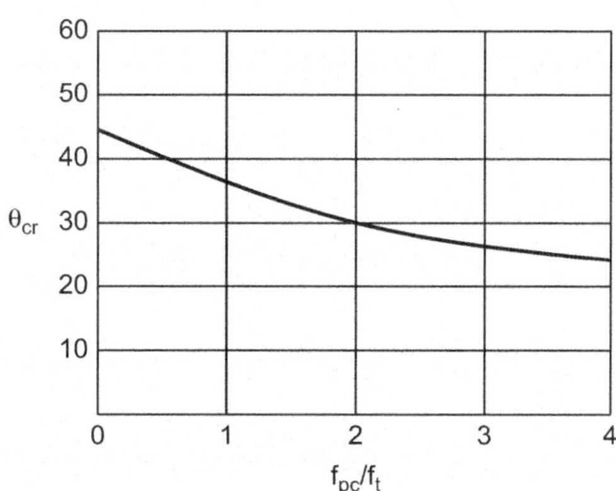

Fig. 7.4 Variation in principal angle with applied prestressing

to result in the principal angle as a function of the prestress and principal tensile stress as shown in the equation below.

$$v^2 = \left(\frac{f_{pc}}{2} + f_1\right)^2 - \left(\frac{f_{pc}}{2}\right)^2 \tag{7.5}$$

$$v = \left(\frac{f_{pc}}{2} + f_1\right)\sin 2\Theta \tag{7.6}$$

$$\left(\frac{f_{pc}}{2} + f_1\right)^2 - \left(\frac{f_{pc}}{2}\right)^2 = \left(\frac{f_{pc}}{2} + f_1\right)^2 \sin^2 2\Theta \tag{7.7}$$

Fig. 7.5 Example of
web-shear cracking

$$1 - \frac{\left(\frac{f_{pc}}{2}\right)^2}{\left(\frac{f_{pc}}{2} + f_1\right)^2} = \sin^2 2\Theta \qquad (7.8)$$

$$\cos^2 2\Theta = \left[\frac{\frac{f_{pc}}{2}}{\frac{f_{pc}}{2} + f_1}\right]^2 \qquad (7.9)$$

$$\Theta = \tfrac{1}{2}\cos^{-1}\left[\frac{\frac{f_{pc}}{f_1}}{\frac{f_{pc}}{f_1} + 2}\right] \qquad (7.10)$$

The plot of this equation illustrates the effect of prestressing on the principal angle as the prestress force is increased. When the section is nonprestressed, a web crack theoretically forms at a 45° angle. When a prestressing is added that is twice the tensile strength of the concrete, then the theoretical crack is at 30° from the longitudinal beam axis. Figure 7.5 shows a beam test in which a web crack has formed in a prestressed concrete girder that is approximately 27°, indicating that the prestress is nearly three times the tensile strength of the concrete.

Adding prestressing increases the shear force required to cause cracking and provides an overall marginal improvement in shear strength. This improvement is reflected in the differences in approach for determining shear strength of prestressed and nonprestressed sections in the ACI Building Code. In addition, prestressing generally reduces the angle (relative to the beam longitudinal axis) of the diagonal cracks.

Example 7.1: Calculate Principal Stresses Due to Shear

Given the 90-ft-long PCI BT-54 bridge girder, calculate the direction and magnitude of the service load principal stresses at the neutral axis of the girder at 10 ft from the face of the support using transformed section properties. Gross section properties are given in Appendix C and the section dimensions in Fig. 7.6. The girder contains 38 0.6-in. diameter prestressing strands placed in a 2-in. grid with an effective prestress after losses of 148.5 ksi. The center of gravity of the tendon (cgs) is located at

Fig. 7.6 PCI BT-54 section dimensions

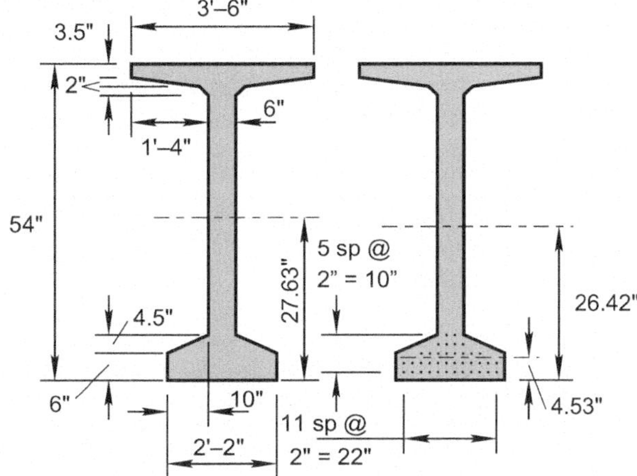

PCI BT-54 section dimensions

Table 7.1 Use parallel axis to calculate the transformed moment of inertia

	A	Ay	I	$\bar{y}$	$A\bar{y}^2$	I_t
Tendon	36.47	165	–	26.42–4.53	17,475	17,475
Concrete	659	18,208	268,056	27.63–26.42	965	269,020
Sum	695.5	18,373	268,056	–	–	286,496

4.53 in. from the bottom of the girder. The section is constructed with $f'_c = 8500\,\text{psi}$ concrete. Use parallel axis approach to determine the location of the centroid of the transformed section and compute the shear stress using VQ/Ib. The beam self-weight is 686 plf and carries a superimposed load of 2600 plf.

Solution: Table 7.1 shows the moment of inertia calculations. The contribution of the strands to the transformed section properties is considered. Modular ratio is needed to transform the section properties to the reference material, which is concrete:

$$E_c = 57,000\sqrt{f'_c} = 57,000\sqrt{8500\,\text{psi}} = 5255\ \text{ksi}$$

$$n = \frac{E_{ps}}{E_c} = \frac{28,500\,\text{ksi}}{5255\,\text{ksi}} = 5.42$$

Calculate the total area of strands:

$$A_{ps} = 38 \cdot 0.216\,\text{in.}^2 = 8.25\,\text{in.}^2$$

And their transformed area (Table 7.1):

$$(n-1)A_{ps} = (5.42 - 1) \cdot 8.25\,\text{in.}^2 = 36.47\,\text{in.}^2$$

Determine the location of centroid of the transformed section (y_{bt}):

$$y_{bt} = \frac{18,373\,\text{in.}^3}{695.5\,\text{in.}^2} = 26.42\,\text{in.}$$

The transformed moment of inertia is then calculated using parallel axis theory. The increase in I_t for this section is approximately 7%.

Q is the first moment of area above the transformed neutral axis and is equal to the web, flange and tapered portions of the flange.

$$Q = b_w \frac{y_{tt}^2}{2} + 3.5 \cdot 36 \left(y_{tt} - \frac{3.5}{2} \right) + 32 \left(y_{tt} - 3.5 - \frac{2}{3} \right)$$

$$+ 4 \left(y_{tt} - 5.5 - \frac{2}{3} \right) + 8(y_{tt} - 3.5 - 1)$$

$$= 6556\,\text{in.}^3$$

The shear at 10 ft in from the end is $(w_g + w_s)l/2 - (w_g + w_l) \cdot 10\,\text{ft} = 115.0\,\text{kips}$. The effective prestress force is $A_{ps}\,f_{se} = 8.25\,\text{in.}^2 \cdot 148.5\,\text{ksi} = 1225\,\text{kips}$. The axial stress at the neutral axis is $f_{pc} = P_e/A_{tr} = 1225/695.5 = 1762\,\text{psi}$. The concrete shear at the neutral axis is

$$v_{max} = VQ/I_{tr}b_w = 115.0 \cdot 6556/(286,496 \cdot 6) = 439\,\text{psi}.$$

Determine the principal stresses and angle using Mohr's circle:
The principal compressive stress is

$$f_2 = \sqrt{v_{max}^2 + \left(\frac{f_{pc}}{2} \right)^2} + \frac{f_{pc}}{2} = \sqrt{439^2 + \left(\frac{1762}{2} \right)^2} + \frac{1762}{2} = 1865\,\text{psi}$$

the principal tensile stress is

$$f_1 = \sqrt{v_{max}^2 + \left(\frac{f_{pc}}{2} \right)^2} - \frac{f_{pc}}{2} = \sqrt{439^2 + \left(\frac{1762}{2} \right)^2} - \frac{1762}{2} = 103\,\text{psi}$$

The angle of the principal stress is

$$\theta = \arctan \left(\frac{v_{max}}{\frac{f_{pc}}{2}} \right) = \arctan \left(\frac{439}{\frac{1762}{2}} \right) = 26.5^\circ$$

Comment: The principal compression stress is about 22% of the concrete compression strength. The principal tensile stress is about 19% of the cracking strength of 8500 psi concrete based on a direct tensile strength of $6\sqrt{f'_c}$. The angle to the principal stresses is less than 45° due to the prestress compression.

7.3   Shear Cracking

Figure 7.7 shows a prestressed concrete beam with idealized cracking that may occur in this type of element. Depending on the beam geometry, loading conditions, and quantity of prestressing, shear cracking patterns vary; consideration of principal tensile stresses that generate this cracking are covered in the previous section.

In general, crack patterns form as illustrated in Fig. 7.7. Consider the beam with the applied load placed at midspan. As the load is increased, the resulting flexural tensile stresses in the bottom of the beam overcome the prestressing and concrete tensile strength to cause the formation of a flexural crack. Flexural cracking is perpendicular to the beam axis without the influence of shear stresses, which theoretically occurs under the load point. In the region with both shear and flexure, the cracks initiate in flexure and transition to shear as the shear stresses exert more influence closer to the neutral axis. These are flexure–shear cracks, Fig. 7.8 and in the right of Fig. 7.7.

When the load is moved closer to the support, the shear-span-to-depth ratio is such that the portion of the beam between the load and support have high shear stresses and low flexural stresses. As both the shear and flexural stresses increase, the diagonal shear stresses eventually exceed the cracking strength, resulting in a diagonal crack in the web that terminates before reaching the top or bottom face of the beam. This is a web-shear crack, Fig. 7.5. The crack does not extend into the bottom flange because the prestressing force is sufficiently high to overcome the flexural tensile stresses imposed by the load.

It is these two classifications of cracking on which the concrete contribution to shear is based. Depending on the location within the length of the beam the concrete contribution (V_c) may either be dominated by flexure–shear cracking (V_{ci}) or web-shear cracking (V_{cw}). The following sections provide the theoretical and empirical basis for the use of these terms in the overall contribution to the shear strength of prestressed concrete beams.

Fig. 7.7 Shear cracking

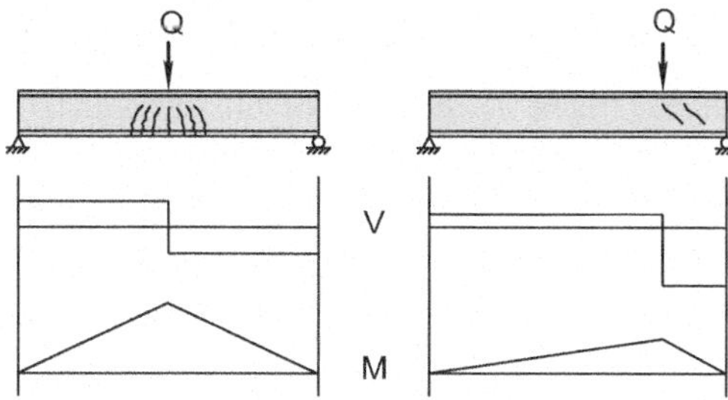

Fig. 7.8 Flexure–shear cracks

7.4 Shear Design Approach

The application of principles learned in mechanics of materials with some modification for nonlinear material behavior is suitable to obtain a reasonable solution for flexure. Shear behavior in reinforced and prestressed concrete, however, requires consideration at both the sectional and member level.

Before cracking, the element can be analyzed using first principles. After cracking, though, flexure can still be analyzed on a sectional basis, but shear design is based on a modified truss analogy Fig. 7.9. The contribution of the concrete tensile strength has traditionally been ignored, so the method underestimates the actual shear strength of the section. To correct for the underestimation of the shear strength, a concrete contribution is added to that of the truss. The general application of the truss model requires that the shear strength for a reinforced concrete section V_n be characterized as the sum of the strength contributed by the concrete V_c and the transverse reinforcement V_s.

$$V_\mathrm{n} = V_\mathrm{c} + V_\mathrm{s} \tag{7.11}$$

The concrete contribution depends on the concrete strength f'_c, axial loading N on the section A, prestress f_pc, and crack width. Transverse reinforcing bars, which are typically termed **stirrups**, provide the steel contribution to shear strength. The key parameters are cross-sectional area of the stirrup, A_v, the strength of the transverse reinforcement f_yt, the depth of the section d, and the spacing of the stirrups s. Thus,

Fig. 7.9 Plastic truss model of reinforced concrete beam

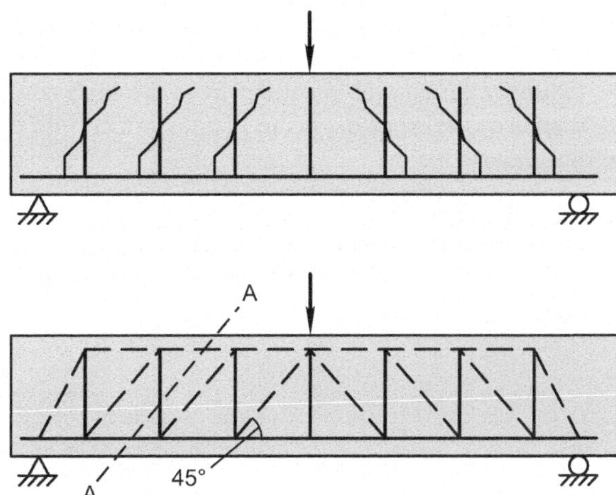

shear strength can be formulated into a general expression, which captures the contribution of the concrete and the steel:

$$V_n = \left[\xi\sqrt{f'_c} + \chi(N/A) + f_{pc}\right] b_w d + \frac{A_v f_{yt} d}{s}$$

where ξ and χ are determined based on the best fit to strength test data. This format is consistent with the ACI Building Code, where Greek characters are used to identify variables to represent best fit data and Latin characters are used to define physical variables. The square root of f'_c is selected because shear is more closely tied to tensile strength than compressive strength in prestressed concrete. Axial load on the section would be negative for tension. Prestress reduces cracking and reduces the width of any cracks that form. The equation could be further expanded to include the reinforcement ratio ρ because a higher ρ value reduces crack width and increases shear capacity. Equations specific to the ACI Building Code are developed, using this basic format.

7.5   Web-Shear Cracking V_{cw}

Web-shear cracking typically occurs in locations where the applied shear-to-moment ratio is high and the web is thin. Web-shear cracking typically initiates in the web suddenly when the principal tensile (diagonal) stresses overcome the tensile strength of the concrete. If web reinforcement is not present the following failure modes are possible:

- Separation of the tension flange from the web as inclined cracks extend horizontally toward the supports.
- Crushing of the web resulting from the high compression acting parallel to the diagonal cracks, as the beam is transformed into the equivalent of a tied arch.
- Secondary inclined tension cracking near the supports, which separates the compression flange from the web.

The concrete contribution to nominal shear strength, when the behavior is dominated by web-shear cracking, is based on the beam theory analysis discussed in Sect. 7.2. If Eq. (7.5) reduces to

$$v_{cw} = f_t \sqrt{1 + \left(\frac{f_{pc}}{f_t}\right)} \tag{7.12}$$

Where f_1 is replaced with the tensile strength of concrete (f_t), v_{cw} is substituted for v_{max} to indicate web shear strength, and f_{pc} is the compressive stress from prestressing. Substituting a conservative estimate of concrete tensile strength for f_t of $3.5\sqrt{f'_c}$

$$v_{cw} = 3.5\sqrt{f'_c} \sqrt{1 + \left(\frac{f_{pc}}{3.5\sqrt{f'_c}}\right)} \tag{7.13}$$

This equation can be conservatively simplified to

$$v_{cw} = 3.5\sqrt{f'_c} + 0.3f_{pc} \tag{7.14}$$

Multiply by the web area to convert shear stress to shear force

$$V_{cw} = \left(3.5\sqrt{f'_c} + 0.3f_{pc}\right) b_w d_p \tag{7.15}$$

where b_w is the width of the web, and d_p is the effective depth of the cgs of the tendon. When the prestressing tendon is not parallel to the beam axis, the vertical component of the tendon force V_d contributes to the shear strength of the section, Fig. 7.10. This component is added directly to the force required to cause web-shear

Fig. 7.10 Contribution of vertical component of prestressing force

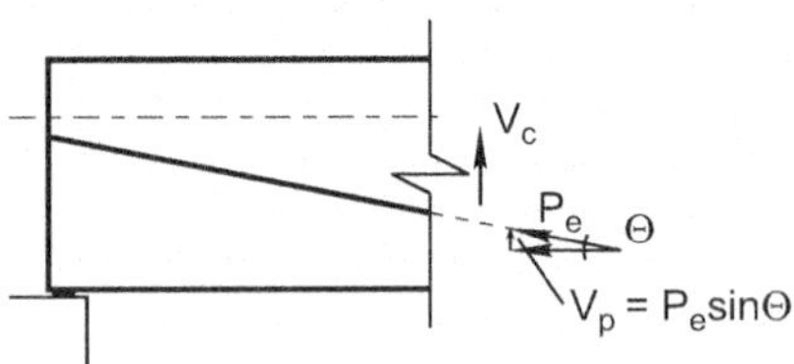

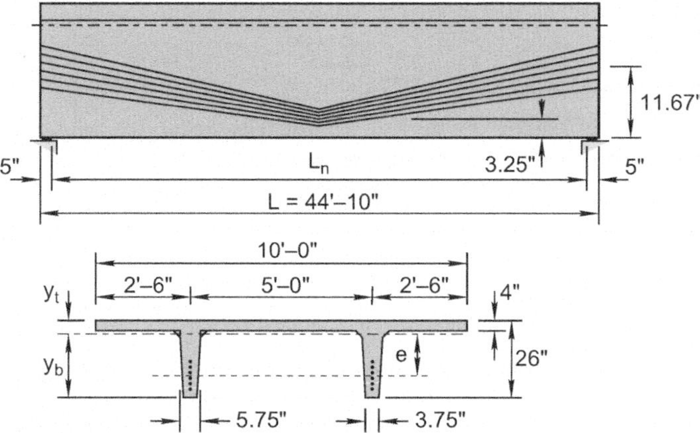

Fig. 7.11 Beam section and strand distribution

cracking. To account for the use of lightweight concrete a factor (λ) is applied, resulting in the ACI Building Code equation for concrete contribution from web-shear cracking:

$$V_{cw} = \left(3.5\lambda\sqrt{f_c'} + 0.3f_{pc}\right)b_w d_p + V_p \tag{7.16}$$

Equation 7.16 of ACI 318-14 indicates that d_p should be taken as 80% of the total beam depth or as the actual effective depth of the prestressing tendon, whichever is greater. This is considered valid based on studies by MacGregor and Hanson (1969). Even so, the tests on which these conclusions are based had either straight prestressed reinforcement or reinforcing bars at the bottom of the section with stirrups that enclosed this longitudinal reinforcement. The basic concept is that because the section is uncracked, the location of the bottom steel does not have a direct effect on the nominal shear stress.

Example 7.2: Calculate V_{cw} of Pretopped Double Tee with Harped Tendon

Determine V_{cw} at a distance of 4 ft from the center of the support in the pretopped double tee section shown in Fig. 7.11. Determine the factored shear at this location and if shear reinforcement is required for strength. Superimposed dead load is 15 psf and live load is 150 psf. The double tee contains 12 0.5-in. dia. 7-wire low relaxation prestressing strands ($f_{pu} = 270$ ksi). The strands are divided evenly between the webs and have a single harp point with the eccentricities shown in Fig. 7.11. The effective prestress immediately after prestress transfer is 182 ksi and effective prestress after all losses is 173 ksi. Concrete compressive strength is $f_c' = 5500$ psi.

Solution: Calculate the factored shear force at the section of interest assuming that the double tee spans center-to-center of support

$$w_u = 1.2w_{DL} + 1.6w_{LL} = 1.2(0.015 \cdot 10 + 0.718) + 1.6 \cdot 0.150 \cdot 10 = 3.44 \, \text{klf}$$

$$V_u = 3.44 \cdot (0.5 \cdot 44.42 - 4) = 62.6 \, \text{kip}$$

Determine the effective depth and eccentricity used to calculate the shear strength. The tendon eccentricity varies from $e_c = 20.29 - 11.67 = 8.62$ in. at the end to $e_m = 20.29 - 3.25 = 17.04$ in. at midspan. Eccentricity at the section of interest is

$$e = 8.62 + (17.04 - 8.62)\frac{4.21}{0.5 \cdot 44.83} = 10.20 \, \text{in.}$$

$$d_p = 10.20 + 5.71 = 15.91 \, \text{in.}$$

But no less than

$$0.8h = 0.8 \cdot 26 = 20.8 \, \text{in.}$$

Effective prestress force in both webs:

$$P_e = 173 \cdot 12 \cdot 0.153 = 317.6 \, \text{kip}$$

Determine shear strength contributed by concrete considering web-shear cracking using Eq. (7.16). Start by calculating f_{pc}. If the centroid was located in the flange, then the compressive stress in the concrete would be calculated at the junction of the web and flange. In this case, however, the depth of the centroid, $y_t = 5.71$ in., is greater than the deck thickness of 4 in. Thus, calculate the compressive stress in the concrete at the centroid of cross section resisting externally applied loads due to both prestress and moments resisted by precast member acting alone. These stresses are to be calculated using unfactored loads, Fig. 7.12. Since there is no flexural stress at the centroid of the section, the stress is due to the axial component of prestress only.

Fig. 7.12 Beam shear and moment distribution

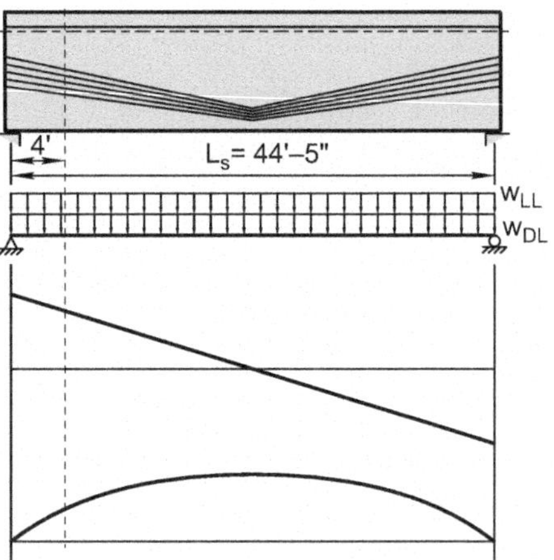

$$f_{pc} = \frac{316.7}{689} = 460\,\text{psi}$$

Vertical component of the prestress force assuming small angles is

$$V_p = \frac{17.04 - 8.62}{0.5 \cdot 44.83 \cdot 12} 316.7 = 9.91\,\text{kip}$$

Web-shear strength contribution is

$$V_{cw} = \left(3.5\lambda\sqrt{5500} + 0.3 \cdot 460\right)9.5 \cdot 20.8/1000 + 9.91 = 88.5\,\text{kip}$$

This value is compared to the factored shear V_u to determine that odd shear reinforcement is not required for strength.

7.6 Flexure–Shear Cracking V_{ci}

Flexure–shear cracks occur as a continuation of flexural cracks that form in the tension face under bending. The flexural crack initiates with an angle that is nearly perpendicular to the beam axis. As the crack propagates, shear stresses change the angle of principal tensile stress causing an angle change in the crack. In prestressed beams, this angle can become relatively flat, nearly parallel with the top of the beam. If shear reinforcement is not present, the crack will widen and extend into the compression zone, potentially causing a shear-compression failure. This type of failure results in the loss of the compression zone due to the combined effect of shear and compression on the compression zone.

For flexure–shear cracking, tests have shown that the critical inclined crack has a horizontal projection at least equal to the effective beam depth. Therefore, it is the flexural cracking within a distance, d, measured in the direction of decreasing moment from the section under consideration that is associated with flexure–shear cracking.

Figure 7.13 shows a beam with flexure–shear cracking. The shear and moment diagrams for the self-weight (V_{DL} and M_{DL}) and superimposed loads (V_{ue} and M_{ue}) are separated. A potential diagonal tension crack is assumed to initiate as a flexural crack at a distance d (Section A) from the critical section (Section C) for shear. Tests indicated that it is the formation of an additional flexural crack, midway between previously formed flexural cracks, which is the event that triggers actual loss of capacity (assuming that the section is not reinforced). The shear and moment at section C are V_i and M_d, respectively, and the shear and moment at section B are V_{cre} and M_{cre}, respectively. The shears and moments are those produced by superimposed dead and live loads, and act in addition to those produced by the member self-weight and the prestressing force.

Fig. 7.13 Development of
flexure–shear cracking
strength equation

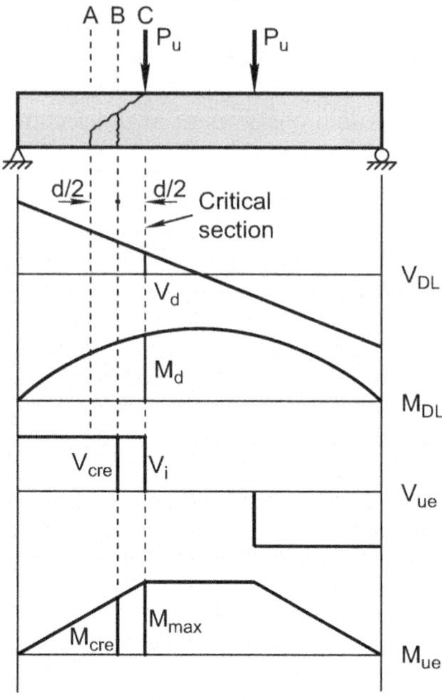

Determine shear due to applied loads ($V_i = V_u - V_d$) at the critical section
required to cause the section to crack at a location shown in the figure. Start with

$$dM = Vdx \tag{7.17}$$

$$M_{max} - M_{cre} = \tfrac{1}{2}(V_i + V_{cre})\tfrac{d}{2} \tag{7.18}$$

Assume that the change in shear between B and C is small ($V_i = V_{cre}$)

$$M_{max} - M_{cre} = V_i \tfrac{d}{2} \tag{7.19}$$

$$V_i = \frac{M_{cre}}{M_{max}/V_i - d/2} \tag{7.20}$$

Further simplification is possible by conservatively assuming that $d/2$ is small
relative to the moment-shear ratio. Shear required to cause flexural cracking is

$$V_i = \frac{V_i M_{cre}}{M_{max}} \tag{7.21}$$

The cracking moment caused by external loads is then

$$M_{cre} = I/y_t \left(6\lambda\sqrt{f'_c} + f_{pe} - f_d \right)$$ (7.22)

where f_d is the stress due to the unfactored dead load.

The total shear force at the critical section is then

$$V_{ci} = 0.6\lambda\sqrt{f'_c} b_w d_p + V_d + \frac{V_i M_{cre}}{M_{max}}$$ (7.23)

where d_p need not be taken less than $0.8h$. The first term is the additional shear force required to fully develop the crack, the second term is the unfactored shear due to all dead load, and the final term is the shear from externally applied loads at the critical section where the flexural crack is initiated at $d/2$ away. The derived equation gives the shear strength at the critical section C, due to the superimposed dead and live loads, when the moment at section B due to these same loads is M_{cre}. In other words, when the shear at section C reaches this value, a flexural crack forms in the beam at location B.

When calculating the shear using Eq. (7.23), it is important to distinguish between composite and noncomposite sections. For noncomposite beams, the staging that occurs with composite construction does not affect the flexural stresses in the section and thus, does not affect the flexural cracking term in V_{ci}. Indeed, if a noncomposite section carries only uniform loading, then Eq. (7.23) reduces to

$$V_{ci} = 0.6\lambda\sqrt{f'_c} b_w d_p + \frac{V_u M_{ct}}{M_u}$$ (7.24)

where M_u is the total factored moment at the section of interest and V_u is the total factored shear associated with M_u. The V_d term has not been eliminated, but rather incorporated into the second term because the same section properties apply to both dead and live load flexural stresses. Consequently, dead load flexural stresses and shear do not need to be separated. As such, the moment required to cause cracking at section B can be expressed as

$$M_{cr} = I/y_t \left(6\lambda\sqrt{f'_c} + f_{pe} \right)$$ (7.25)

where M_{cr} is the total moment required to cause flexural cracking. M_{cr} reflects the total stress change from effective prestress to a tension of $6\lambda\sqrt{f'_c}$, assumed to cause flexural cracking.

Incorporating composite construction into the shear strength presents some difficulties. Composite members are generally composed of a prestressed, often precast,

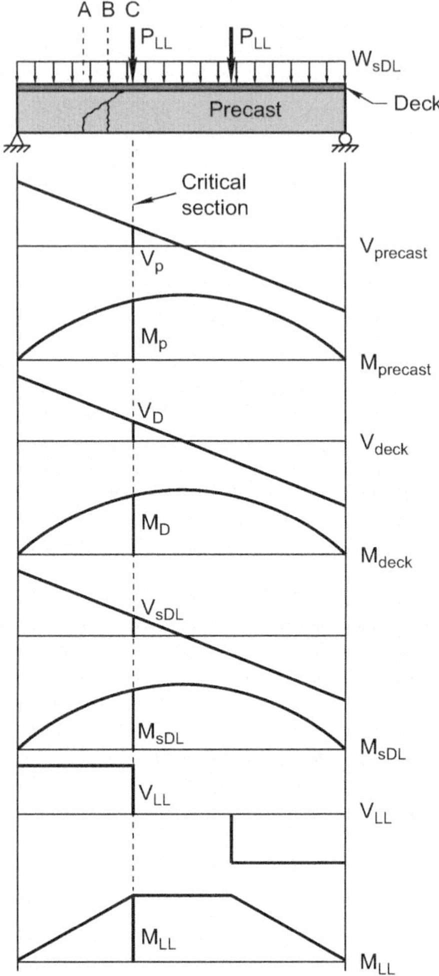

Fig. 7.14 Breakdown of shear strength based on flexure–shear cracking

member topped with a cast-in-place component, which is usually a deck, Fig. 7.14. While the entire section resists shear caused by superimposed loads, the prestressing is only applied to the precast section. Because Eq. (7.23) is based partially on the initiation of flexural cracking, this must be incorporated into the calculation of the concrete contribution. Depending on characteristics of the construction, the two components resist different proportions of the total applied shear. If the member is unshored, then the precast portion carries both its self-weight and that of the deck. If shored, then the precast member carries its own self-weight, but the composite section carries the deck load. As such, appropriate section properties should be used to calculate f_d. Consider the composite beam shown in Fig. 7.14. Since the purpose of the third term in Eq. (7.23) is to estimate the point at which a flexural crack is initiated, composite section properties and shoring must be considered when it is calculated. The dead load stress that should be discounted in the cracking moment due to externally applied loads when shoring is used becomes

$$f_d = {}^{M_{precast}}\!\big/_{S_b} + {}^{M_{deck}}\!\big/_{S_b} \tag{7.26}$$

where the individual service moments are shown in the figure and S_b is the section modulus for extreme fiber stress in the bottom of the precast section. If shoring is used then the deck is carried by the composite section rather than just the precast section. Thus, the dead load stress is

$$f_d = {}^{M_{precast}}\!\big/_{S_b} + {}^{M_{deck}}\!\big/_{S_{bc}} \tag{7.27}$$

where S_{bc} is the bottom section modulus of the composite section.

The appropriate stress is then used in Eq. (7.22) with the composite section properties to compute cracking moment

$$M_{cre} = S_{bc}\left(6\lambda\sqrt{f'_c} + f_{pe} - f_d \right) \tag{7.28}$$

Because of the distinction required by the composite construction, V_i and M_{max} must also account for the dead load separately. For both shored and unshored cases, the moment due to the unfactored dead load is

$$M_d = M_{precast} + M_{deck} \tag{7.29}$$

The maximum factored moment at a section due to externally applied loads is then

$$M_{max} = \text{factored}\left(M_{precast} + M_{deck} + M_{sDL} + M_{LL} \right) - M_d \tag{7.30}$$

The unfactored shear used to calculate V_i is

$$V_d = V_{precast} + V_{deck} \tag{7.31}$$

So the factored shear occurring with M_{max} is

$$V_i = \text{factored}\left(V_{precast} + V_{deck} + V_{sDL} + V_{LL} \right) - V_d \tag{7.32}$$

The lower bound for Eq. (7.23) is

$$V_{ci} = 1.7\lambda\sqrt{f'_c}b_w d \tag{7.33}$$

where d need not be taken less than $0.8h$.

Example 7.3: Calculate V_{ci} of Pretopped Double Tee with Harped Tendon
For the double tee in Example 7.2, calculate V_{ci} at a distance of 4 ft from the center of the support in the pretopped double tee section shown in the figure.

Solution: Determine shear strength contributed by concrete considering flexure–shear cracking using Eq. (7.23). Use the non-composite equations to solve for the shear strength. Start with the precompression in the concrete

$$f_{pe} = 317.6\left(\frac{1}{689} + \frac{10.20}{1514}\right) \cdot 1000 = 2600\,\text{psi}$$

Calculate the cracking moment due to externally applied loads using Eq. (7.22)

$$M_{cr} = 1514 \cdot \left(6\sqrt{5500} + 2600\right)/12{,}000 = 384.2 \ \text{kip ft}$$

Maximum factored moment due to externally applied loads is

$$M_u = 0.5 \cdot 3.44 \cdot 44.42 \cdot 4 - 0.5 \cdot 3.44 \cdot 4^2 = 278.1\,\text{kip ft}$$

Factored shear force is

$$V_u = 3.44 \cdot (0.5 \cdot 44.42 - 4) = 62.6\,\text{kip}$$

Contribution to shear strength when considering flexure–shear cracking is

$$V_{ci} = 0.6\sqrt{5500} \cdot 9.5 \cdot 20.8/1000 + \frac{384.2}{278.1} \cdot 62.6 = 95.3\,\text{kip}$$

The ACI Building Code requires that both V_{ci} and V_{cw} be checked and that the lowest value is used to calculate the total shear resistance.

7.7 Critical Sections

Shear capacity is checked at each "critical section;" however, the ACI Building Code does not provide complete definition of what constitutes a critical section. Some critical sections, such as the change in spacing of stirrups, are a function of design decisions and others are a function of the loading. Beams loaded on the top flange generate an internal compression strut bringing the load to the support as seen in

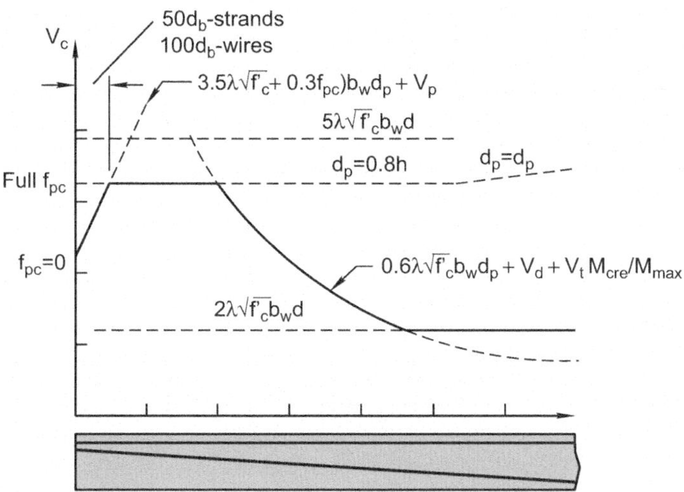

Fig. 7.15 Characteristic parts of the V_c diagram for prestressed concrete

Fig. 7.9. This strut prevents a crack from forming at the support and, consequently, the ACI Building Code allows the critical section to be located a distance d for reinforced beams or $h/2$ for prestressed beams in from the face of the support. In cases where the load is applied to the bottom of the section, as occurs with inverted-T beams, the critical section is at the face of the support. Shear calculations for beams with concentrated loads applied near the end of the beam should consider the face of the support as the critical section. Additional critical sections occur adjacent to concentrated loads and at locations of discontinuity in the member.

The critical section at the end of a pretensioned beam is complicated by the transfer length of the prestressing strand, Fig. 7.15. The prestressing force is assumed to be zero at the end of the beam and increases to the full effective prestress at the end of the transfer length. Further complicating factor is that the shear equations for prestressed concrete are only valid where

$$A_{ps}f_{se} \geq 0.4\left(A_{ps}f_{pu} + A_sf_y\right) \qquad (7.34)$$

Consequently, the shear strength at the end of the member will be the minimum value of

$$V_{c,min} = 2\lambda\sqrt{f'_c}b_wd \qquad (7.35)$$

until the prestress level increases sufficiently to satisfy Eq. (7.34). At that point, V_{cw} equation is valid and can be used to calculate V_c. This equation plateaus at the end of the transfer length where full prestress is transferred. The following equation may also limit V_c in this region of the beam

$$V_{c,\max} = 5\lambda\sqrt{f'_c}\,b_w d \qquad\qquad (7.36)$$

Away from the support, flexure–shear cracking is the dominate cracking mode where Eq. (7.23) controls the concrete contribution. Near midspan, the minimum shear strength may control. Excel spreadsheets are commonly used to determine the shear capacity as seen in Fig. 7.15.

7.8 Shear Reinforcement V_s

The previous sections described the cracking shear strength of prestressed concrete members. In general, the equations are derived to provide an estimate of the shear required to fully develop either a web-shear crack (V_{cw}) or flexure–shear crack (V_{ci}), depending on the location of interest. If reinforcement is not included in the section to prevent the unstable growth of these shear cracks, then very brittle and unpredictable failure would follow. Consequently, transverse reinforcement is used in prestressed concrete as it is in nonprestressed concrete. Smaller bars (#3, #4, and #5) are bent and placed transversely in the beam to ensure that the shear cracks are controlled and that, should an extreme loading occur, the failure mode is relatively ductile.

Both the ACI and AASHTO shear design provisions are based on plastic truss models. The ACI Building Code uses a fixed-angle (45°) model and AASHTO uses a variable angle truss model that requires the angle of the struts be determined to calculate the shear strength of the section. The particular approach used in AASHTO has traditionally been referred to as the modified compression field theory (MCFT).

In The ACI Building Code the 45° truss model is applied by assigning a portion of the total shear strength to the concrete and the remainder to the shear reinforcement. While beams do not actually behave as if the two contributions are separate and distinct, the approach provides a computationally convenient method to determine the amount and distribution of shear reinforcement required to provide an adequate total nominal shear strength. The strength contributed by the concrete is assumed to be that required to cause web or flexure shear cracking and is the same regardless of the amount of shear reinforcement present. Once diagonal cracking has occurred, whether initiated in the web or at the tension face, the concrete contribution is ascribed to aggregate interlock (V_a), dowel action (V_d), shear transmitted across the concrete compression zone (V_{cz}), and the vertical component of the effective prestress force (V_p), Fig. 7.16. The concrete contribution due to these various components are combined into a single term based on an average shear stress over the effective cross section, $b_w d$.

The contribution of the transverse reinforcement to the shear strength is then a matter of summing forces vertically on the section assuming that the stirrups have all yielded. For the free body diagram shown in the figure, the number of stirrups crossed by a 45° diagonal crack is

Fig. 7.16 Free body
diagram illustrating the
various contributions to
shear strength at a section

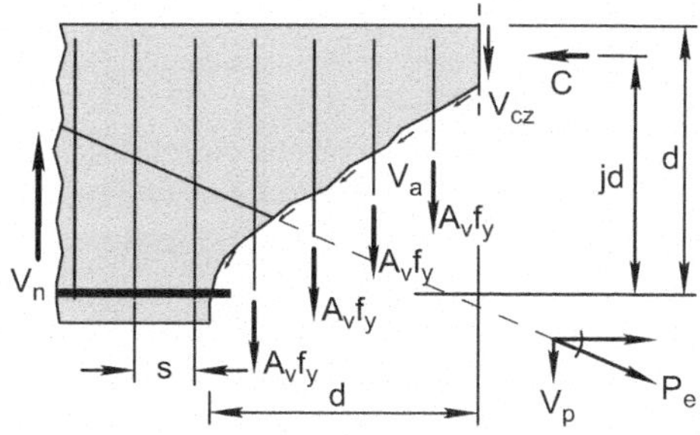

$$n_s = \frac{jd}{s \times \tan 45} \tag{7.37}$$

If the difference between jd and d is ignored, then the contribution to shear strength is

$$V_s = \frac{A_v f_{yt} d}{s} \tag{7.38}$$

This steel contribution is typically accomplished using deformed bars, deformed wires, or welded wire reinforcement placed perpendicular to the axis of the beam. Reinforcing bars are bent into shapes that are convenient to integrate into a reinforcing **cage** in which the longitudinal and transverse reinforcement are tied together and placed into the forms. A tied reinforcing cage along with the support provided by the formwork fixes the reinforcing bars into place during the rather rough placement of fresh concrete.

Figure 7.17 shows some of the more common shapes of stirrups used with both non-prestressed and prestressed concrete sections. Stirrups are commonly formed with two legs because they can be conveniently incorporated into a box-shaped reinforcing cage. To ensure that the plastic truss fully develops at nominal shear strength, the vertical legs of the stirrups must fully develop their yield strength on either side of the diagonal cracks. This is ensured by anchoring the top and bottom end of the stirrup either through the bends at the bottom or the standard hook in the top. In addition, stirrups are required to reach as close to the compression and tension faces as cover requirements allow. They are also required to extend a distance d from extreme compression fiber. Chapter 25 of ACI 318-14 provides detailing requirements for bends or hooks at the ends of stirrups. The hooks or bends are typically required to enclose the longitudinal reinforcement to assist with the anchorage.

In members with thin webs, stirrups can be formed with a single leg and using hooks at both ends to ensure they are fully developed. Reinforcement for torsion must form a closed loop to be effective. Another technique to address shear reinforcement in thin members is the use of welded wire reinforcement (WWR), which is common in the precast, prestressed concrete industry, Fig. 7.17. To ensure anchorage of the vertical elements, the ACI Building Code requires that longitudinal

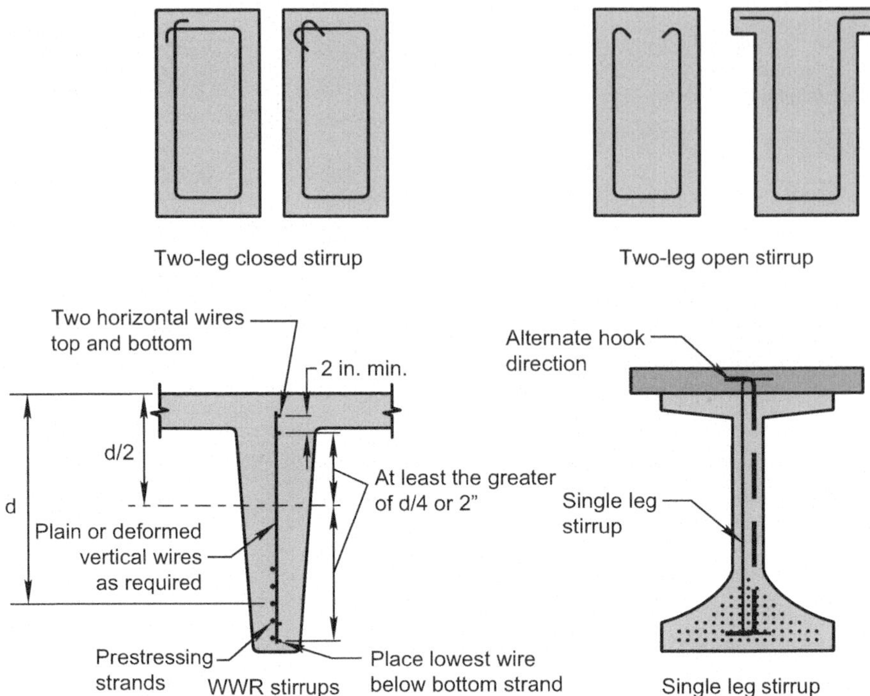

Fig. 7.17 Configurations of shear reinforcement

wires be located at the same depth as the primary flexural reinforcement to avoid a splitting problem at the level of the tension reinforcement. In thin web sections that do not allow for stirrup hooks to be developed at the top and bottom of the member, such as precast double tee sections, welded wire reinforcement (WWR) is commonly used.

7.9 Design of Shear Reinforcement

ACI and AASHTO both have similar provisions for the design of shear reinforcement in prestressed concrete members. Nominal one-way shear strength for beams and slabs is calculated by summing the individual contributions to shear strength from concrete and steel:

$$V_\mathrm{n} = V_\mathrm{c} + V_\mathrm{s} \tag{7.39}$$

where V_c for prestressed members is the lesser of V_{cw} (Eq. 7.16) and V_{ci} (Eq. 7.23) and V_s is calculated using Eq. (7.31). To satisfy the shear strength limit state the design shear strength must be greater than or equal to the factored shear

$$\phi V_n \geq V_u \tag{7.40}$$

The design strength includes both concrete section and transverse reinforcement. Once the section dimensions have been selected, the shear design is primarily the selection of stirrup spacing. As this stirrup spacing is decreased, the design strength is increased. Closely spaced stirrups create reinforcement congestion. In addition, the excessive transverse reinforcement may result in a shift in failure mode from yielding of stirrups to compression failure in the diagonal struts of the plastic truss, which is commensurate with a brittle failure mode. To avoid this situation, Section 25.5.1.2 of ACI 318-14 limits the design strength of a section to

$$V_u \leq \phi \left(V_c + 8\lambda \sqrt{f'_c} b_w d \right) \tag{7.41}$$

If this limit is exceeded, then the cross-sectional dimensions must be increased to limit cracking and the prospect of diagonal compression failure in the concrete.

Given a section of sufficient size, the reinforcement can then be sized using

$$V_{s,req} = \frac{V_u}{\phi} - V_c \tag{7.42}$$

where $V_{s,req}$ is the strength required of the stirrups based on the factored shear and concrete contribution. Combine Eqs. (7.31) and (7.42) to provide the maximum stirrup spacing required by strength:

$$s = \frac{A_v f_{yt} d}{V_{s,req}} \tag{7.43}$$

based on a given stirrup configuration, bar area, and yield strength.

To ensure a ductile failure mode, the ACI Building Code requires the use of a minimum amount of shear reinforcement. The limitation is in terms of both the area of transverse reinforcement and the spacing of the reinforcement. Having a minimum amount of transverse reinforcement present ensures that if a diagonal crack forms, then reinforcement is present to restrain the crack growth that might occur due to overload. Other potential causes of crack initiation include restrained shrinkage and temperature movement. Minimum shear reinforcement is required where $V_u > 0.5\phi V_c$ except for the cases show in Table 7.2 in which minimum reinforcement is required where $V_u > \phi V_c$. This is valid for prestressed beams and slabs. Where shear reinforcement is required and torsional effects are negligible, the requirements in Table 7.3 apply.

Table 7.2 Cases where $A_{v,min}$ is not required (where $0.5\phi V_c \leq V_u \leq \phi V_c$)

Beam type	Conditions
Shallow depth	$h \leq 10\,\text{in.}$
Integral with slab	$h \leq$ greater of $2.5t_f$ or $0.5b_w$ and $h \leq 24\,\text{in.}$
Constructed with steel fiber-reinforced normalweight concrete conforming to Section 26.4.1.5 of ACI 318-14	$h \leq$ greater of $2.5t_f$ or $0.5b_w$ and $V_u \leq \phi 2\sqrt{f_c'}\,b_w d$
One-way joist system	In accordance with ACI 318-14

Table 7.3 Minimum shear reinforcement for slabs and beams ($A_{v,min}$) from Section 7.9.6.3.3 of ACI 318-14

Beam type	$\dfrac{A_{v,\min}}{s}$			Eq. (7.44)
Nonprestressed and prestressed with $A_{ps}f_{se} < 0.4$ $(A_{ps}f_{pu} + A_s f_y)$	Greater of:		$0.75\sqrt{f_c'}\dfrac{b_w}{f_{yt}}$	a
			$50\dfrac{b_w}{f_{yt}}$	b
Prestressed with $A_{ps}f_{se} \geq 0.4(A_{ps}f_{pu} + A_s f_y)$	Lesser of:	Greater of:	$0.75\sqrt{f_c'}\dfrac{b_w}{f_{yt}}$	c
			$50\dfrac{b_w}{f_{yt}}$	d
		$\dfrac{A_{ps}f_{pu}}{80 f_{yt}d}\sqrt{\dfrac{d}{b_w}}$		e

Table 7.4 Maximum stirrup spacing for prestressed beams (in.) from ACI 318-14

$V_s \leq 4\sqrt{f_c'}\,b_w d$	Lesser of:	$3h/4$
		24
$V_s > 4\sqrt{f_c'}\,b_w d$	Lesser of:	$3h/8$
		12

These requirements ensure that a sufficient amount of reinforcement is available to resist the applied factored shear. To ensure that a sufficient number of stirrups are in place to cross every potential diagonal crack, a maximum stirrup spacing in Table 7.4 is specified by ACI 318-14.

For thin-web, post-tensioned members such as those used in joists and waffle slabs, the ACI Building Code commentary recommends the use of shear reinforcement even when V_u is less than $0.5\phi V_c$. Ducts for post-tensioned tendons are placed without the tendon being tensioned, which means that some deviation from the theoretical profile (wobble) occurs due to placement tolerances. The use of some web reinforcement is recommended to help secure the duct in place during concrete placement. During tendon stressing the unintended curvature of the tendon may result in lateral forces on the webs. The presence of web reinforcement resists tensile stresses generated in the webs under such forces. ACI Building Code commentary recommends a maximum spacing of stirrups used for this purpose as the smaller of $1.5h$ or 4 ft.

Example 7.4: Calculate V_c and Shear Reinforcement

For the BT-63 girder shown in Fig. 7.18, determine the concrete contribution to shear at 20 ft from the left end of the member. The section contains 40 prestressing strands of which 12 are harped. Determine the stirrup size and spacing required to resist the given load case at 20 ft from the end of the member. The girder is constructed such that composite action can be assumed with the deck. Girder spacing is 9 ft and deck thickness is 8 in. V_u at the section of interest is 159 kip.

The effective prestress immediately after prestress transfer is 182 ksi and effective prestress after all losses is 172.5 ksi. Girder concrete compressive strength is $f'_c = 6500\,\text{psi}\ f'_{ci} = 5500\,\text{psi}$. Deck compressive strength is $f'_c = 5500\,\text{psi}$.

Precast section properties.

$$A = 713\,\text{in.}^2$$
$$I = 392,638\,\text{in.}^4$$
$$y_b = 32.12\,\text{in.}$$
$$S_b = 12,224\,\text{in.}^3$$
$$y_t = 30.88\,\text{in.}$$

Composite section properties:

$$I_c = 819,070\,\text{in.}^4$$
$$b_{\text{eff}} = 108\,\text{in.}$$
$$y_{bp} = 49.16\,\text{in.}$$
$$y_{td} = 21.84\,\text{in.}$$
$$y_{bd} = y_{tp} = 13.84\,\text{in.}$$

Solution: Determine the effective depth and eccentricity used to calculate the shear strength. Calculate the average tendon depth considering both the straight and harped tendons. The tendon eccentricities of the straight tendon and top and bottom of the harped tendon:

$$e_{\text{straight}} = y_b - \frac{20 \cdot 3 + 6 \cdot 6 + 2 \cdot 8}{28} = 28.12\,\text{in.}$$

$$e_{\text{harped}} = y_b - \left[\frac{55 - 7}{35} \cdot (35 - 20) + 7\right] = 4.55\,\text{in.}$$

$$d_p = \left[28\left(e_{\text{straight}}\right) + 12\left(e_{\text{harped}}\right)\right]\frac{1}{40} + t_d + y_t = 59.93\,\text{in.}$$

But no less than

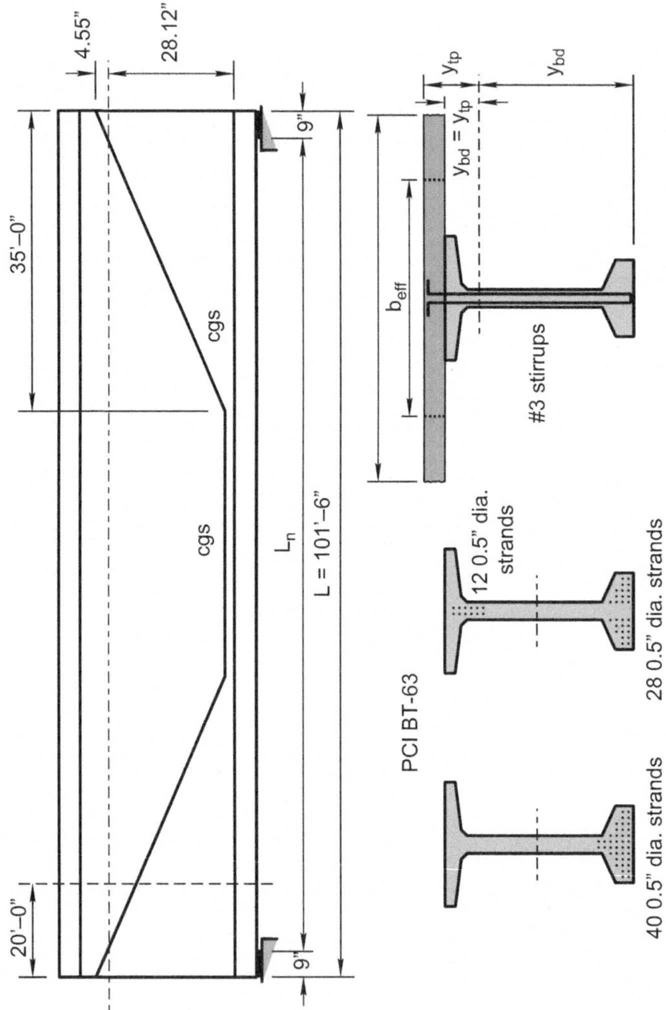

Fig. 7.18 Beam section and tendon location

$$0.8h = 0.8 \cdot (63 + 8) = 56.8 \text{ in.}$$

Use the actual depth for shear strength of 59.93 in.
Effective prestress force subdivided into straight and harped strands are

$$P_{e\_harped} = 12 \cdot 0.153 \cdot 172.5 = 316.7 \text{ kip}$$

$$P_{e\_straight} = 28 \cdot 0.153 \cdot 172.5 = 739.0 \text{ kip}$$

Calculate shear strength contributed by concrete. Unfactored dead load moment including precast and deck load is

$$w_d = 0.743 + 0.15 \cdot 9 \cdot 0.667 = 1.64 \text{ klf}$$

Assume that the reaction is at the center of bearing.

$$L = 101.5 - 0.75 = 100.75 \text{ ft}$$

Dead load moment at section of interest is

$$M_d = 0.5 w_d \left(Lx - x^2\right) = 0.5 \cdot 1.64 \left(100.75 \cdot 19.63 - 19.63^2\right) = 1306 \text{ kip ft}$$

At the extreme fiber of the section where tensile stress is caused by externally applied loads, stress due to unfactored dead load is

$$f_d = \frac{M_d y_b}{I} = {}^{1306 \cdot 32.12 \cdot 12,000}\!/\!{}_{392,638} = 1282 \text{ psi}$$

Compressive stress at the extreme tensile fiber due to effective prestress force alone is:

$$f_{pe} = \left(\frac{316.7 + 739.0}{713} + \frac{316.7 \cdot 4.55 + 739.0 \cdot 28.12}{12,224}\right) \cdot 1000 = 3299 \text{ psi}$$

Cracking moment due to externally applied loads using Eq. (7.22) is

$$M_{cre} = 12,224 \cdot \left(6\sqrt{6500} + 3299 - 1282\right)/12,000 = 2547 \text{ kip ft}$$

Shear force due to unfactored dead load is

$$V_d = w_d(0.5L - x) = 1.64 \cdot (0.5 \cdot 100.75 - 19.63) = 50.42 \text{ kip}$$

Maximum factored moment due to externally applied loads is

$$M_{\max} = M_{\mathrm{u}} - M_{\mathrm{d}}$$

$$= 1.2 \cdot 1306 + 1.6 \cdot 0.5 \cdot 2.0\left(100.75 \cdot 19.63 - 19.63^2\right) - 1306$$

$$= 2809\,\mathrm{kip\ ft}$$

Factored shear force due to externally applied loads occurring simultaneously with $M_{\max}$ is

$$V_{\mathrm{i}} = V_{\mathrm{u}} - V_{\mathrm{d}}$$
$$= 1.2 \cdot 50.42 + 1.6 \cdot 2.0 \cdot (0.5 \cdot 100.75 - 19.63) - 50.42$$
$$= 108.5\,\mathrm{kip}$$

Nominal shear strength provided by the concrete according to Eq. (7.23) is

$$V_{\mathrm{ci}} = 0.6\sqrt{6500} \cdot 6 \cdot 59.93/1000 + 50.42 + \frac{108.5}{2809} \cdot 2547 = 166.2\,\mathrm{kip}$$

Determine the nominal shear strength when considering web-shear cracking. In a composite member, f_{pc} is the resultant compressive stress at the centroid of the composite section (or at junction of web and flange where the centroid lies within the flange) due to both prestress and moments resisted by precast member acting alone. Assuming that the girder remains unshored during deck placement, the precast girder carries self-weight and deck.

$$f_{\mathrm{pc}} = \frac{P_{\mathrm{e}}}{A} + \frac{P_{\mathrm{e}} \cdot e\left(y_{\mathrm{bp}} - y_{\mathrm{b}}\right)}{I} + \frac{M_{\mathrm{d}}\left(y_{\mathrm{bp}} - y_{\mathrm{b}}\right)}{I}$$

$$f_{\mathrm{pc}} = \left(\frac{316.7 + 739.0}{713} - \frac{(316.7 \cdot 4.55 + 739.0 \cdot 28.12)(49.16 - 32.12)}{392,638}\right.$$

$$\left. + \frac{1306(49.16 - 32.12)12}{392,638}\right) \cdot 1000 = 1194\,\mathrm{psi}$$

The vertical component of the prestress force assuming small angles is

$$V_{\mathrm{p}} = \frac{55 - 5}{35 \cdot 12}\,316.7 = 37.7\,\mathrm{kip}$$

Web-shear strength contribution is

$$V_{\mathrm{cw}} = \left(3.5\lambda\sqrt{6500} + 0.3 \cdot 1194\right)6 \cdot 59.93/1000 + 37.7 = 268.0\,\mathrm{kip}$$

V_{ci} controls concrete contribution to shear strength. The required shear reinforcement using Eq. (7.42) at the section of interest is

$$V_{s,\text{req}} = \frac{180.2}{0.85} - 166.2 = 45.8\,\text{kip}$$

The maximum contribution of shear reinforcement according to Eq. (7.41) is

$$V_{s,\text{max}} = 8\sqrt{6500} \cdot 6 \cdot 59.93/1000 = 231.9\,\text{kip} > 45.80\,\text{kip}$$

Section is adequately sized for this condition.

Typically, when designing a beam, the minimum stirrup requirement is calculated and a reasonable spacing is selected. Stirrup spacing in areas of high demand are then reduced to complete the design. For this example, calculate the minimum requirements first, then determine the required stirrup spacing for strength at the section under consideration. Depending on the application, the stirrup size may already be selected. In general, for beams of this size, stirrup size would range from #3 to #5. In this case, 2-legged #3 stirrups are selected.

$$A_v = 0.11 \cdot 2 = 0.22\,\text{in.}^2$$

Rearrange Eqs. (7.44c) and (7.44d) to determine the maximum spacing of stirrups.

$$\frac{A_v f_{yt}}{0.75\sqrt{f'_c}\,b_w} = \frac{0.22 \cdot 60{,}000}{0.75\sqrt{6500} \cdot 6} = 36.4\,\text{in.}$$

$$\frac{A_v f_{yt}}{50 b_w} = \frac{0.22 \cdot 60{,}000}{50 \cdot 6} = 44.0\,\text{in.}$$

Smaller value of 36.4 in. controls, which corresponds to a larger transvers reinforcement ratio (A_v/s).

Spacing may also be limited by Eq. (7.44e).

$$\frac{80 A_v f_{yt} d}{A_{ps} f_{pu}} \sqrt{\frac{b_w}{d}} = \frac{80 \cdot 0.22 \cdot 60{,}000 \cdot 59.93}{40 \cdot 0.153 \cdot 270{,}000} \sqrt{\frac{6}{59.93}} = 12.1\,\text{in.}$$

Use the greater of 36.4 and 12.1 for the maximum stirrup spacing that satisfies the minimum stirrup area requirement when $V_u > 0.5\phi V_c$

$$s_{\text{max}} = 36.4\,\text{in.}$$

Stirrup spacing is further limited by ACI Building Code detailing requirements. The maximum spacing of stirrups when

$$V_s \le 4\sqrt{f'_c}\,b_w d = 4\sqrt{6500} \cdot 6 \cdot 59.93/1000 = 116.0\,\text{kip}$$

$$\frac{3h}{4} = \frac{3 \cdot (63 + 8)}{4} = 53.3\,\text{in.}$$

or 24 in., whichever is less.

and when $V_s > 4\sqrt{f'_c}b_w d$

$$\frac{3h}{8} = \frac{3 \cdot (63 + 8)}{8} = 26.6\,\text{in.}$$

or 12 in., whichever is less.

In a typical design approach, V_s associated with the minimum spacing determined above would be calculated and used to determine stirrup spacing. In this example, we calculate the stirrup spacing required at the section of interest and use the least spacing. Stirrup spacing at the section of interest based on required strength from Eq. (7.43) is

$$\frac{2 \cdot 0.11 \cdot 60,000 \cdot 59.93}{180.2/0.85 - 166.2} = 17.3\,\text{in.}$$

$$\frac{180.2}{0.85} - 166.2 = 45.8\,\text{kip} < 116.0\,\text{kip}$$

Stirrup spacing requirements for strength (17.3 in.) controls over maximum spacing from detailing requirements (24 in.) and maximum spacing based on minimum area of transverse reinforcement (36.4 in.). Stirrup spacing can be rounded down to a whole even number (16 in.).

Example 7.5: Design Shear Reinforcement Using Shear Diagram
For the pretopped double T in Examples 7.2 and 7.3, plot the concrete contribution to the shear strength. Software programs are available that automate this process and allow the designer to visualize the areas where shear reinforcement is needed so that the transverse reinforcement layout can be customized to more efficiently utilize the shear reinforcement. For illustrative purposes, the shear reinforcement requirements are calculated at strategic locations and plotted with the concrete contribution to form the design shear strength plot.

In Fig. 7.19, V_c is plotted, which is a composite of the minimum values of V_{cw} and V_{ci} over the full length of the member. With the factored shear diagram superimposed, the areas where shear reinforcement is required become apparent. This diagram illustrates that for typical beams V_{cw} controls near the support and V_{ci} controls the concrete contribution closer to midspan.

For double T sections, welded wire reinforcement (WWR) is commonly used for shear reinforcement. For this problem, WWR ASTM A1064, Gr 70 deformed. Start by calculating the maximum spacing required for the D5 wire. WWR will be placed in each stem so the area of shear reinforcement is

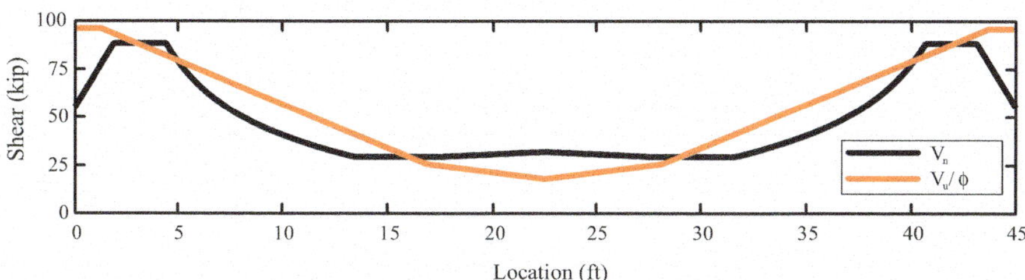

Fig. 7.19 Values for V_{cw} and V_{ci}

$$A_{\mathrm{v}} = 0.055 \cdot 2 = 0.11\,\mathrm{in.}^2$$
$$f_{\mathrm{yt}} = 70\,\mathrm{ksi}$$

Maximum spacing of stirrups is

$$\frac{A_{\mathrm{v}}f_{\mathrm{yt}}}{0.75\sqrt{f_{\mathrm{c}}'}\,b_{\mathrm{w}}} = \frac{0.11 \cdot 70,000}{0.75\sqrt{5500} \cdot 9.5} = 14.5\,\mathrm{in.}$$

$$\frac{A_{\mathrm{v}}f_{\mathrm{yt}}}{50b_{\mathrm{w}}} = \frac{0.11 \cdot 70,000}{50 \cdot 9.5} = 16.3\,\mathrm{in.}$$

14.5 in. controls.

Minimum effective depth is $0.8h = 0.8 \cdot 26 = 20.8$ in. Use this to determine spacing limitations of Eq. (7.44e)

$$\frac{80A_{\mathrm{v}}f_{\mathrm{yt}}d}{A_{\mathrm{ps}}f_{\mathrm{pu}}}\sqrt{\frac{b_{\mathrm{w}}}{d}} = \frac{80 \cdot 0.11 \cdot 70,000 \cdot 20.8}{12 \cdot 0.153 \cdot 270,000}\sqrt{\frac{9.5}{20.8}} = 17.4\,\mathrm{in.}$$

Use the greater of 14.5 in. and 17.4 in. for the maximum stirrup spacing that satisfies the minimum stirrup area requirement where $V_{\mathrm{u}} > 0.5\phi V_{\mathrm{c}}$. To provide even spacing increments use $s_{\max} = 17$ in..

Stirrup spacing is further limited by the ACI Building Code detailing requirements. The maximum spacing of stirrups where

$$V_{\mathrm{s}} \le 4\sqrt{f_{\mathrm{c}}'}\,b_{\mathrm{w}}d = 4\sqrt{5500} \cdot 9.5 \cdot 20.8/1000 = 58.6\,\mathrm{kip}$$

$$\frac{3h}{4} = \frac{3 \cdot (26)}{4} = 19.5\,\mathrm{in.}$$

or 24 in., whichever is less.

and where $V_{\mathrm{s}} > 4\sqrt{f_{\mathrm{c}}'}\,b_{\mathrm{w}}d$

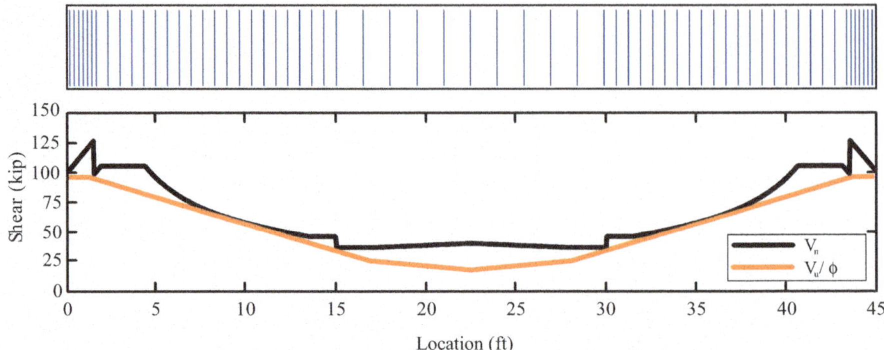

Fig. 7.20 Shear reinforcement distribution

$$\frac{3h}{8} = \frac{3 \cdot (26)}{8} = 9.75 \,\text{in.}$$

or 12 in., whichever is less.

Determine the maximum spacing at the end of the beam where shear is maximum. Try a spacing of 3 in.

$$\phi V_s = \frac{\phi A_v f_y d}{s} = \frac{0.75 \cdot 0.11 \cdot 70 \cdot 20.8}{3} = 45.3 \,\text{kip}$$

$$\phi V_n = \phi V_s + \phi V_c = 40.0 + 40.9 = 80.9 \,\text{kip} > V_u = 72.0 \,\text{kip} \ \ \text{OK}$$

Maximum stirrup requirement is less than $4\sqrt{f'_c}b_w d$ so maximum spacing is 17 in. WWR leg spacing is increased to 10 in. at 2 ft from the support. The spacing is further increased to 17 in. at 16 ft from the support. The stirrup spacing is symmetric about the midspan, Fig. 7.20.

Comment: this example demonstrates how software can be used for design of shear reinforcement. Spot checks of the software were calculated as a check, which the engineer should do to ensure that the software program is performing as expected.

7.10 Causes of Torsion

Torsion occurs when a flexural member carries loads that are offset from the centerline of the member. This situation occurs commonly in spandrel beams that act as a boundary element along the edge of a floor slab or when supporting floor joists are placed perpendicular to the beam, Fig. 7.21. The unit moment (m) applied along the slab connection to the spandrel beam results in a twisting deformation. To maintain stability under this loading, the spandrel beam must be fixed against twist at

Fig. 7.21 Torsion in a
spandrel beam supporting
a slab

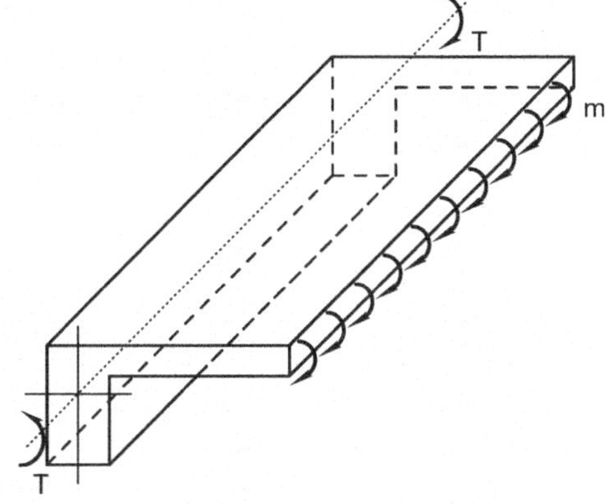

Fig. 7.22 Torsion in curved
box girder

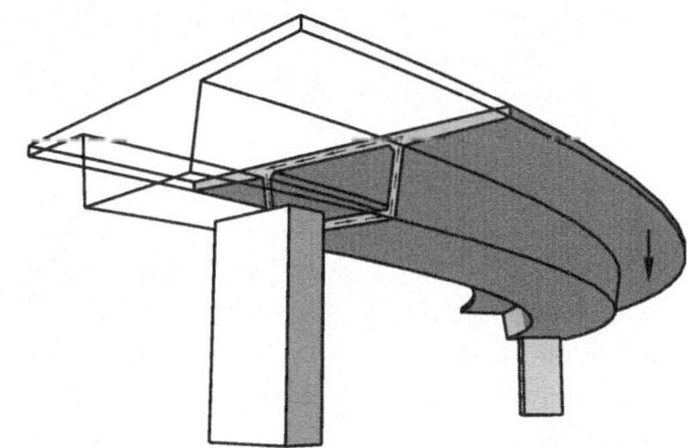

its connection to the supporting column. This fixity results in a torsional moment (T)
applied in opposite direction at each end. Similar to shear, torsion in concrete beams
results in diagonal tension, which, if large enough, causes inclined cracking in the
beam.

The most efficient shape to provide torsional stiffness and strength is a hollow
section in which the material is placed as far as possible from the centerline of the
member. Curved girders are typically used in elevated highway interchanges. Sometimes these girders must be constructed with small horizontal radii, which can result
in a significant torsional component (Fig. 7.22).

In the design of concrete for torsion, the nature of the applied loads and the
boundary conditions of the member dictate whether the member is considered to
carry compatibility torsion or equilibrium torsion. **Compatibility torsion** occurs
when an element is displaced through the action of adjacent elements such as seen in
Fig. 7.21. The structural configuration of compatibility torsion allows loads to be
redistributed to another part of the structure if the member cracks in torsion.
Equilibrium torsion occurs when the stability of the member or system is solely
dependent on the torsional strength, such as that shown in Fig. 7.22. In conditions
where the member is statically determinate, or where the member or structure relies

on torsion for stability, the section must be designed to carry the full factored torsion obtained from the structural analysis.

7.11 Torsional Strength

ACI 318-14 commentary R22.7.6 states "The torsional design strength ϕT_n must equal or exceed the torsional moment T_u due to factored loads. In the calculation of T_n, all the torsion is assumed to be resisted by stirrups and longitudinal reinforcement, neglecting any concrete contribution to torsional strength." This is the basis for the development of the torsional resistance of concrete sections.

When torque is applied to an elastic member with circular cross section, the shear strains vary linearly from zero at the centroidal axis to a maximum at perimeter of the section. In rectangular sections, the shear strains vary non-linearly from the center to the perimeter. Furthermore, the shear stresses vary around the perimeter with the maximum occurring at the mid-height of the widest face. Analytical solutions for the maximum shear stress have been proposed for solid rectangular sections. Instead of a solid section, the ACI Building Code approach to calculating the torsional cracking strength is to utilize a thin-walled tube analogy. A thin outer layer is assumed to provide the majority of the torsional resistance for the member, whether solid or hollow, and this layer is assumed to be centered on the closed stirrups.

In the design for torsion, the cracking strength and the nominal strength of the section are important aspects of design and the thin-wall analogy is used for both. For a section in which the cracking strength is needed, the element is idealized as a thin-walled tube in which the core concrete cross section is ignored, Fig. 7.23. The equivalent wall thickness prior to cracking is

$$t = 0.75 \frac{A_{cp}}{p_{cp}} \tag{7.45}$$

where A_{cp} is area enclosed by outside perimeter of concrete cross section, and p_{cp} is outside perimeter of concrete cross section. The area enclosed by the thin wall centerline is also defined as the gross area enclosed by torsional shear flow path:

$$A_o = \frac{2}{3} A_{cp} \tag{7.46}$$

The principal tensile stress caused by the applied torque in the thin-walled analogy is

$$\tau = \frac{T}{2A_o t} \tag{7.47}$$

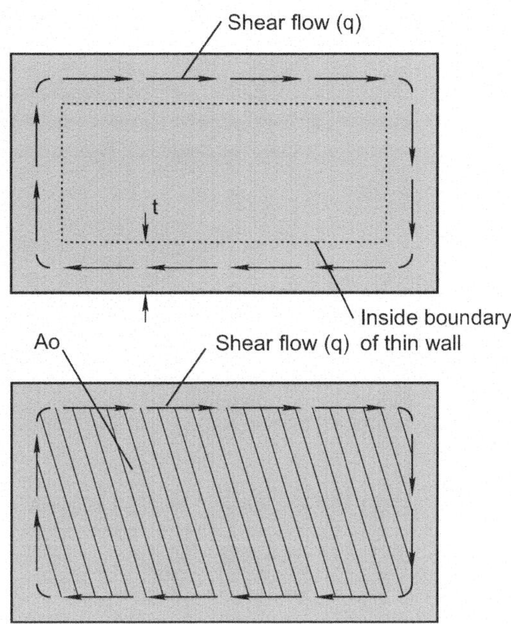

Fig. 7.23 Thin-walled model used to calculate torsional strength in solid sections

Table 7.5 Cracking torsion

Member type	T_{cr}
Nonprestressed	$4\lambda\sqrt{f'_c}\left(\dfrac{A_{cp}^2}{p_{cp}}\right)$
Prestressed	$4\lambda\sqrt{f'_c}\left(\dfrac{A_{cp}^2}{p_{cp}}\right)\sqrt{1+\dfrac{f_{pc}}{4\lambda\sqrt{f'_c}}}$

Based on an assumed lower bound cracking strength of $4\lambda\sqrt{f'_c}$ the cracking torque for nonprestressed section is shown in Table 7.5. Prestressing can be incorporated into the cracking equation by using a Mohr's circle analysis of the average stresses.

Once a reinforced concrete beam has cracked in torsion, the ACI Building Code considers the torsional strength provided solely by the plastic truss action of the reinforcement made up of closed stirrups, longitudinal bars, and inclined concrete struts, Fig. 7.24. Because the truss action depends on the resistance of both the transverse reinforcement and longitudinal reinforcement, the torsional strength of the section is taken as the lesser of

$$T_n = \frac{2A_oA_t f_{yt}}{s}\cot\theta \qquad (7.48)$$

or

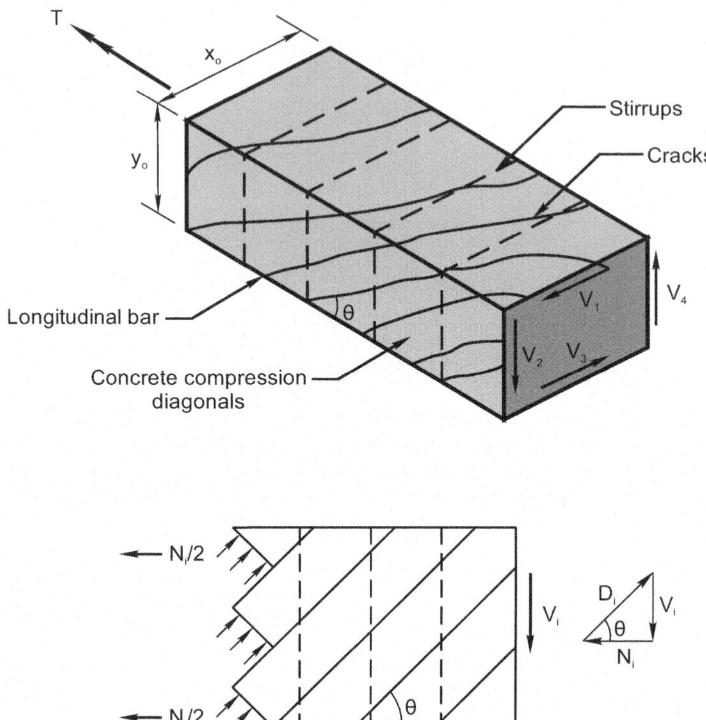

Fig. 7.24 Plastic space truss model to calculate torsional strength

$$T_n = \frac{2A_oA_lf_{yt}}{p_h}\cot\theta \tag{7.49}$$

where A_t is the area of one leg of a closed stirrup resisting torsion; A_l is the area of longitudinal torsional reinforcement; p_h is the perimeter of the centerline of the outermost closed stirrup; and A_o is either determined by analysis or may be taken as $0.85A_{oh}$, where A_{oh} is the sectional area enclosed by the centerline of the outermost closed transverse torsional reinforcement, Fig. 7.25. As indicated in Fig. 7.24, θ is the angle between the inclined concrete strut and is to be between 30 and 60°. This angle can be determined either by analysis or as indicated in Table 7.6.

While the use of the plastic truss model is similar to the approach for shear, torsional strength does not include a concrete contribution. In cases of combined shear and torsion, the concrete contribution to shear strength does not need to be reduced.

Fig. 7.25 Examples of
closed stirrups

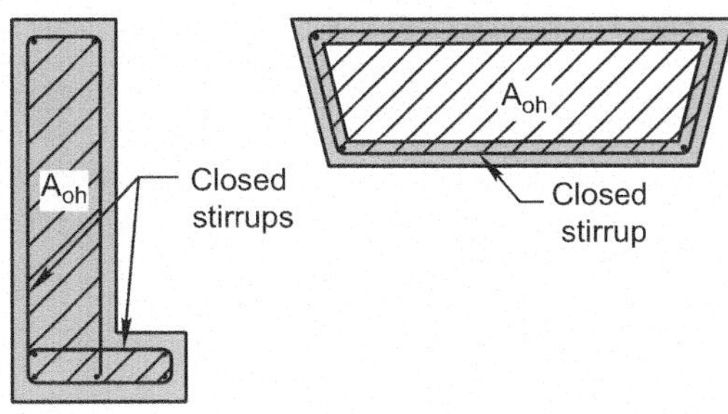

Table 7.6 Strut angle for
torsional strength based on
effective prestress

$A_{ps}f_{se}$	θ
$<0.4(A_{ps}f_{pu} + A_s f_y)$	45°
$\geq 0.4(A_{ps}f_{pu} + A_s f_y)$	37.5°

7.12 Design for Torsion

Not all torsion must be considered in design. The factored torsion at a particular
section is calculated from the structural analysis. That torsion is compared to a
limiting torsion to determine if the torsion can be ignored; the ACI Building Code
defines this limit as **threshold torsion**. If $T_u < \phi T_{th}$, then the effects of torsion can be
neglected. Threshold torsion is essentially one-fourth the cracking torsional moment
T_{cr} and is calculated using the applicable equation in Table 7.7. In developing the
equations for torsional cracking, the interaction between inclined cracking caused by
combined shear and torsion was considered. An elliptical function is suitable to
represent the interaction of solid members. Using this approach along with the A_{cp} in
the equations in Table 7.5, results in a negligible reduction shear cracking strength.
For hollow sections, however, the relationship transitions from elliptical for sections
with small voids, to straight line for sections with large voids. The expressions for
hollow sections are derived from the solid section relationships modified by the
factor $(A_g/A_{cp})^2$; this accounts for the reduction of cracking shear resulting from the
interaction between torsional and shear stresses in hollow sections. A_g for hollow
sections is the area of concrete only and does not include voids. The ACI Building
Code does not explicitly define the hollow sections, but the commentary suggests
that beams may be considered solid where $A_g/A_{cp} \geq 0.95$ for determining T_{th}.

If the factored torsion is equal to or greater than the threshold torsion, then torsion
must be considered in the member design. In indeterminate structures where the
torsion can be redistributed, the factored torsion (T_u) can be reduced to the cracking
torsion (ϕT_{cr}) determined using the applicable equation in Table 7.5. Adjoining
members, however, must be designed to accommodate the redistribution of torsion.
It follows that this provision is applicable to typical framing systems. Unusual
framing plans or disproportionate distribution of uncracked member stiffness may
result in significant torsional rotations which should be investigated in more detail
before reducing the design torsion.

Table 7.7 Threshold torsion for solid and hollow sections

$A_{ps}f_{se}$	Cross section	T_{th}
$< 0.4(A_{ps}f_{pu} + A_s f_y)$	Solid	$\lambda \sqrt{f'_c} \left(\dfrac{A_{cp}^2}{p_{cp}} \right)$
	Hollow	$\lambda \sqrt{f'_c} \left(\dfrac{A_g^2}{p_{cp}} \right)$
$\geq 0.4(A_{ps}f_{pu} + A_s f_y)$	Solid	$\lambda \sqrt{f'_c} \left(\dfrac{A_{cp}^2}{p_{cp}} \right) \sqrt{1 + \dfrac{f_{pc}}{4\lambda \sqrt{f'_c}}}$
	Hollow	$\lambda \sqrt{f'_c} \left(\dfrac{A_g^2}{p_{cp}} \right) \sqrt{1 + \dfrac{f_{pc}}{4\lambda \sqrt{f'_c}}}$

7.13 Shear and Torsion Interaction

Torsion usually occurs in combination with shear. Because both actions result in diagonal tension, ignoring one effect in the design of the other is likely unconservative. When considering the individual effects of shear and torsion on uncracked concrete, shear produces diagonal tension on each side-face of a member that are oriented in the same direction, resulting in a diagonal crack that passes through the member. Torsion produces diagonal tensile stresses on all four faces of a rectangular member with the stresses on opposing faces oriented in opposing directions. This results in cracking that follows a spiral path around the member. Under most conditions, the shear and torsion stresses are additive on one face and subtractive on the opposite face. This results in the reduction of shear cracking strength when torsion is present and vice versa, and hence the need to consider interaction.

If shear and torsion must be considered, then section size is limited by the applicable equation from Table 7.8, which are in terms of shear and torsional stresses. The limitation on section size provides control on crack width at service load levels and reduces the potential of a diagonal strut compression failure. The applied stress terms on the left side of the equation for solid sections utilizes a circular interaction relationship while a linear relationship is used for the hollow section limitations. The limiting stress term is the same for both equations and is similar to that of shear alone, Eq. (7.41), and is expressed using V_c to allow its use with either nonprestressed or prestressed concrete.

As with shear, for prestressed members, the value of d used in the equations from Table 7.8 need not be taken less than $0.8h$. Some hollow sections may have thin top or bottom sections in which the torsional stresses are larger than those in the webs. In these sections where the wall thickness varies, the equation for hollow sections from Table 7.8 should be evaluated at the location where

Table 7.8 Limits on cross-sectional dimensions for solid and hollow sections

Cross section	Limiting condition
Solid	$\sqrt{\left(\dfrac{V_\mathrm{u}}{b_\mathrm{w}d}\right)^2 + \left(\dfrac{T_\mathrm{u}p_\mathrm{h}}{1.7A_\mathrm{oh}^2}\right)^2} \leq \phi\left(\dfrac{V_\mathrm{c}}{b_\mathrm{w}d} + 8\sqrt{f_\mathrm{c}'}\right)$
Hollow	$\left(\dfrac{V_\mathrm{u}}{b_\mathrm{w}d}\right) + \left(\dfrac{T_\mathrm{u}p_\mathrm{h}}{1.7A_\mathrm{oh}^2}\right) \leq \phi\left(\dfrac{V_\mathrm{c}}{b_\mathrm{w}d} + 8\sqrt{f_\mathrm{c}'}\right)$

$$\left(\frac{V_\mathrm{u}}{b_\mathrm{w}d}\right) + \left(\frac{T_\mathrm{u}p_\mathrm{h}}{1.7A_\mathrm{oh}^2}\right) \tag{7.50}$$

is maximum.

7.14   Flexure, Shear, and Torsion Reinforcement

When a design requires the use of torsional reinforcement, the ACI Building Code requires that the longitudinal (A_l) and transverse (A_t) reinforcement requirements to be added to the reinforcement required for flexure, shear, and axial actions that are acting in combination with the torsion.

The combination of torsional and shear reinforcement is considered on a section-by-section basis. When stirrups are required for torsional strength, then the transverse reinforcement must be composed of closed stirrups to ensure that each leg of the closed stirrup can be developed to resist the forces V_1 through V_4 (Fig. 7.24) that develop under torsion. Torsional reinforcement is defined based on the area of the stirrup leg in the face of the section (A_t), which is typically the area of a single leg of the stirrup. Shear reinforcement is defined based on the number of legs that cross the potential diagonal crack that forms from the diagonal shear stresses (A_v), which is twice the area of the leg of the two-leg stirrup. To satisfy both the torsion and shear strength requirement, the area of steel required for each is summed. One alternative for summing the reinforcement requirements is provided by ACI 318 commentary as

$$\text{Total}\left(\frac{A_\mathrm{v+t}}{s}\right) = \frac{A_\mathrm{v}}{s} + 2\frac{A_\mathrm{t}}{s} \tag{7.51}$$

Only closed stirrup legs on the exterior of the section are effective at resisting torsion. Consequently, any interior legs should be ignored in this equation.

In practical terms, unless the torsional demand is unusually high, shear design is conducted first to ensure that the section is adequately sized and determine a preliminary stirrup bar size and configuration. Once the section is sized for shear, the threshold torsion is checked to determine if torsion can be ignored. If not, then torsional reinforcement must also be sized. Once the two demands are determined, then the final spacing is determined. When torsional reinforcement is required, the

Table 7.9 Minimum transverse reinforcement requirements for beams and slabs when torsional reinforcement is required

$\dfrac{(A_v + 2A_t)_{\min}}{s} \geq$	greater of	$0.75\sqrt{f_c'}\dfrac{b_w}{f_{yt}}$
		$50\dfrac{b_w}{f_{yt}}$

Table 7.10 Minimum longitudinal reinforcement requirements for beams and slabs when torsional reinforcement is required

$A_{l,\,\min} \geq$	lesser of	$\dfrac{5\sqrt{f_c'}A_{cp}}{f_y} - \left(\dfrac{A_t}{s}\right)p_h\dfrac{f_{yt}}{f_y}$
		$\dfrac{5\sqrt{f_c'}A_{cp}}{f_y} - \left(\dfrac{25b_w}{f_{yt}}\right)p_h\dfrac{f_{yt}}{f_y}$

ACI Building Code also requires that minimum transverse (Table 7.9) and longitidunal (Table 7.10) reinforcement be included.

The combination of longitudinal reinforcement for flexure, axial, and torsional loads should also be considered on a section-by-section basis. One example is that of a beam in which the maximum reinforcement for flexure occurs at midspan. But due to the support conditions imposed, the maximum longitudinal reinforcement for torsion occurs at the end of the member where the flexural demand is small. Reinforcement requirements for torsion and flexure are evaluated and summed individually at these key locations. Practical considerations such as bar selection and reinforcement cutoffs must be considered (Table 7.10).

7.15   Alternative Design Approach for Shear and Torsion

The previous approach for shear and torsion is suitable if there are a small number of sections to be designed. The problem is compounded if several sections along the beam or multiple beams must be designed or checked. One design approach is to construct a shear-torsion interaction diagram. The basis for the diagram is the fact that shear and torsion interact as second order functions. Hollow sections lose some of the second order effect but still display conservative linear behavior. Therefore, a linear interpolation between the shear and the torsion is conservative, Fig. 7.26.

With this background, an interaction diagram is constructed for a member subjected to shear and torsion. Values for the design shear capacity ϕV_n includes the contribution of the concrete ϕV_c and the contribution of stirrups at prescribed spacing ϕV_s. These values are added and plotted on the shear axis. If desired, the location where no stirrups are required could be added; however, if the torsion threshold is exceeded, minimum stirrups are used regardless of the magnitude of the forces. Similarly, values for design torsion capacity based on the stirrup spacing are calculated and plotted on the torsion axis. The resulting interaction diagram is

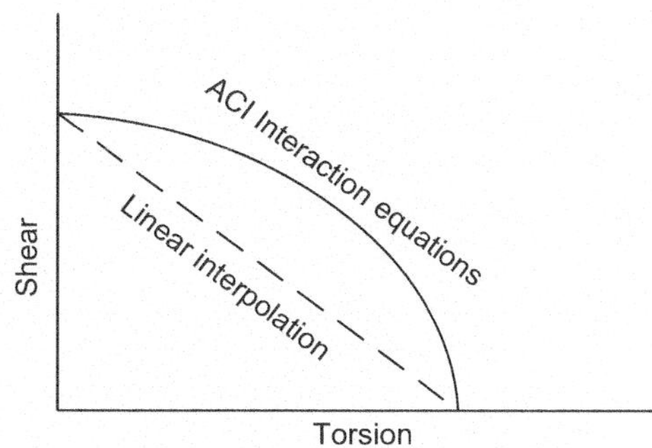

Fig. 7.26 Shear and torsion interaction

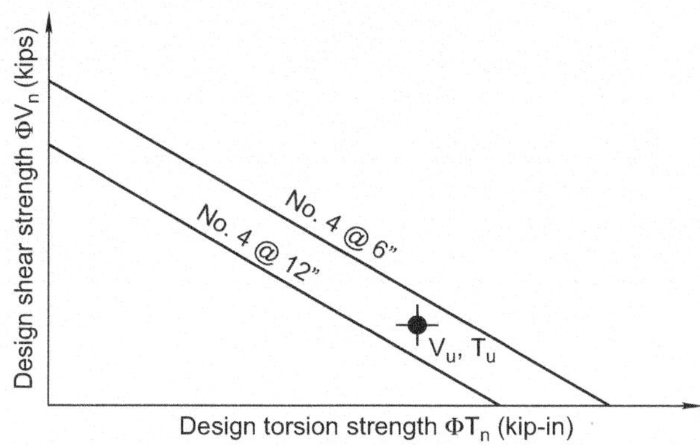

Fig. 7.27 Shear torsion design diagram

used to select stirrup spacing based on the factored loads V_u and T_u. Figure 7.27 is a schematic solution using No. 4 stirrups spaced at 12 and 6 in. For this example, the point V_u, T_u would require stirrups at 6 in. on centers to meet the demand.

Construction of the interaction diagram is complicated by the variable concrete contribution to shear. Three alternatives are possible. First, the interaction diagram is conservatively constructed using a shear contribution of $2\sqrt{f'_c}b_w d$. Second, the interaction diagram is constructed using a lower bound of local concrete contributions, usually V_{cw} near the end and V_{ci} at the interior of the beam. Lastly, the interaction diagram is constructed using the detailed concrete shear values.

7.16 Shear and Torsion Design Example

Structures, such as those used for transit systems, require attention to shear and torsion resulting from environmental and operational loads. The Detroit People Mover, for example, required substantial torsional reinforcement for the 350-ft radius horizontal curves, Fig. 7.28. The analysis includes the variation in vehicle weights, variation in vehicle location, determination of vehicle centroids, centripetal

Fig. 7.28 Detroit downtown people mover guideway

forces of the vehicle moving around curves, wind, and emergency braking. The example that follows examines a similar guideway structure. The vehicles are smaller than those in Detroit and torsional effects are due to wind and vehicle eccentricity on a straight guideway section. The simplifications focus on the shear and torsion design rather than the complexity of the full structural analysis.

This example is based on the original automated guideway transit system at the Dallas-Fort Worth regional airport. The guideway consists of straight beams that are 70, 80, and 90 ft long. The end 70-ft long simple span beam allows for motion between at grade sections and the fixed structure. The remaining structure is 80-ft long end spans and 90-ft long interior spans. The beams have identical prestress with the tendon eccentricity varying with the span length. The beam is a box girder with an extended flange to support the vehicle, Fig. 7.29. Continuity for live load is provided by reinforcement in the cast-in-place reinforcement placed in the parapet walls. A three-car train is subjected to wind loads and minor vehicle lateral movement on the guideway resulting in shear, torsion, and moments at the beam end and quarter points summarized in Fig. 7.30. The final design is based on the shear and moment envelopes for the moving vehicles. The ACI Building Code equations for shear require that the moment and shear are for the same loading condition. Figure 7.30 is a set of consistent factored loads. The beam dimensions and prestressing properties are provided in Table 7.11.

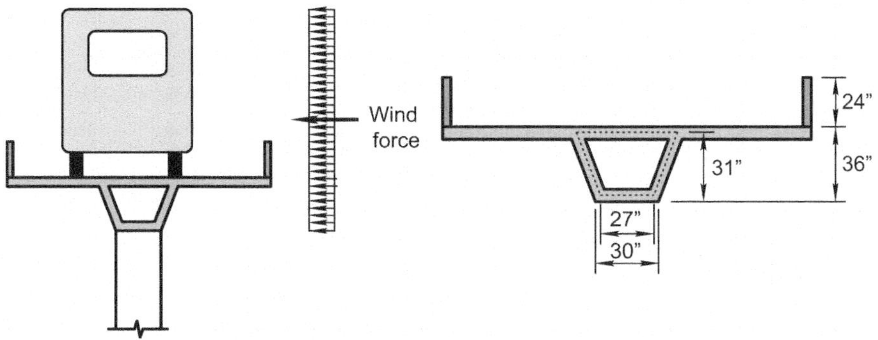

Fig. 7.29 Wind force, vehicle eccentricity, and section details

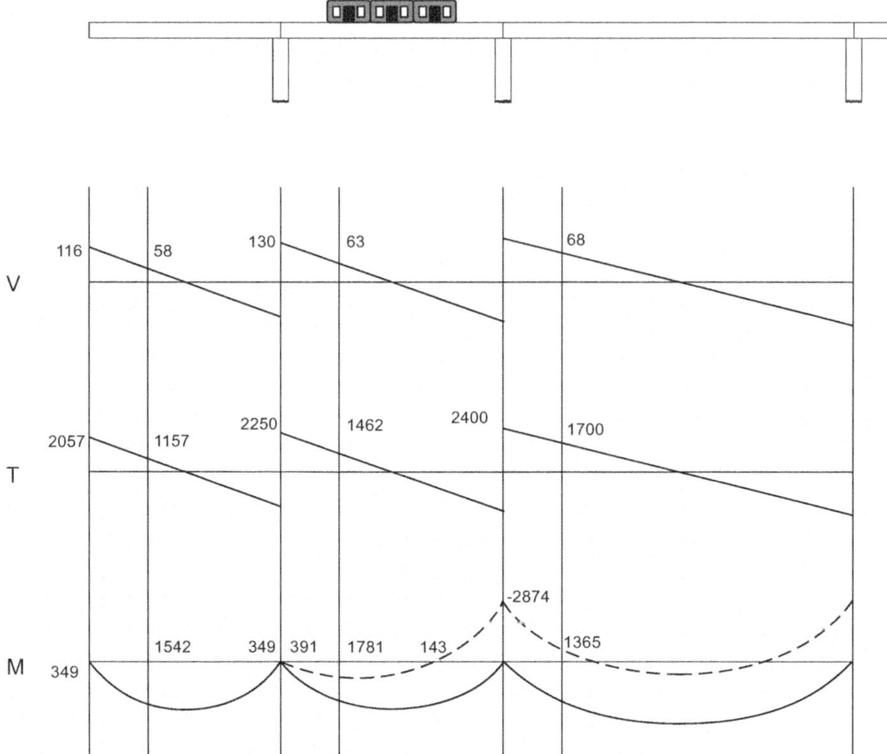

Fig. 7.30 Shear, torque and moment on the guideway

The beams are fabricated with normal weight concrete and pretensioned with the same number of tendons, and thus all have an area of prestress of 5.508 in.^{2}, an effective prestress after losses of 154 ksi, and an average precompression in the concrete of $f_{\mathrm{pc}} = 799$ psi. The eccentricity of the tendons is set to approximately load

Table 7.11 Material and sectional properties

Material properties	Sectional properties	Weights
$f'_c = 5000\,\text{psi}$ $f_{yt} = 60{,}000\text{ psi}$ $f_{pu} = 270\text{ ksi}$ $f_{ps} = 154\text{ ksi}$	$A_g = 1061\text{ in.}^2$ $I_g = 160{,}970\text{ in.}^4$ $b_w = 5\text{ in.}$ $x_1 = 28.5\text{ in.}$ $y_1 = 31\text{ in.}$	$w_g = 1.179\text{ kip/ft}$ $w_v = 0.975\text{ kip/ft}$
$A_v = 0.62\text{ in.}^2$ $A_t = 0.31\text{ in.}^2$ $A_{ps} = 5.508\text{ in.}^2$	$A_o = 883.5\text{ in.}^2$ $A_{oh} = 846.7\text{ in.}^2$ $p_{cp} = 119\text{ in.}$	

balance the beam deal load and the harp point set at one quarter of the span length. For the 70-ft beam, the eccentricity is

$$e = \frac{w_g l^2}{8 P_e} = \frac{1.179 \cdot 70^2 \cdot 12}{8 \cdot 848} = 10.2\,\text{in.}$$

Correspondingly, the eccentricity for the 80-ft beam is 13.3 in. and for the 90-ft beam is 16.9 in.

The threshold torque for the section is

$$T_{th} = \lambda \sqrt{f'_c} \frac{A_g}{p_{cp}} \sqrt{1 + \frac{f_{pc}}{4\lambda\sqrt{f'_c}}} = 1.0 \cdot \sqrt{5000}\,\frac{1061}{119}\sqrt{1 + \frac{799}{4 \cdot 1\sqrt{5000}}} = 1308\,\text{in.-kip}$$

Applying a phi factor of 0.75 gives $\phi T_{th} = 0.75 \cdot 1308 = 981$ in.-kip. This is less than the torsion values in Fig. 7.29, requiring torsion to be considered at each section.

Based on initial trials, No. 5 stirrups are selected for the design. Stirrup spacing at 3, 4, 6, 8, and 12 in. are used to develop the interaction diagrams. The torsional design strength for stirrups at 3 in. spacing is

$$\phi T_n = \phi \frac{2 A_{oh} A_t f_{yt}}{s} = 0.75\,\frac{2 \cdot 846.7 \cdot 0.31 \cdot 60}{3} = 7874\,\text{in.-kip}$$

In a similar manner, the design torque capacity spacing of 4, 6, 8, and 12 in. are 5906, 3937, 2953, and 1969 in.-kip respectively. These torque capacities are independent of beam length.

7.16.1   Solution Using V_c of $2\sqrt{f'_c}b_w d$

For the simplest solution, the concrete shear is based on using a concrete contribution of $2\sqrt{f'_c}b_w d$ gives $V_c = 2\sqrt{5000} \cdot 2 \cdot 5 \cdot 28.2 = 41\,\text{kip}$. The value of 2 before the 5 in. thick web is because the box has two webs for shear. In this case the structural depth is taken as $0.80h = 28.2$ in. As shown later, this is a conservative value. The design shear capacity for stirrups at 3 in. spacing is

$$\phi V_n = \phi V_c + \phi\frac{A_v \cdot f_{yt} \cdot d}{s} = 0.75 \cdot 41 + 0.75\frac{0.62 \cdot 60 \cdot 28.2}{3} = 298\,\text{kip}$$

Using the same method, the shear capacity for stirrups at 4, 6, 8, and 12 in. spacing are 231, 164, 131, and 98 kips respectively. An interaction diagram is constructed and the shear and torque values from Fig. 7.29 are plotted, Fig. 7.31. From the interaction diagram, the 90-ft beam requires No. 5 closed stirrups at 3 in. near the end transitioning to No. 5 stirrups at 6 in. near the quarter point. The spacing requirements for the 70 and 80-ft beams are read from the chart accordingly.

A 3-in. spacing is fairly tight and the selection based on the simple value for V_c is conservative. This solution is adequate if only a few beams are required for a project. If there are many beams repeated with the same length, a more refined solution based on V_{cw} and V_{ci} is justified.

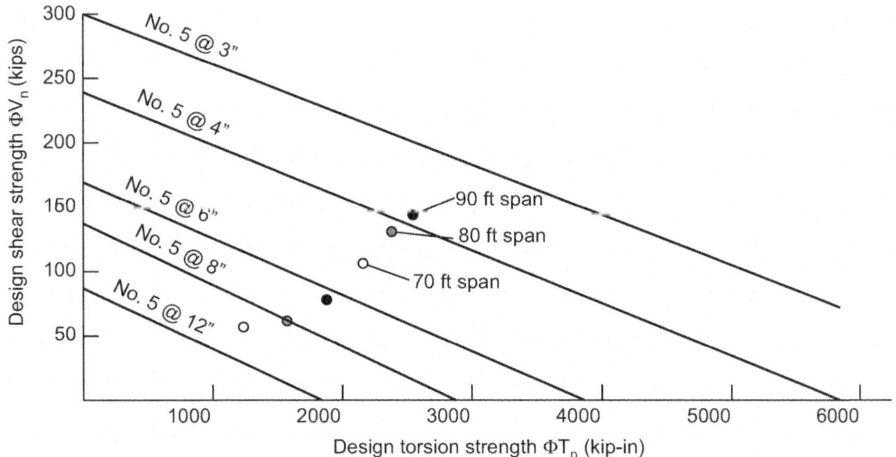

Fig. 7.31 Interaction diagram for solution with simple V_c

7.16.2 Refined Shear and Torsion Solution

The concrete shear contribution varies along the beam length requiring the lesser of V_{cw} or V_{ci} be selected. This example looks at the shear and torsion at h, in this case 3 ft, in from the beam end and at the beam quarter point. Based on experience, V_{cw} is the governing condition at the beam end and V_{ci} governs at the quarter point. The design shear values are summarized in Table 7.12.

The torsion design capacities are unchanged from the simplified approach. For ease of presentation and design, a value of $V_{cw} = 182$ kips and $V_{ci} = 80$ kips is used. The results are then plotted and presented in Table 7.12, Fig. 7.32 for the beam end, and Fig. 7.33 for the quarter point.

Table 7.12 Beam design shear capacities

Beam length (ft)	Design shear capacity at beam end $-\phi V_{cw}$ (kips) Source Eq. (7.15)	Design shear capacity at quarter point $-\phi V_{ci}$ (kips) Source Eq. (7.22)
70	181.6	78.5
80	187.5	92.4
90	193.4	89.1

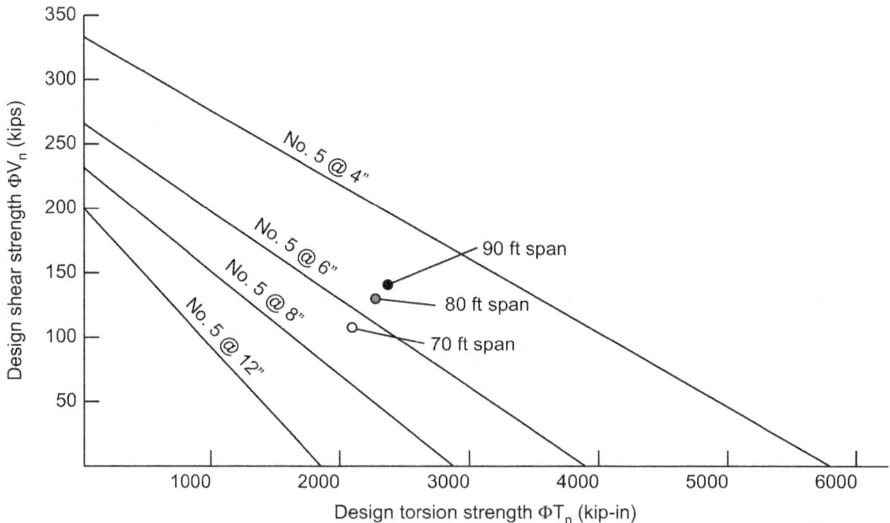

Fig. 7.32 Design shear and torsion capacity at beam end

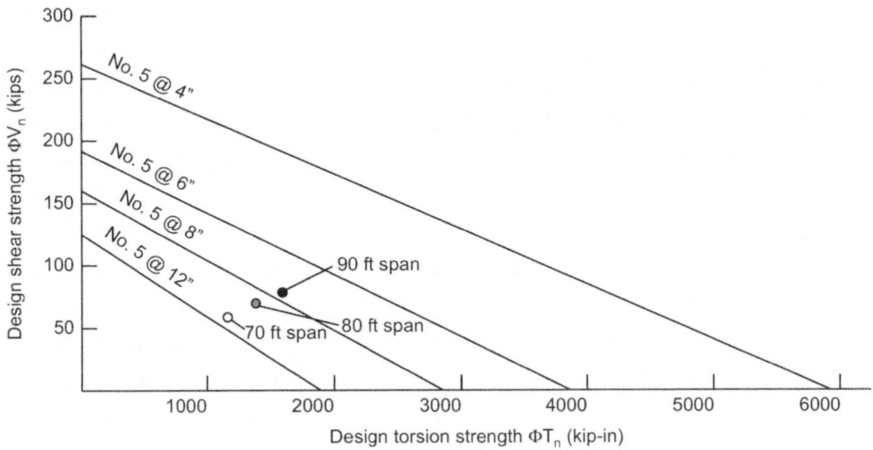

Fig. 7.33 Design shear and torsion capacity at quarter point

7.16.3 *Observations on Combined Shear and Torsion Design Solutions*

Table 7.12 requires an extensive calculation effort to determine the intermediate values used in Eqs. (7.15) and (7.22). Even after the calculation effort, a lower bound value for the shear capacity is selected to allow the data to be reduced onto two figures. Examination of Figs. 7.31, 7.32, and 7.33 indicates that the stirrup spacing at the end of the 90-ft beam is increased from 3 to 4 in. with the refined calculations. The 70 and 80-ft beam stirrup spacings are unchanged. At the quarter point, the stirrup spacing in the 80-ft beam is on the line for No. 5 at 8 in. with the simple solution and below the 8-in. line with the refined solution. Knowing that the simple solution is conservative, the engineer would likely select the greater spacing and the results would be the same with either approach.

Only two locations in the beam are selected for this example. A computer analysis of the beams would provide the shear and torsion values at other locations in the structure. The simplified interaction diagram provides a ready method to determine the change in stirrup spacing along the beam.

Problem

7.1. Recalculate the principal stresses in Example 7.1 using the gross section properties and a concrete shear stress of $v_c = V/b_w d$. Use $d = 51.5$ in. Comment on the comparison of the solutions.

References

AASHTO LRFD Bridge Design Specification (8th Ed.). (2017). Washington, DC: American Institute of State Highway and Transportation Officials (AASHTO).

ACI 318-14. (2014). *Building Code Requirements for Structural Concrete* (ACI Committee 318-14, pp. 519). Farmington Hills, MI: American Concrete Institute.

MacGregor, J. G., & Ghoneim, M. G. (1995). Design for torsion. *ACI Structural Journal, 92*(2).

Hassan, T., Lucier, G., Rizkalla, S., & Zia, P. (2007). Modeling of L-shaped, precast, prestressed concrete spandrels. *PCI Journal, 52*(2), 78–92.

MacGregor, J. G., & Hanson, J. M. (1969). Proposed changes in shear provisions for reinforced and prestressed concrete beams. *ACI Journal Proceedings, 66*(4), 276–288.

Chapter 8
Camber and Deflections

8.1 Introduction

The primary purpose of prestressing concrete is to increase the load at which the concrete cracks. The consequence is that, depending on the choices made by the engineer, the section remains uncracked over a much larger proportion of the total load than does nonprestressed concrete. This allows longer spans or shallower sections, or both, thus making a more efficient use of the material.

When designing a prestressed concrete beam for a given set of loads the engineer must select the section geometry and arrange the prestressing force and eccentricity to ensure both strength and serviceability. Serviceability is subdivided into stresses, addressed in Chap. 5, and deflections.

For a given set of loads, and by adjusting the section geometry, prestressing force, and eccentricity, the engineer dictates the load at which the element cracks. When prestressed concrete was first used in the United States, prestressing was applied such that the section remained in compression, and therefore uncracked, under full dead and live load. This approach was soon determined to be excessive and engineers began to design beams with some net tension in the section under full load. Even though the section may not crack under service loads, sustained gravity loads on prestressed concrete results in time-dependent deflections, which can be significant.

8.2 Controlling Deflections

Reinforced concrete beam design is not typically controlled by deflections. For nonprestressed slabs and beams, deflection calculations can be avoided by using a minimum thickness element according to ACI 318-14. Estimating deflections in prestressed concrete, however, is particularly important because of the slender

© Springer Nature Switzerland AG 2019
C. W. Dolan, H. R. Hamilton, *Prestressed Concrete*,
https://doi.org/10.1007/978-3-319-97882-6_8

members and longer spans that are possible. Deflections must be controlled so that serviceability is maintained when service loads are in place. Excessive deflections can cause cracking of nonstructural elements or make doors and windows difficult to open and close. Flat roof members may collect rainwater, thus increasing the deflection and retaining yet more rainwater or snow melt.

For Class U members, immediate deflections can be calculated using the gross section properties. Class T and C members, however, must be calculated assuming a cracked section. Section R24.2.4.2 of ACI 318-14 states: "Any suitable method for calculating time-dependent deflections of prestressed members may be used, provided all effects are considered." Further, the ACI Building Code requires deflection calculations for all prestressed members.

8.3   Deflections in Nonprestressed Concrete

It is not possible, nor is it necessarily desirable, to calculate deflections to a high degree of accuracy. For example, deflections are directly dependent on modulus of elasticity, which is notoriously difficult to ascertain and has a natural variations even within the same mixture. In addition, structural calculations are typically simplified and conservative such that some of the effects of moment restraint at supports are reduced to pins and point supports. Creep, shrinkage, and temperature effects further complicate the calculation. Refined accuracy, however, is not typically required. Deflections are calculated, not to ensure safety, but rather to ensure that the structure meets its intended use adequately and that occupants of the structure are comfortable.

Deflections, then, are estimates and should be treated commensurately with the number of significant figures used in the calculations. Determining the deflection of a beam to the nearest 1/1000th inch is nonsensical. The calculation of time-dependent deflections in structural concrete is further affected by the contribution of creep and shrinkage.

When analyzing a structure for deflections it is important to understand the reason for limiting deflections, and it is important to strike a balance with the complexity. Spending time incorporating all aspects of the structure into the calculation may not be time well spent. Conversely, too many simplifying conservative assumptions may lead to excessive size requirements that may not, in reality, be warranted.

8.4   Effect of Prestressing on Section Properties

Before deflections and camber can be computed, the section properties must be determined. As detailed in Chap. 6, ACI 318-14 classifies beam elements by the tension stress at the extreme tension fiber under full service loads, Table 8.1. This is computed at the precompressed tensile zone. If the net flexural tension stress under

Table 8.1 Section properties to be used in computing deflections based on member classification

Condition	Classification			Nonprestressed
	Class U	Class T	Class C	
Tensile stress at service loads	$f_t < 7.5\sqrt{f'_c}$	$7.5\sqrt{f'_c} < f_t < 12\sqrt{f'_c}$	$f_t < 12\sqrt{f'_c}$	–
Assumed behavior	Uncracked	Transition between uncracked and cracked	Cracked	Cracked
Section properties for deflections	Gross	Cracked	Cracked	Effective moment of inertia

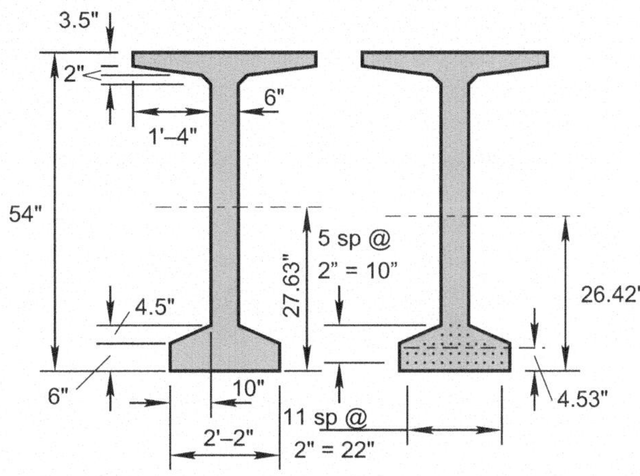

Fig. 8.1 Beam section dimensions

service load is less than the concrete cracking stress, then the gross section properties should be used to compute deflections. If the flexural stresses are greater than the cracking stress, then a cracked section analysis must be conducted. The cracking moment is computed as follows:

$$M_{cr} = \frac{\left(f_r + f_{pe}\right)I_g}{y_t} \tag{8.1}$$

where f_r is the tensile strength of concrete, f_{pe} is the effective prestress, I_g is the gross moment of inertia, and y_t is the distance to the extreme tension fiber of the precompressed tensile zone. The engineer selects and controls this classification by their choice of the section geometry, prestress force, and prestress eccentricity (Table 8.1).

Example 8.1: Calculate Transformed Section Properties Affecting Deflections
Given the PCI bridge section shown in Fig. 8.1, calculate the increase in moment of inertia when comparing gross and transformed section properties. Gross section properties are given in the appendix. The girder contains 38 0.6-in. diameter prestressing strands placed in a 2-in. grid. The center of gravity of the tendon (cgs)

Table 8.2 Use parallel axis to calculate the transformed moment of inertia

	A	Ay	I	$\bar{y}$	$A\bar{y}^2$	I_{t}
Tendon	36.47	165	–	26.42–4.53	17,475	17,475
Concrete	659	18,208	268,056	27.63–26.42	965	269,020
Sum	695.5	18,373	268,056			286,496

is located at 4.53 in. from the bottom of the girder. The section is constructed with $f'_{\mathrm{c}} = 8500$ psi concrete. Use parallel axis approach to determine the location of the centroid of the transformed section. This is done in the table shown below.

Solution: Because the accuracy of deflection calculations is not that high, in many prestressed concrete applications deflections calculated using the gross section properties is sufficient. When the section contains a large amount of prestressing steel, however, such as in a bridge girder, then it may be necessary to calculate the transformed section properties. Assume that the gross sections properties are given (reference table in which the sections are given) and it is desired to calculate the transformed section properties based on an arbitrary quantity of prestressing steel. Table 8.2 shows the calculation of the transformed section properties. Modular ratio is needed to transform the section properties to the reference material, which is concrete (Table 8.2):

$$E_{\mathrm{c}} = 57,000\sqrt{f'_{\mathrm{c}}} = 57,000\sqrt{8500}\ \mathrm{psi} = 5255\ \mathrm{ksi}$$

$$n = \frac{E_{\mathrm{ps}}}{E_{\mathrm{c}}} = \frac{28,500\ \mathrm{ksi}}{5255\ \mathrm{ksi}} = 5.42$$

Calculate the total area of strands:

$$A_{\mathrm{ps}} = 38 \times 0.216\,\mathrm{in.}^2 = 8.25\ \mathrm{in.}^2$$

And their transformed area:

$$(n - 1)A_{\mathrm{ps}} = (5.42 - 1) \times 8.25\ \mathrm{in.}^2 = 36.47\ \mathrm{in.}^2$$

Determine the location of centroid of the transformed section y_{bt}:

$$y_{\mathrm{bt}} = \frac{18,373\ \mathrm{in.}^3}{695.5\ \mathrm{in.}^2} = 26.42\ \mathrm{in.}$$

The transformed moment of inertia is then calculated using parallel axis theory. The increase in I_{t} for this section is approximately 7%. This reflects both the low modular ratio and the low percent of reinforcement compared to a conventionally reinforced beam.

Comment: Most deflection calculations are made with the gross section. In terms of precision, a 7% difference in deflections is immediately possible by ignoring the transformed section.

8.5 Camber

Curvature induced by prestressing causes the member to deform resulting in **camber,** which is the upward displacement of a flexural member resulting from an eccentric prestress force. If the member is prevented from deforming, then secondary moments are generated as discussed in Chap. 9. Generally, induced curvature due to prestress is in a direction that opposes the curvature caused by gravity load and offsets gravity load in terms of stress and deflections. For instance, a beam designed to support a load such as that shown in Fig. 8.2 naturally deflects downward in the direction of the load. If the prestressing tendon is placed below the section centroid, then the prestressing force causes an upward deflection in the opposite direction.

Precast or cast-in-place beams that are prestressed are constructed on forms and cured sufficiently before prestressing. For beams that are simply supported, prestress transfer causes the beam to camber upward. Sufficient prestress can lift the beam off the formwork and cause the beam to be supported at its end. Consequently, camber and self-weight deflection necessarily exist simultaneously.

Example 8.2: Camber after Transfer
Given the rectangular section shown in Fig. 8.3, calculate the estimated camber and deflections upon prestress transfer. The girder contains 2 7-strand tendons using 0.6-in. diameter prestressing strands placed in a single duct. The center of gravity of the tendon (cgs) is located at 3.5 in. from the bottom of the girder and at the beam centroid at the ends. The effective prestress immediately after prestress transfer is 189 ksi. The section is constructed with $f'_c = 5500$ psi $f'_{ci} = 4500$ psi.

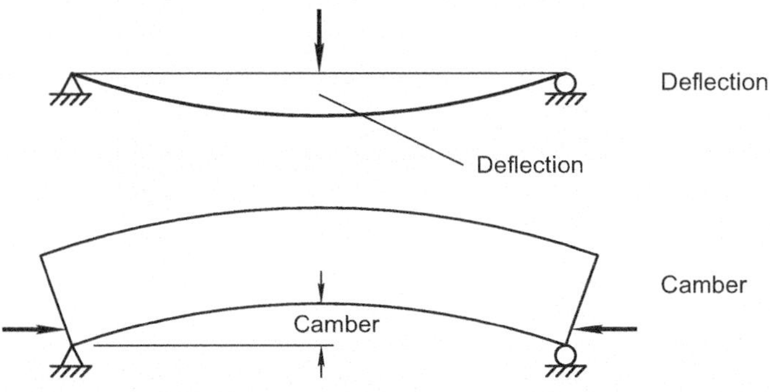

Fig. 8.2 Deflection and camber in a beam

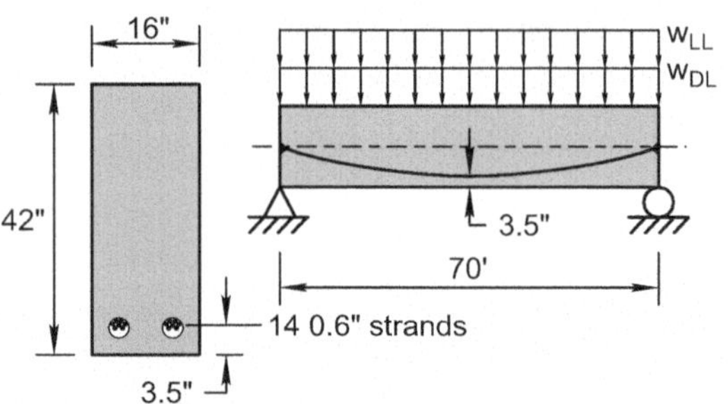

Fig. 8.3 Beam section and geometry

Solution: Calculate the camber of the beam under self-weight immediately following prestress transfer. The modulus of elasticity must be calculated based on the concrete strength at time of prestress transfer:

$$E_c = 57,000\sqrt{f'_{ci}} = 57,000\sqrt{4500}\ \text{psi} = 3824\ \text{ksi}$$

$$I = 16 \times 42^3/12 = 98,784\ \text{in.}^4$$

Prestressing force causes an upward displacement. Calculate the initial prestress force at release:

$$P_{pi} = f_{pi}A_{ps} = 189 \times 14 \times 0.217$$

$$P_{pi} = 574\ \text{kips}$$

Calculate the displacement caused by prestress force:

$$\Delta_{pi} = \frac{5P_i eL^2}{48E_{ci}I} = \frac{5(574)(0.5 \times 42 - 3.5)(70 \times 12)^2}{48 \times 3824 \times 98,784}$$

$$\Delta_{pi} = 1.96\ \text{in.}\ \uparrow$$

Calculate the displacement caused by self-weight:

$$\Delta_{pi} = \frac{5w_{sw}L^4}{384E_{ci}I} = \frac{5(0.70/12)(70 \times 12)^4}{384 \times 3824 \times 98,784}$$

$$\Delta_{sw} = 1.00\ \text{in.}\ \downarrow$$

In this example, prestress force and resulting camber lifts the beam from the formwork. This results in the downward self-weight displacement and upward deformation due to prestressing being superimposed into a net upward displacement:

$$\Delta_{ci} = 1.96\ \text{in.} - 1.00\ \text{in.} = 0.96\ \text{in.}\ \uparrow$$

8.6 Control of Deflections

Deflections under service loads are controlled to ensure that the structure is serviceable in terms of both human comfort and system functionality. Deflections are limited by the governing building code requirements. The 2015 International Building Code limits deflections by prescription as does the ACI Building Code. Table 8.3 displays the deflection limitations prescribed by ACI 318-14. The table is divided into two parts. The upper portion of the table addresses immediate deflections due to elastic deformations while the lower portion of the table covers the combined effects of elastic and time-dependent deflections. In each part of the table, both roof and floor elements are included. Deflection limitations are presented as a fraction of the element span l. For example, a deflection limit of $l/400$ for a 50-foot-long beam would be $(50 \cdot 12 \text{ in./ft})/400 = 1.5$ in.

Deflections limits are based on the elements being supported and their function. The deflection limitation is not intended to safeguard flat roof elements against ponding. If roof drainage is susceptible to ponding of rainwater or snowmelt, then ponding should be anticipated and accounted for in the design and checked during the deflection calculations. In addition to the immediate elastic deflections due to ponding, time-dependent deflections should be considered if appropriate.

Time-dependent deflections are compared to the limits in Tables 8.3 and 8.4 and may be reduced by amount of deflection calculated to occur before attachment of nonstructural elements. This adjustment is determined on basis of accepted engineering data relating to time-deflection characteristics of members being supported and the planned construction sequence. The limits in Tables 8.3 and 8.4 may be exceeded if adequate measures are taken to prevent damage to supported or attached elements. If camber is present the total deflection minus camber should not exceed the prescribed limits. Lastly, deflection limits should not be greater than tolerance provided for nonstructural elements.

Table 8.3 Deflection limitations according to ACI 318-14

Member	Condition		Deflection to be considered	Deflection limitation
Flat roofs	**Not supporting** or attached to nonstructural elements likely to be damaged by large deflections		**Immediate deflection** due to maximum of L_r, S, and R	$l/180$
Floors			**Immediate deflection** due to L	$l/360$
Roof or floors	**Supporting** or attached to nonstructural elements	**Likely** to be damaged by large deflections	That part of the total deflection **occurring after attachment of nonstructural elements**, which is the sum of the time-dependent deflection due to all sustained loads and the immediate deflection due to any additional live load	$l/480$
		Not likely to be damaged by large deflections		$l/240$

Table 8.4 IBC deflection limitations

Construction	L	S or W	D + L
Roof members:			
Supporting plaster or stucco ceiling	$l/360$	$l/360$	$l/360$
Supporting non-plaster ceiling	$l/240$	$l/240$	$l/240$
Not supporting ceiling	$l/180$	$l/180$	$l/180$
Floors	$l/360$		$l/240$
Exterior walls:			
With plaster with stucco finishes		$l/360$	
With other brittle finishes		$l/240$	
With flexible finishes		$l/120$	
Interior partitions:			
With plaster with stucco finishes	$l/360$		
With other brittle finishes	$l/240$		
With flexible finishes	$l/120$		

Adapted from IBC 2015, Table 1604.3

Table 8.5 Bridge deflection limits

In absence of other criteria, the following deflection limits may be considered for concrete vehicular bridges	
Vehicular load—general	$l/800$
Vehicular and pedestrian loads	$l/1000$
Vehicular load on cantilever arms	$l/300$
Vehicular and pedestrian loads on cantilever arms	$l/375$

Adapted from Ontario Highway Bridge Design Code (1991)

The AASHTO LRFD Bridge Specification deflection limits are based on the perception of a pedestrian or vehicle occupant to accelerations of the bridge due to a passing vehicle. Extensive research on human response to motion indicates that the primary factor affecting human sensitivity is acceleration, rather than deflection, velocity, or the rate of change of acceleration for bridge structures. Correlation of sensitivity and acceleration remains somewhat subjective. Thus, there are as yet no simple definitive guidelines for the limits of tolerable static deflection or dynamic motion. Among current specifications, the Ontario Highway Bridge Design Code (1991) contains the most comprehensive provisions regarding vibrations tolerable to humans, Table 8.5.

Example 8.3: Immediate Deflections, Class U

Using the beam from the previous example determine the classification and calculate the immediate deflections for comparison with the ACI Building Code limitations. Superimposed dead load is 15 psf and live load is 40 psf with a tributary width of 15 ft.

Solution: Calculate the camber after losses requires the use of the modulus of elasticity based on f'_c:

$$E_c = 57,000\sqrt{f'_c} = 57,000\sqrt{5500} \text{ psi} = 4227 \text{ ksi}$$

$$I = 16 \times 42^3/_{12} = 98,784 \text{ in.}^4$$

Prestressing force causes an upward displacement. Calculate the initial prestress force at release:

$$P_e = f_{pe}A_{ps} = 175 \times 14 \times 0.217$$

$$P_e = 532 \text{ kips}$$

Calculate the displacement caused by prestress force:

$$\Delta_p = \frac{5P_e e L^2}{48E_c I} = \frac{5(532)(0.5 \times 42 - 3.5)(70 \times 12)^2}{48 \times 4227 \times 98,784}$$

$$\Delta_p = 1.64 \text{ in.} \quad \uparrow$$

Calculate the displacement caused by self-weight:

$$\Delta_{sw} = \frac{5w_{sw}L^4}{384E_c I} = \frac{5(0.70/12)(70 \times 12)^4}{384 \times 4227 \times 98,784}$$

$$\Delta_{sw} = 0.91 \text{ in.} \quad \downarrow$$

$$\Delta_c = 1.64 \text{ in.} - 0.91 \text{ in.} = 0.73 \text{ in.} \quad \uparrow$$

Calculate the instantaneous deflection due to superimposed loads:

$$\Delta_D = \frac{5w_D L^4}{384E_c I} = \frac{5(15 \times 15/12,000)(70 \times 12)^4}{384 \times 4227 \times 98,784}$$

$$\Delta_D = 0.29 \text{ in.} \quad \downarrow$$

$$\Delta_L = \frac{5(15 \times 40/12,000)(70 \times 12)^4}{384 \times 4227 \times 98,784}$$

$$\Delta_L = 0.78 \text{ in.} \quad \downarrow$$

$$\Delta_T = 0.29 \text{ in.} + 0.78 \text{ in.} - 0.73 = 0.34 \text{ in.} \quad \downarrow$$

Calculate the deflection due to superimposed loads:

Compare instantaneous live load deflection to Table 8.3 to determine if deflections are within permissible limits. For flat roofs "not supporting or attached to nonstructural elements likely to be damaged by large deflections" the limit is:

$$\frac{L}{180} = \frac{70 \times 12}{180} = 4.67 \text{ in.}$$

For floors "not supporting or attached to nonstructural elements likely to be damaged by large deflections" the limit is:

$$\frac{L}{360} = \frac{70 \times 12}{360} = 2.33 \text{ in.}$$

Comparing either of these to the live load deflection Δ_L indicates that this beam would be adequate as either a roof or floor element.

8.7 Effect of Cracking on Deflections

Prestressed concrete is typically designed to avoid cracking under most of the applied load. Members designed so that no tension occurs under full dead and live load require an excessive amount of prestressing, which may cause excessive elastic camber and potentially camber growth over the life of the structure. Some tension allowed in the section at full live load is a prudent design approach from a serviceability perspective and also results in economic savings in the cost of prestressing. As long as the section remains uncracked, deflection calculations are straightforward because the member is prismatic. Once the section cracks, however, estimating the stiffness becomes more complicated. Figure 8.4 illustrates this concept with a simply supported beam loaded at midspan. In the nonprestressed member, service loads cause significant cracking, which results in a variation of section properties along the length of the member. Between cracks, the section retains the full gross section properties (I_g). At the crack, moment of inertia decreases sharply to the cracked moment of inertia (I_{cr}). Effective moment of inertia (I_e) is a weighted average that can be used to represent the overall stiffness of the member in a single value even though the member is far from prismatic. This facilitates the use of traditional structural analysis techniques to analyze the member or structure. The effective moment of inertia is calculated based on the actual moment and the cracking moment, where M_{cr} is given in Eq. (8.1).

$$I_e = \left[\frac{M_{cr}}{M_a}\right]^3 I_g + \left[1 - \left(\frac{M_{cr}}{M_a}\right)^3\right] I_{cr} \tag{8.2}$$

The ACI Building Code is considering an alternative form of Eq. (8.2) for reinforced concrete; however, Eq. (8.2) is shown to better represent deflections in prestressed concrete members.

Prestressed concrete, however, delays the cracking until most of the service load is in place. If the cracking load is exceeded and the intermittent load is removed,

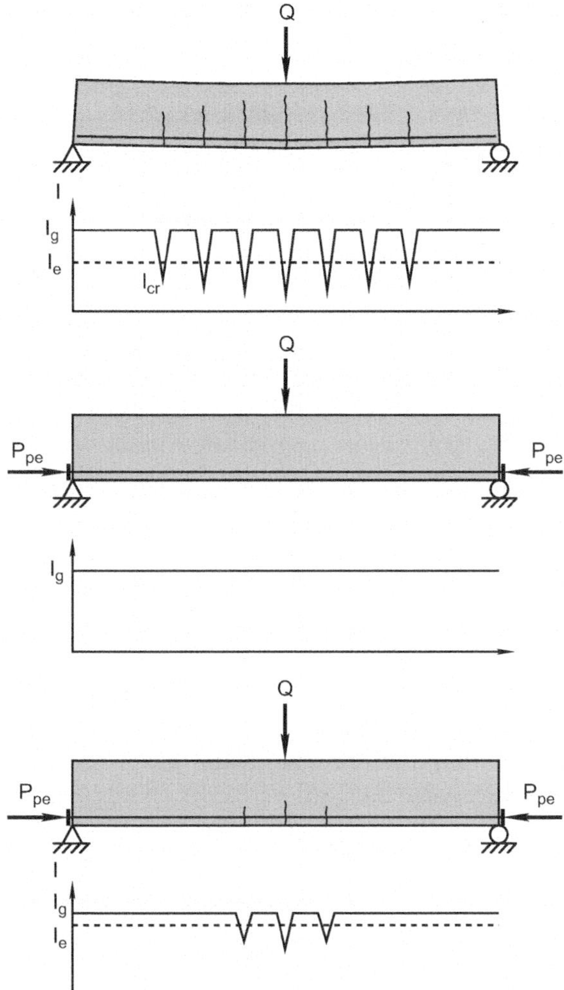

Fig. 8.4 Variation in moment of inertia in nonprestressed and prestressed concrete beams

however, the cracks close and the section returns to its full gross moment of inertia. Even if the member is designed to allow the live load to cause cracking, the extent and depth of cracking is much less for a prestressed member than that of a nonprestressed member. Practically, most prestressed concrete is prestressed sufficiently so that the sum of the dead load tensile stresses will be less than f_r. Determining the cracked section properties of prestressed concrete is not trivial as is shown for a rectangular cross section in Sect. 5.5.

The ACI Building Code requires the effect of cracking to be considered in Class T and C members. The member is analyzed using a cracked transformed section and one of the two following methods. The first method uses the effective moment of inertia. The second method uses a bilinear moment-deflection relationship in which the gross moment of inertia is used before cracking and the cracked moment of inertia is used for load beyond cracking. The PCI manual suggests the following approximation for the cracked moment of inertia in lieu of the approach in Sect. 5.5:

$$I_{cr} = nA_{ps}d_p^2\left(1 - 1.6\sqrt{n\rho_p}\right) \tag{8.3}$$

where ρ_p is the ratio of A_{ps} to bd_p. Unlike the effective moment of inertia approach for nonprestressed concrete, the bilinear approach assumes prismatic section properties after cracking. Consequently, this approach likely to be more conservative than other methods that seek to model the cracked behavior more closely. The design emphasis, though, should be on the primary advantage of prestressed concrete, which is improved serviceability by the avoidance of cracking except in the relatively rare cases when the design full live load is in place. If the member is designed such that a relatively small percentage of the live load is carried by the cracked section, then the conservative nature of the deflection calculations should not be an issue.

Example 8.4: Immediate Deflections, Class T or C

Determine the live load deflection for pretopped 10 foot wide double-T 10DT26 and compare to ACI 318 deflection limitations. DT has a span of 60 ft and 12 ½" dia. prestressing strands with six placed in each web with the cgs at 7 in. from the bottom of the DT giving $d_t = 19$ in. Effective prestress is 175 ksi. Self-weight is 718 plf, superimposed dead load is 5 psf, and live load is 40 psf. The section is constructed with $f'_c = 5000$ psi, $I_g = 30{,}716$ in.^{4}, and $\rho_p = 12 \times 0.153/689 = 6.98 \times 10^{-4}$.

Solution: Calculate the camber after losses requires the use of the modulus of elasticity based on f'_c:

$$E_c = 57{,}000\sqrt{f'_c} = 57{,}000\sqrt{5000} = 4031 \text{ ksi}$$

and $n = E_s/E_c = 28{,}500/4031 = 7.1$.

Effective prestressing force is

$$\begin{aligned}P_e &= f_{pe}A_{ps} = 175 \times 12 \times 0.153 \\ P_e &= 321 \text{ kip}\end{aligned}$$

Determine classification of the DT. Self-weight moment is

$$M_g = \frac{1}{8}\left(\frac{718}{1000}\right)(60)^2 = 323.1 \text{ kip ft}$$

Dead load moment is

$$M_{\mathrm{DL}} = \frac{1}{8}\left(\frac{10\times 5}{1000}\right)(60)^2 = 22.5 \ \mathrm{kip \ ft}$$

Live load moment is

$$M_{\mathrm{LL}} = \frac{1}{8}\left(\frac{10\times 40}{1000}\right)(60)^2 = 180.0 \ \mathrm{kip \ ft}$$

Total moment is

$$M_{\mathrm{TL}} = 323.1 + 22.5 + 180.0 = 525.6 \ \mathrm{kip \ ft}$$

Section modulus for calculating the bottom stress is

$$S_{\mathrm{b}} = \frac{I}{y_{\mathrm{b}}} = \frac{30,716}{20.29} = 1514$$

Maximum tensile stress is

$$f_{\mathrm{b}} = \frac{P_{\mathrm{e}}}{A_{\mathrm{g}}} + \frac{P_{\mathrm{e}}e}{S_{\mathrm{b}}} - \frac{M_{\mathrm{TL}}}{S_{\mathrm{b}}} = \frac{321}{689} + \frac{321\times(20.29-7)}{1514} - \frac{525.6\times 12}{1514} = -0.882 \ \mathrm{ksi}$$

Tensile stress that defines class C is

$$12\sqrt{f_{\mathrm{c}}'} = 12\sqrt{5000} = 848 \ \mathrm{psi}$$

Therefore, the member is Class C. Since the tensile stress is not far beyond the transition, it is necessary to use the cracked section properties only for some percentage of the live load. Solving the following equation for alpha provides the exact amount.

$$-7.5\sqrt{f_{\mathrm{c}}'} - \frac{P_{\mathrm{e}}}{A_{\mathrm{g}}} + \frac{P_{\mathrm{e}}e}{S_{\mathrm{b}}} - \frac{M_{\mathrm{SW}}}{S_{\mathrm{b}}} - \frac{M_{\mathrm{DL}}}{S_{\mathrm{b}}} - \frac{\alpha M_{\mathrm{LL}}}{S_{\mathrm{b}}}$$

$$\alpha = \left(\frac{P_{\mathrm{e}}}{A_{\mathrm{g}}} + 7.5\sqrt{f_{\mathrm{c}}'}\right)\frac{S_{\mathrm{b}}}{M_{\mathrm{LL}}} + \frac{P_{\mathrm{e}}e - (M_{\mathrm{SW}} + M_{\mathrm{DL}})}{M_{\mathrm{LL}}}$$

$$= \left(\frac{321}{689} + \frac{7.5\sqrt{5000}}{1000}\right)\frac{1514}{180.0\times 12} + \frac{321\times(20.29-7){}^{1}/{}_{12} - (323.1+22.5)}{180.0}$$

$$\alpha = 0.753$$

Which means that the gross section properties carry all the dead load and 75.3% of the live load before cracking.

The cracked moment of inertia is

$$I_{cr} = nA_{ps}d_p^2\left(1 - 1.6\sqrt{n\rho_p}\right) = 7.1 \cdot 12 \cdot 0.153 \cdot 19^2\left(1 - 1.6\sqrt{7.1 \cdot 6.98 \times 10^{-4}}\right)$$

$$= 4121 \ \text{in.}^4$$

which is 13 percent of I_g. Before cracking the uncracked section properties carry part of the live load.

$$\Delta_g = \frac{5\alpha w_{LL}L^4}{384E_cI_g} = \frac{5(0.753 \times 10 \times 40)(60)^4 \times 1728}{384 \times 4031 \times 30,716 \times 1000}$$

$$\Delta_g = 0.71 \ \text{in.} \ \downarrow$$

Calculate the incremental displacement for the remainder of the live load on the cracked moment of inertia section

$$\Delta_{cr} = \frac{5(1-\alpha)w_{LL}L^4}{384E_cI_{cr}} = \frac{5(1-0.753)(10 \times 40)(60)^4 \times 1728}{384 \times 4031 \times 4121 \times 1000}$$

$$\Delta_{cr} = 1.75 \ \text{in.} \ \downarrow$$

The total live load displacement is then

$$\Delta_{LL} = 0.71 \ \text{in.} + 1.75 \ \text{in.} = 2.46 \ \text{in.} \ \downarrow$$

Compare instantaneous live load deflection to Table 8.3 of ACI 318-14 to determine if deflections are within permissible limits. For flat roofs "not supporting or attached to nonstructural elements likely to be damaged by large deflections" the limit is:

$$\frac{L}{180} = \frac{70 \times 12}{180} = 4.67 \ \text{in.}$$

By inspection this section meets the limit on deflection. For floors "not supporting or attached to nonstructural elements likely to be damaged by large deflections" the limit is:

$$\frac{L}{360} = \frac{70 \times 12}{360} = 2.33 \ \text{in.}$$

This element exceeds the limit for floor systems. Because of the substantial reduction in stiffness caused by the cracking in this method, most of the live load deflection occurs after cracking. This is illustrated by the load-displacement plot for this problem, Fig. 8.5.

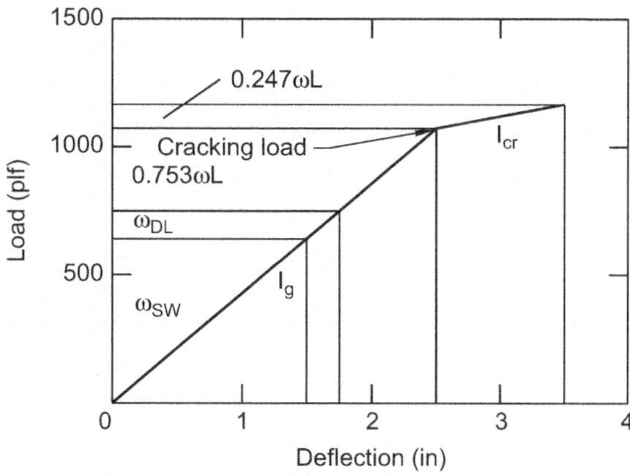

Fig. 8.5 Bilinear approach to determine deflections

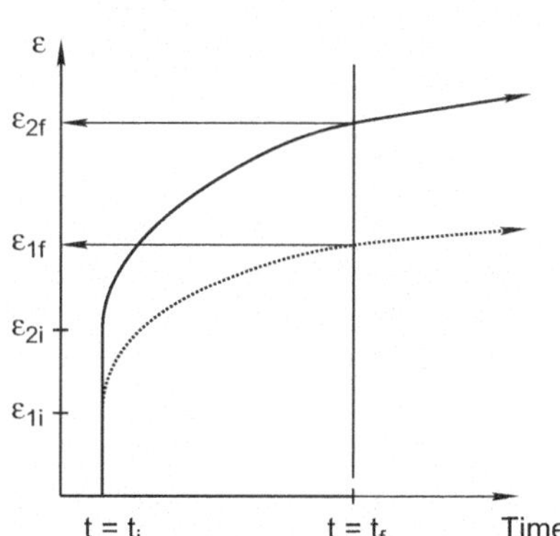

Fig. 8.6 Time-dependent deformation of concrete in compression

8.8 Time-Dependent Deflections

Structural concrete members are susceptible to time-dependent deformations as a result of the creep properties of concrete. As illustrated in Fig. 8.6, an applied compressive stress results in an immediate deformation that occurs at time t_i followed by continued deformations over the life of the structure. A lower initial stress (and accompanying strain) results in a lower elastic strain and a lower total creep strain at the end of the design life span, t_f.

In reinforced concrete, these deformations are manifested in increased curvature resulting from the deformations in the compression zone with time (Fig. 8.7). Over time, the portion of the concrete section in compression due to flexural stresses continues to deform. Reinforcing steel has virtually no time-dependent deformation compared to the concrete, which results in a change in curvature without a change in internal forces. In prestressed concrete with the cgs placed eccentrically to the centroid of the concrete section, compressive stresses are greater in the bottom of the section than in the top. Because the initial stress is larger in the bottom of the

Fig. 8.7 Time-dependent
(**a**) deflections in a
nonprestressed beam and (**b**)
camber growth due to
eccentric prestressing

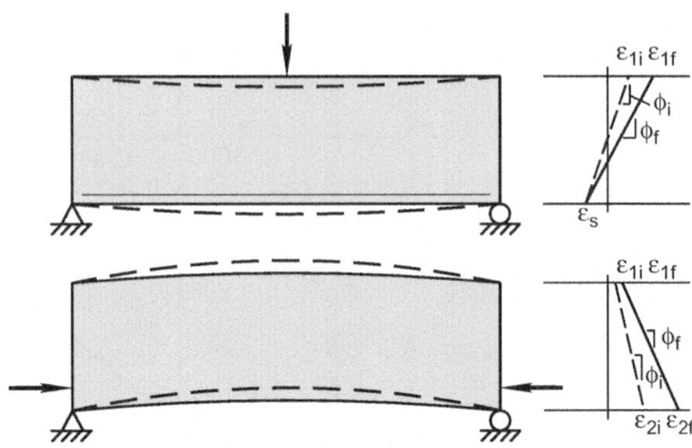

section, the creep deformations are larger, resulting in a change in curvature in time, which can result in **camber growth**. Superimposed on the curvature caused by the prestressing is the curvature caused by the member loads, which is in the opposite direction and results in a balance of competing deformations. The engineer must then select the prestressing force and eccentricity to provide a serviceable balance between camber and deflections, both for instantaneous and long-term effects.

The ACI Building Code addresses this by use of a multiplier λ_Δ that addresses long-term deflections in nonprestressed members. ACI 318 further requires that the long-term effects be considered, but no specific requirements beyond those in Table 8.3 are given. Instead, the commentary indicates that any suitable method may be used, provided all effects are considered. Some of the methods mentioned can be found in ACI 209R, ACI Committee 435 (1963), Branson et al. (1970), and Ghali and Favre (1986).

Although not specifically mentioned in ACI 318 commentary, Martin (1977) proposed a method for estimating camber and deflections of prestressed members using multipliers at specific critical points in the life of the structure. Figure 8.8 illustrates the method by tracking each component of camber and deflection for prestress release, assembly, and final. The multipliers are assembled in Table 8.6. In developing these multipliers, Martin (1977) seeks to temper the expectation that deflections and camber can be calculated with a high degree of accuracy:

> "It should be noted that because of the inherent variables that affect camber and deflection, such as concrete mix, storage method, time of release of prestress, time of erection and placement of superimposed loads, relative humidity, etc., and the data scatter under the most closely controlled tests, calculated long-time values should never be considered any better than estimates."

Example 8.5: Time-Dependent Deflections Using Multipliers

Using the beam from Example 8.4, determine the time-dependent deflections using the PCI factors. Assume that the superimposed dead load is applied at the time of assembly approximately 60 days after fabrication and prestressing the element. Compare the results with the ACI limitations. Assume that nonstructural elements were attached to the structure at the time of assembly.

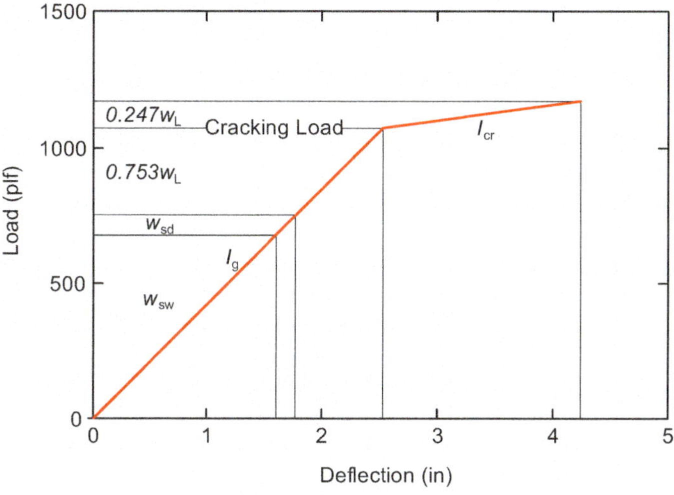

Fig. 8.8 Time-dependent increases in deflection illustrating simplified long-term deflection calculations (Adapted from Martin 1977)

Table 8.6 Multipliers for determining long-term camber and deflections in prestressed members

Incremental deflection	Stage	Noncomposite	Composite
δ_c, δ_{sw}	Prestress transfer	n/a	n/a
Assembly			
δ_{sw}	(1) Deflection (downward) component—apply to the elastic deflection due to the component weight at release of prestress	1.85	1.85
δ_c	(2) Camber (upward) component—apply to the elastic camber due to prestress at the time of release of prestress	1.80	1.80
Final			
δ_{sw}	(3) Deflection (downward) component—apply to the elastic deflection due to the component weight at release of prestress	2.70	2.40
δ_c	(4) Camber (upward) component—apply to the elastic camber due to prestress at the time of release of prestress	2.45	2.20
δ_{SDL}	(5) Deflection (downward)—apply to elastic deflection due to superimposed dead load only	3.00	3.00
δ_{CD}	(6) Deflection (downward)—apply to elastic deflection caused by the composite topping	–	2.30

Table 8.7 Time-dependent deflections

	Prestress transfer	LT factor	1–2 months
Prestress	1.95 in. ↑	1.80	3.51 in. ↑
Self-weight	1.00 in. ↓	1.85	1.85 in. ↓
			1.66 in. ↑
SDL			0.26 in. ↓
Total	**0.95 in. ↑**		**1.40 in. ↑**

Table 8.8 Time-dependent and live load deflections

	Prestress transfer	LT factor	Final
Prestress	1.95 in. ↑	2.45	4.78 in. ↑
Self-weight	1.00 in. ↓	2.70	2.70 in. ↓
SDL	0.26 in. ↓	3.00	0.78 in. ↓
Total			1.30 in. ↑
LL			1.55 in. ↓

Solution: Adjust the previously calculated immediate deflections using the time-dependent multipliers to compute the estimated deflections at the key times, Table 8.7:

Initial deflection due to prestress and self-weight are factored individually before combining for the camber at assembly, which is assumed to be at 1–2 months following element fabrication. This camber is then combined with the superimposed dead load, which is also assumed to be applied at this point in time. Furthermore, any partitions considered in the time-dependent deflections are assumed to be installed at this point in time. Table 8.8 applies the factors for the element after all time-dependent movement is assumed to have occurred. Again, the factors are applied to the initial prestress and self-weight components of camber and combined with the superimposed dead load for a final camber.

To determine the total deflection absorbed by the partitions, the change in camber from the time of partition installation (assembly) is subtracted from the final camber. This change is then added to the live load deflection and compared to the ACI limits:

$$\Delta_f = 1.55 + (1.40 \text{ in.} - 1.30 \text{ in.}) = 1.65 \text{ in.} \quad \downarrow$$

Compare this to the allowable deflections in Table 8.3 to determine if deflections are within permissible limits. For roof or floor "supporting or attached to partitions not likely to be damaged by large deflections" the limit is:

$$\frac{L}{240} = \frac{70 \times 12}{240} = 3.50 \text{ in.}$$

For roof or floor "not supporting or attached to nonstructural elements likely to be damaged by large deflections" the limit is:

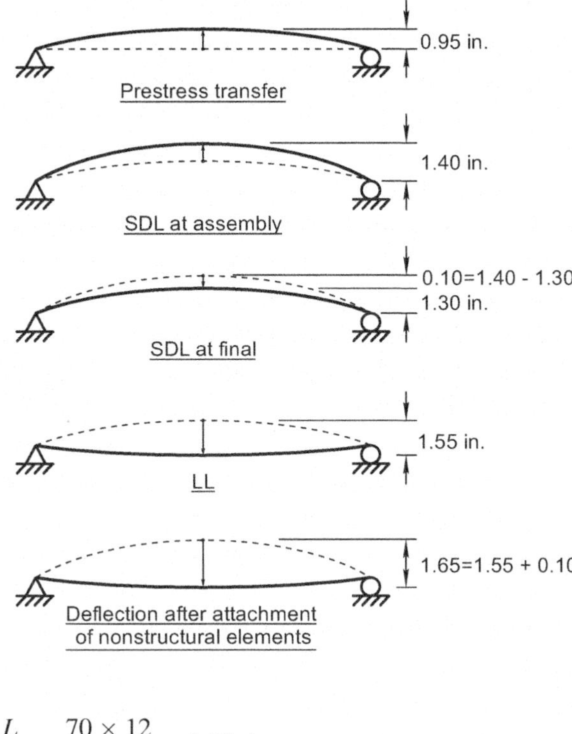

Fig. 8.9 Schematic summary of beam deflections

$$\frac{L}{480} = \frac{70 \times 12}{480} = 1.75 \ \text{in.}$$

Comparing either of these to the that part of the total deflection occurring after attachment of nonstructural elements live load deflection Δ_f indicates that this beam would be adequate as either a roof or floor element, Fig. 8.9.

8.9   Deflections in Composite Members

Composite members should be designed so that the horizontal shear strength requirements of the ACI Building Code are satisfied as well as the flexural strength requirements. Typically, whether the member is shored or unshored does not affect the design of the member for these actions. For deflections, however, the construction method and timing may have a significant effect on both the time-dependent deflections. If the member is unshored, then the precast section alone carries the flexural stresses associated with the deck. These stresses have time-dependent components associated with them. Conversely, if the member is unshored, then the entire composite section carries the deck load and thus has less initial and time-dependent deflections associated with the deck.

Another consideration in composite sections is the differential shrinkage that can occur between deck and precast. The deck is typically placed in the field, possibly several months after the precast section was fabricated. Consequently, the precast section has relatively little shrinkage movement compared to that of the deck that is placed in the field and subjected to job site conditions. Deck shortening from shrinkage results in curvature that increases the deflection of the member.

8.10   Deflections due to Thermal Gradient

Beams with wide flanges and narrow webs in the shadow of the flange or slabs with narrow supporting beams are subject to deflection from differential thermal effects. The expansion of the top flange is restrained by the beam stem and causes a simple span beam to camber upward. Similarly, radiant cooling can cause a beam to sag. Beams with fixed end supports resist the deflection by internal moments. The AASHTO Bridge Specification contains guidance for the calculation of these deflections; however, the methods and temperature distribution are more complex than may be needed for building construction.

The balanced temperature approach and parabolic curve fit evolved from a series of measurements on 110 ft long single-T beams with web thickness of 10 in. and average flange thickness of 2.5 in., Fig. 8.10. A maximum temperature differential of 26 F was recorded and the total daily deflection motion was approximately 1.20 in. The storage of the single-T beams provided maximum flange sun exposure and full shade for the stems. A reference line was run down the neutral axis of the beam and the daily deflections recorded relative to the reference line.

The following two approaches are simplified methods to obtain a quick estimate of the potential deflections. Both are variations on an approach by (Maher and Aust 1970). The first case examines a balanced temperature approach. The second case examines an equivalent parabolic fit to the thermal strains. Both cases assume the single-T beam is subjected to a 20 F temperature rise at the top of the beam flange.

Neither of the following methods is "exact" and both provide a quick estimate of the thermal deflections of a beam or slab exposed to differential temperature. The methods are particularly useful is assessing whether a member has excessive thermal camber at the time of placement. Based on the measurements in Fig. 8.10, the

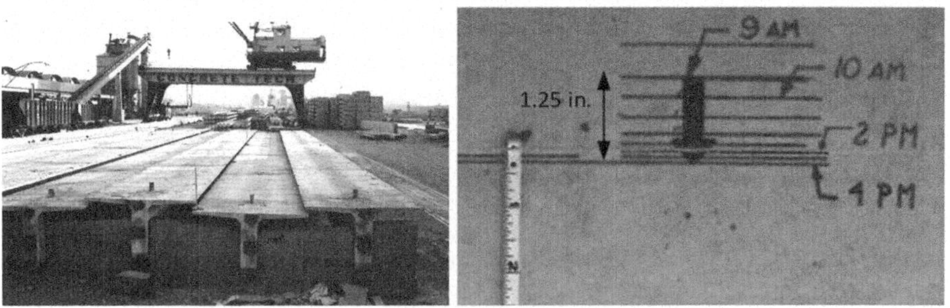

Fig. 8.10 Single-T beam deflections due to differential temperature

parabolic approximation overestimates the deflection. The overestimation occurs because using just the top layer thermal strain does not fully capture the reduction in strain through the flange. The balanced temperature method underestimates the deflection because the idealized stress blocks produce a slightly smaller internal moment arm.

8.10.1 Balanced Temperature Approach

The stem of the beam is assumed to be in the shade. The temperature rise of 20 F is on the top of the flange and the bottom of the flange is assumed to be at the same temperature as the stem. The average temperature rise is applied to the flange of an idealized beam and flange, Fig. 8.11. The procedure balances the temperature differential and creates an equivalent internal moment used to calculate the thermal deflection.

The analysis procedure based on the internal moment:

- Balances the thermal gradient to get the average stem temperature.
- Calculates the thermal strains on the balanced section.
- Calculates the thermal stresses and forces from these strains.
- Sums the moment on the section to calculate the restrained internal moment.

Using the beam in Fig. 8.11 and an average temperature rise of 10°F in the top flange, the average temperature on the structure is

$$T_{av} = \frac{\sum A_c \Delta T}{\sum A_c} = \frac{7 \cdot 96 \cdot 10}{7 \cdot 96 + 8 \cdot 29} = 7.43 \ F$$

Therefore, the stem is subject to 7.43° and the flange effective temperature is $10 - 7.43 = 2.57$ F, Fig. 8.12.

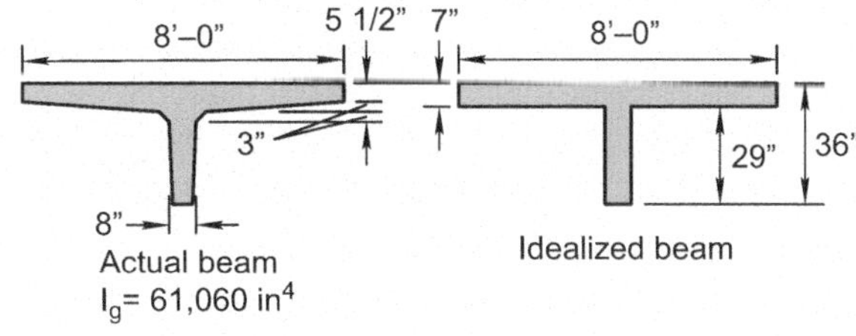

Fig. 8.11 Actual and idealized beam

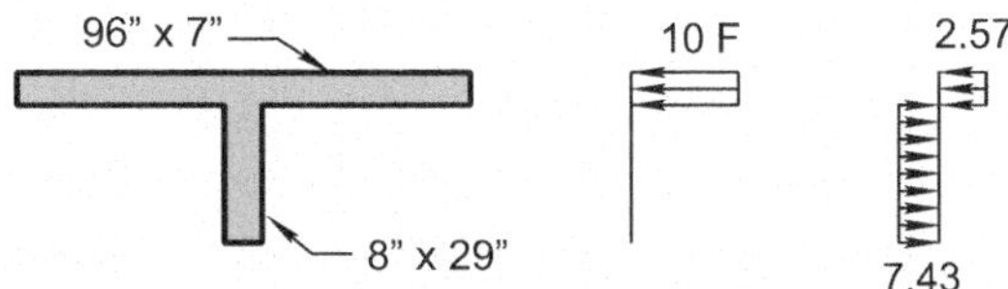

Fig. 8.12 Balanced thermal gradient

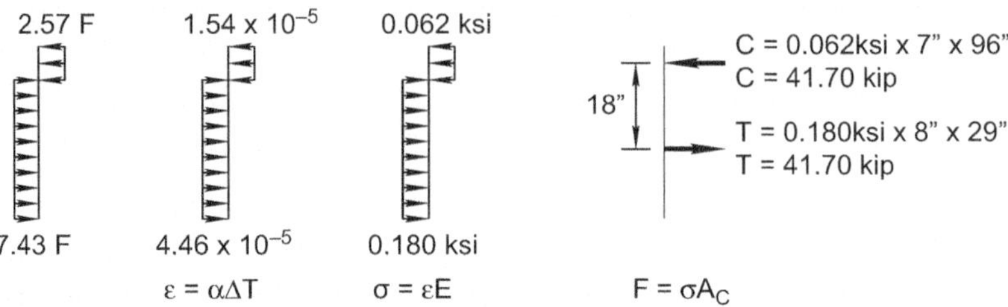

Fig. 8.13 Resultant temperatures, strains, stresses, and forces

The corresponding temperatures, strains, stresses, and forces are given in Fig. 8.13 and use a coefficient of thermal expansion of 6.0×10^{-6}/F, a concrete strength of $f'_c = 5000$ psi, a modulus of elasticity is $57,000 \sqrt{f'_c} = 4030$ ksi. The stress is the strain times the concrete modulus of elasticity.

The moment arm between the centroids of temperature couples is 18 in. giving a constant moment on the section of $M = 41.7$ kip $\times$ 18 in. $= 751$ in.-kip. For a 70-ft long simple span beam, the resulting deflection is

$$\delta = \frac{Ml^2}{8E_c I_g} = \frac{795 \cdot 70^2 \cdot 144}{8 \cdot 4030 \cdot 61,060} = 0.27 \text{ in.}$$

This deflection is approximately $l/3100$, giving an indication of the influence of temperature on the total beam deflection allowed by various codes. A fixed beam is restrained so no end rotations can occur. In such cases, the internal moment can be used to calculate stresses due to thermal gradients.

8.10.2   *Parabolic Approximation*

For simple span beams an approximate deflection can be calculated by using the radius of curvature and an equivalent parabola. In this case, the maximum differential strain on the top fiber results from the 20 F temperature rise or $6.0 \times 10^{-6} \times 20 = 1.20 \times 10^{-4}$ at the top surface. Assume that the entire strain is at the top of the beam and the bottom of the stem has no differential strain. The radius of curvature is then:

$$\frac{1}{R} = \frac{\epsilon_{top}}{h} = \frac{1.2 \cdot 10^{-4}}{36} = 3.33 \cdot 10^{-6}$$

The middle offset of a parabola is

$$y = \frac{l^2}{8R} = \frac{70^2 \cdot 144}{8}\,3.33 \cdot 10^{-6} = 0.29 \ \text{in.}$$

which is close to the first approximation.

Problems

8.1. Compute the camber immediately following prestress transfer for a precast untopped double tee 8DT24.

Load: LL = 280 plf, superimposed DL = 80 plf

Geometry: L = 70 ft assume span extends from end-to-end of beam (bearing length = 0)

Tendon: 12 0.5 in. diameter ASTM A416 low relaxation seven-wire prestressing strands with single-point harp. $e_c = 13.9$ in., $e_e = 5.4$ in.

Concrete: $f'_c = 5000$ psi, $f'_{ci} = 3500$ psi

8.2. For the section in Problem 8.1, compute the immediate deflection due to dead and live loads. Compute the moment of inertia using the bi-linear stiffness approach. Do the deflections satisfy the limits given in ACI 318?

8.3. For the section in Problem 8.1, compute the time-dependent deflections using the deflection multipliers.

References

ACI 209R. (2008). *Prediction of creep, shrinkage, and temperature effects in concrete structures (reapproved 2008)* (p. 47). ACI Farmington Hills, MI.

ACI 435 Subcommittee 5. (1963). Deflections of Prestressed Concrete Members (ACI 435.1R-63). *ACI Journal Proceedings, 60*(12), 1697–1728.

Branson, D. E., Meyers, B. L., & Kripanarayanan, K. M. (1970). Time-dependent deformation of noncomposite and composite prestressed concrete structures. *Symposium on Concrete Deformation, Highway Research Record 324* (pp. 15–13). Highway Research Board.

Ghali, A., & Favre, R. (1986). *Concrete structures: Stresses and deformations* (348p). New York: Chapman and Hall.

Ghali, A. (1993). Deflection of reinforced-concrete members—A critical review. *ACI Structural Journal, 90*(4), 364–373.

Maher, D. R. H., & Aust, M. I. E. (1970). The effects of differential temperature on continuous prestressed beams. *Civil Engineering Transactions*, Australia, pp. 29–42.

Martin, L. D. (1977). A rational method for estimating camber and deflection of precast prestressed members. *PCI Journal, 22*(1), 100–108.

Ontario Ministry of Transportation. 1991. *Ontario Highway Bridge Design Code*, 3rd ed. Ministry of Transportation, Downsview, Ont.

PCI Design Handbook: Precast and Prestressed Concrete (8th edition, MNL-120-04). (2017). Precast/Prestressed Concrete Institute, Chicago, IL, pp. 4–68 to 4–72.

Robertson, I. N. (2005). Prediction of vertical deflections for a long-span prestressed concrete bridge structure. *Engineering Structures, 27*(12), 1820–1827.

Sabnis, G. M., Meyers, B. L., and Roll, F., (1974), *Deflections of concrete structures.* American Concrete Institute SP-43, 637 pp.

Tadros, M. K., Ghali, A., & Meyer, A. W. (1985). Prestressed loss and deflection of precast concrete members. *Journal Prestressed Concrete Institute, 30*(1), 114–141.

Chapter 9
Continuous Slabs and Beams

9.1 Introduction

Continuity provides an economic benefit with shallow sections and longer spans possible. Continuity also inherently improves safety due to the redundancy provided by the continuity. This comes at a price of increased calculation and detailing complexity, increased construction complexity, and increased propensity for cracking and other distress due to the effect of restrained creep and shrinkage.

Although there are exceptions, precast prestressed construction is typically statically determinate, thus increasing the efficiency of assembly process and isolating the effects of creep and shrinkage. Indeed, a good portion of the deformation associated with creep and shrinkage typically occurs by the time the prefabricated elements have been erected.

Cast-in-place post-tensioned construction lends itself to the use of continuity. Continuous slabs and beams are easily placed in continuous concrete pours and prestressed with continuous post-tensioning tendons either single monostrand or multi-strand bundled tendons. This form of construction allows more flexibility in the structural form over that of prefabricated elements. In some cases, the two systems can be combined to provide an elegant solution such as the drop-in girder spans discussed in Chap. 13.

Advantages of continuity include:

1. Reduced moments—shallower members
2. Redundancy
3. Reduced number of anchorages

Disadvantages of continuity include:

1. Higher friction losses
2. Lateral forces from time-dependent axial shortening
3. Secondary moments due to prestress and temperature change
4. Potential moment sign reversal from pattern loading

C. W. Dolan, H. R. Hamilton, *Prestressed Concrete*,
https://doi.org/10.1007/978-3-319-97882-6_9

9.2   Factored and Service Load Analysis

Just as in the simply supported flexural elements discussed in earlier chapters, flexural stresses and strength must be checked in continuous members to ensure compliance with Code requirements for serviceability. An analysis must be conducted on the indeterminate system to ascertain the distribution of moments. To determine moments and shears caused by gravity loads in columns, beams, and slabs, the ACI Building Code permits floor levels, other than the one being considered, to be ignored in the analysis. In analyzing each floor, the columns above and below that floor are included in the analysis with the far ends of columns fixed if they are built integrally with the structure, Fig. 9.1. In addition to isolating the floor system structurally, the engineer may assume that live load is applied only to the level under consideration.

Two analyses are typically required, one to determine service load effects and one to determine factored load effects for strength design. The in-service performance of the structure is evaluated based on the defections, vibrations, and building periods that can be determined from the service load analysis; for prestressed concrete, the service level concrete stresses are checked. The moments of inertia used in the service load analysis should be representative of the degree of cracking that would be expected under the load levels investigated. For prestressed concrete, the degree of cracking used in service stress and deflection calculations is directly related to the classification of the member. Consequently, the initial structural analysis is conducted with gross section properties to determine the service moments. Flexural stresses are then determined by combining stresses imposed by the prestressing force, secondary stresses caused by the redundant restraints, and stresses from load effects into a final set of concrete stresses.

The flexural stresses calculated in the precompressed tensile zone under full service dead and live loads are used to classify the beam. For simply supported

Fig. 9.1 One-way floor system, beam or slab, idealized for analysis

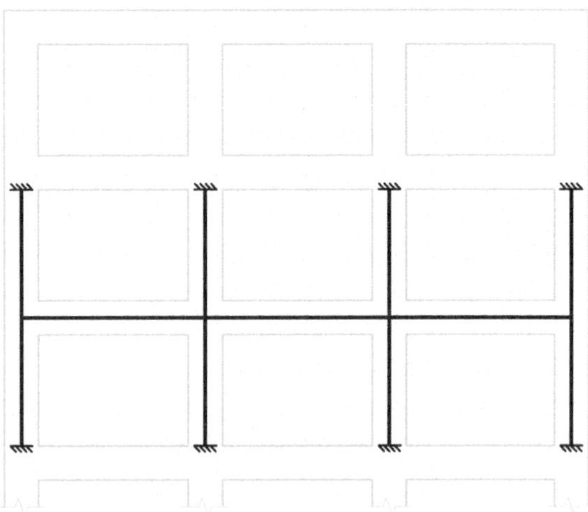

Table 9.1 Reduction in member stiffness for factored load analysis

Member and condition	Moment of inertia	Cross-sectional area
Columns	$0.70I_g$	$1.0A_g$
Walls	$0.70I_g$	
Beams	$0.35I_g$	
	$0.35I_g$	
Flat plates and flat slabs	$0.25I_g$	

beams, this location is usually the bottom of the beam at or near mid-span and does not typically change if pattern loading is applied. For continuous beams and slabs, however, flexural tensile stresses commonly occur in the top of the section near the supports and at the bottom of the section near midspan.

Generally, the prestressing force is adjusted so that the classification is either U or T. Depending on the classification of the beam, member stiffness is then be adjusted accordingly for subsequent analyses to reflect the relative stiffness for the particular design parameters and load case. Final stresses are then checked against ACI Building Code limits.

Structural analyses conducted to determine factored load effects take into account flexural cracking similar to the approach used for nonprestressed concrete because prestressed concrete members crack when they reach their nominal moment strength. The ACI Building Code allows the use of reduction factors for the relative flexural stiffness of members incorporated in the analysis, Table 9.1. The ACI Building Code commentary indicates that these values were derived for nonprestressed members. Although the code does not explicitly address their use for prestressed concrete, the commentary cautions that the stiffness values for prestressed concrete members should include an allowance for the variability of the member stiffnesses.

Pattern loading (or "skip" loading) in which the live load is applied to alternate spans, affects the location of the maximum flexural stresses. The use of a topping or deck on the beam further complicates the stress check. Consequently, all cases of loading are considered to determine the magnitude and location of the peak flexural stresses so that the member can be classified. This ensures that the appropriate section properties are used when calculating the serviceability requirements for concrete stresses and deflections.

Pattern loading must also be addressed in the analysis for factored load effects. In conducting the analysis, the ACI Building Code indicates that pattern loading should be considered for multi-span floor or roof systems, Fig. 9.2. If the arrangement of live load is known and fixed, then the structure may be analyzed for that pattern. If not, then the worst-case loading pattern is considered for all the critical locations, which are typically near midspan and at the face of the columns. For one-way slabs and beams, the maximum factored moment at midspan can be based on the span carrying live load. In addition, live load is placed on alternate spans. Maximum negative moment may be based on live load being placed only on the spans adjacent to the location under consideration. If the live load does not exceed 75% of the dead

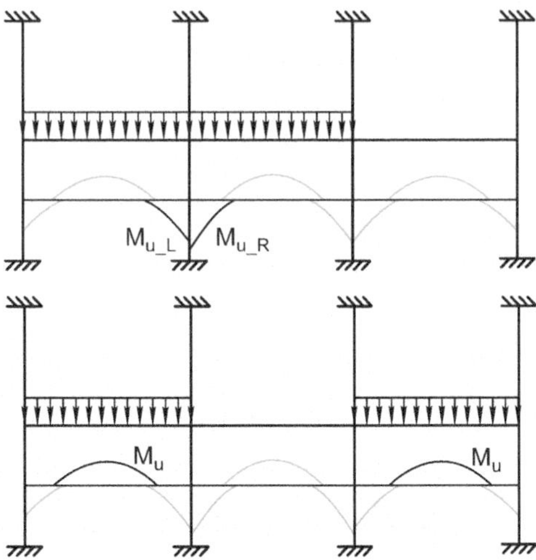

Fig. 9.2 Factored moment envelope development using pattern loading

load, then the ACI Building Code allows pattern affects to be ignored and all factored moments can be determined from the condition where all spans are loaded.

9.3 Tendon Profiles and Stressing

The advantages of post-tensioning allow for creative use of tendon configuration to solve a number of design and construction issues faced by the engineer. Figure 9.3 illustrates a few of the possible tendon layouts that can be used in a multiple span element or structure. These profiles could be used on slabs, beams, or girders. Figure 9.3a shows three spans of continuous concrete and a continuous post-tensioning tendon. This configuration could be used on one- or two-way slab systems using monostrand tendons in which the eccentricities are only a couple of inches. Continuous tendons can be used on deeper cast-in-place transfer girders with multi-strand PT tendons in which the eccentricities are much larger.

When the load balancing method was introduced, the idealized parabolic tendon layout was used to determine the amount of load that could theoretically be balanced by the tendon based on its eccentricity and effective prestress force. The tendon profile contains a sharp break at the support that allows the balancing load to be calculated easily for a single parabolic profile. The actual layout used in construction, however, is actually a series of parabolas that are either concave up or concave down. The concave down parabola over the support replaces the concentrated load from the sharp break in the theoretical layout. Advantage of the Fig. 9.4a is that the tendon is at a relatively constant depth through the support. This is advantageous in slab systems where reinforcement is congested in the column-slab joint and immediately surrounding areas. This configuration allows reinforcement and tendons to be placed at the same elevation to avoid interference. Figure 9.4b is more likely to be

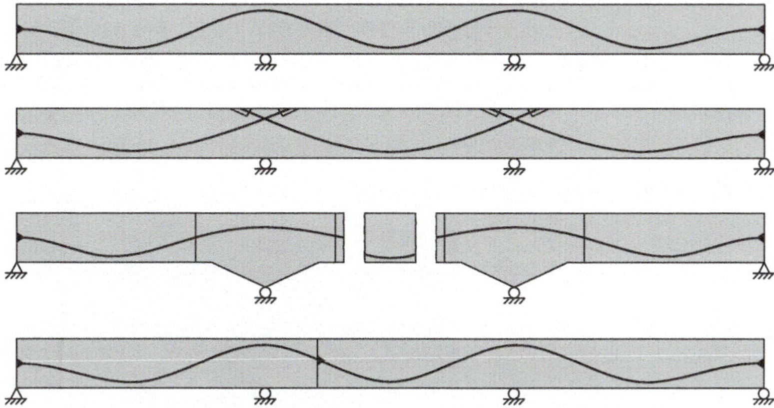

Fig. 9.3 Examples of possible tendon profiles in a three-span configuration (**a**) continuous tendon (**b**) overlapping tendons (**c**) spliced girder with continuous tendon (**d**) discontinuous tendon with coupling

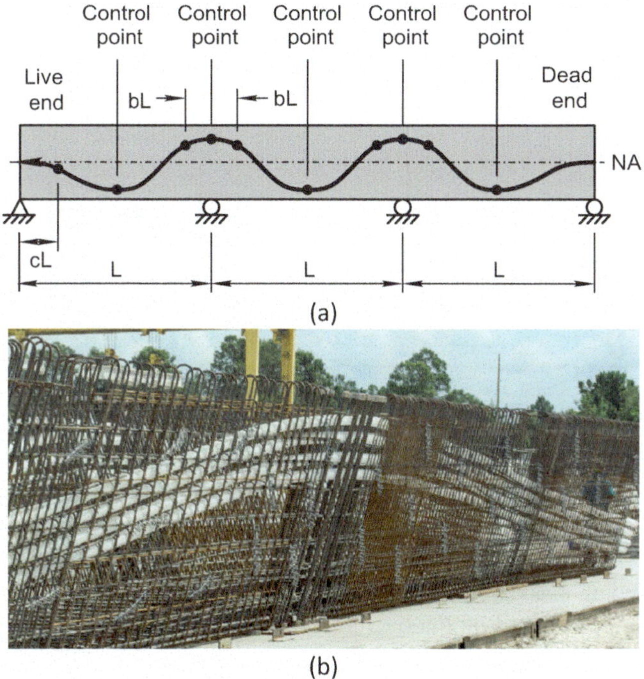

Fig. 9.4 Example of inverse parabolic tendon profile over the support region (**a**) schematic and (**b**) box-girder for bridge element

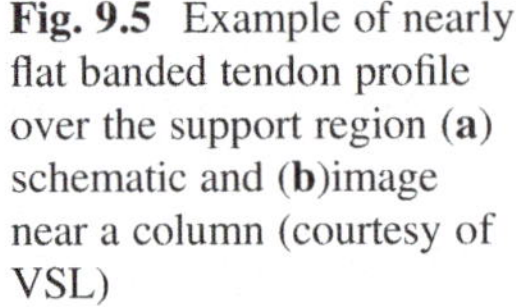

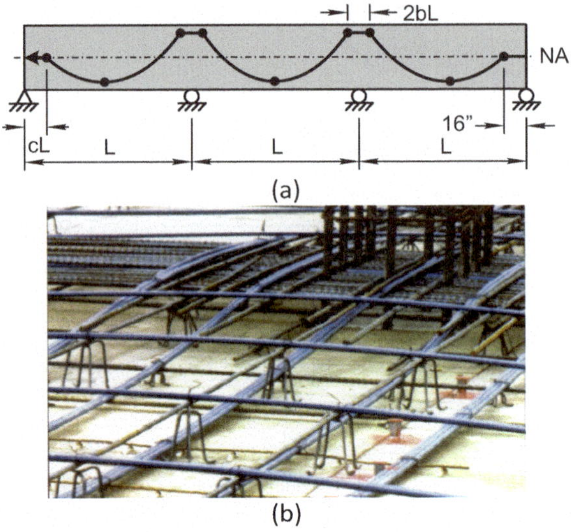

Fig. 9.5 Example of nearly flat banded tendon profile over the support region (**a**) schematic and (**b**)image near a column (courtesy of VSL)

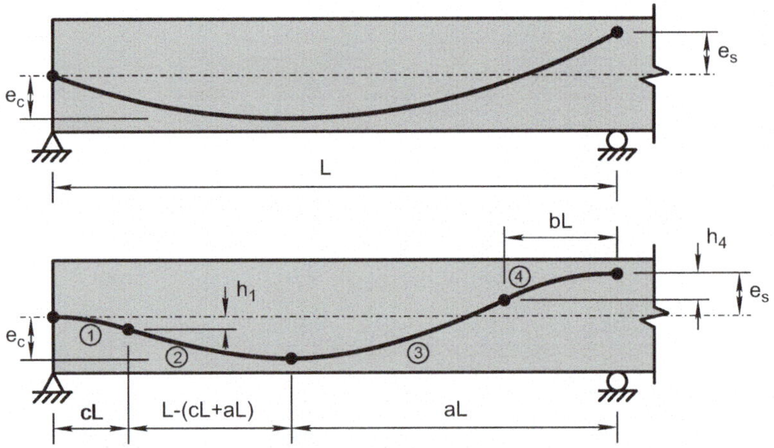

Fig. 9.6 Comparison of idealized and real tendon profile in end span of multi-span element

used in longer span elements that are deeper and have larger and stiffer multi-strand tendons that more naturally conform to the reverse parabolic shape over the support (Fig. 9.5).

The following derivation results in equations that define the eccentricities of the individual tendons based on the overall layout, Fig. 9.6. In addition, the minimum radius of curvature over the support is calculated to ensure that the actual duct and tendon radius is greater than the minimum. Assume that the slope at the node between segments 2 and 3 is zero and that the slope at the right end of segment 4 is also zero. Based on an assumed set of geometric constants a, b, and c, determine

Table 9.2 Coefficients for a typical tendon profile

a	b	cL
0.4–0.5	0.1	16 in.

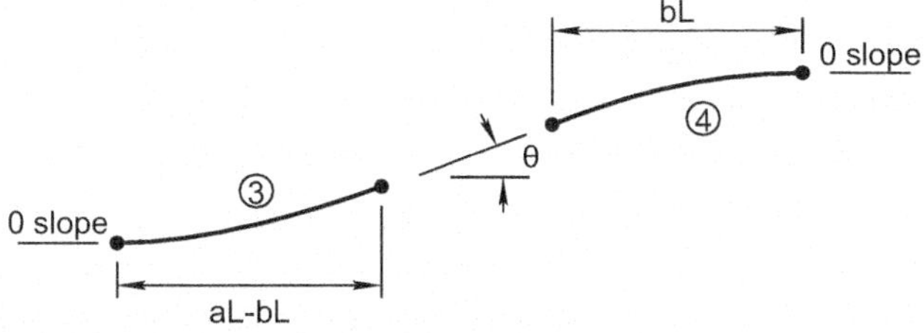

Fig. 9.7 Inflection point between tendon segments 3 and 4

the elevation (h_3) of the node between segments 3 and 4. First, write equations for the slope at the ends of segment 3 (Table 9.2):

$$\theta_3 = \frac{2(e_c + e_s - h_3)}{aL - bL} \tag{9.1}$$

and segment 4:

$$\theta_4 = \frac{2h_3}{bL} \tag{9.2}$$

The slope at the end of each segment is equal (Fig. 9.7):

$$\theta_3 = \theta_4 \tag{9.3}$$

$$\frac{2h_4}{bl} = \frac{2(e_c + e_s - h_4)}{aL - bL} \tag{9.4}$$

Solve for h_4.

$$h_4 = {}^{b}\!/_{a}(e_c + e_s) \tag{9.5}$$

Likewise, for h_1

$$h_1 = \left(\frac{c}{1 - a}\right)e_c \tag{9.6}$$

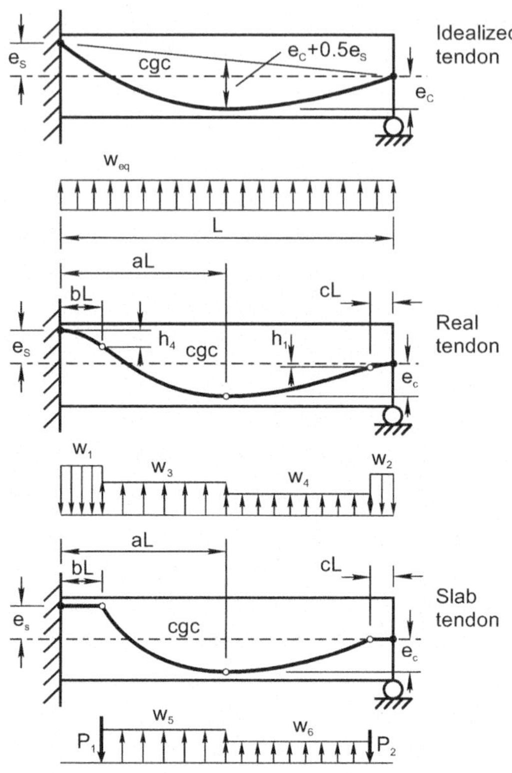

Fig. 9.8 Tendon configurations and equivalent loads

Example 9.1: Comparison of PT Tendon Equivalent Loads

Given the three tendon configurations shown in Fig. 9.1, determine the equivalent loading intensity and distribution applied by each tendon. Use $l = 30$ ft, $a = 0.5$, $b = 0.1$, and $c = 0.05$. Effective prestressing force is $P_e = 270$ kip and eccentricity $e_c = e_s = 12$ in. (Fig. 9.8).

Solution: Calculate the equivalent distributed and concentrated loads away from the supports caused by prestressing. Equivalent load for the idealized tendon profile is

$$w_{eq} = \frac{8P_e(e_c + 0.5e_s)}{L^2} = \frac{8 \times 270 \times (12 + 0.5 \times 12)}{(30)^2 \times 12} = 3.6\,\text{kip/ft}$$

Equivalent loads for the realistic tendon profile must be calculated and applied separately for each tendon segment. The location of both inflection points must be determined to complete the geometry of the parabolic segments. Using Eq. 9.6, PT segment 1 inflection point is located at

$$h_1 = \left(\frac{c}{1-a}\right)e_c = \left(\frac{0.05}{1-0.5}\right)12 = 1.2\,\text{in.}$$

In a beam of this size, a change in tendon elevation of 1.2 in. is not a large portion of the eccentricity. Using Eq. 9.5, PT segment 4 inflection point is located at

$$h_4 = {}^{b}/_{a}(e_c + e_s) = {}^{0.1}/_{0.5}(12 + 12) = 4.8\,\text{in.}$$

The equivalent loads using equations from Chap. 1 are

$$w_1 = \frac{8P_e h_4}{(2bL)^2} = \frac{8 \times 270 \times 4.8}{(2 \times 0.1 \times 30)^2 \times 12} = 24.0\,\text{kip/ft}$$

$$w_2 = \frac{8P_e h_1}{(2cL)^2} = \frac{8 \times 270 \times 1.2}{(2 \times 0.05 \times 30)^2 \times 12} = 24.0\,\text{kip/ft}$$

$$w_3 = \frac{8P_e(e_c + e_s - h_4)}{[2L(a-b)]^2} = \frac{8 \times 270 \times (12 + 12 - 4.8)}{[2 \times 30(0.5 - 0.1)]^2 \times 12} = 6.0\,\text{kip/ft}$$

$$w_4 = \frac{8P_e(e_c - h_1)}{[2L(1-a-c)]^2} = \frac{8 \times 270 \times (12 - 1.2)}{[2 \times 30(1 - 0.5 - 0.05)]^2 \times 12} = 2.67\,\text{kip/ft}$$

For the slab tendon, the equivalent loads for the left parabolic segment are

$$w_5 = \frac{8P_e(e_c + e_s)}{[2L(a-b)]^2} = \frac{8 \times 270 \times (12 + 12)}{[2 \times 30(0.5 - 0.1)]^2 \times 12} = 7.5\,\text{kip/ft}$$

$$P_1 = w_1 L(a - b) = 90\,\text{kip}$$

The right parabolic segment equivalent loads are

$$w_6 = \frac{8P_e e_c}{[2L(1-a-c)]^2} = \frac{8 \times 270 \times 12}{[2 \times 30(1 - 0.5 - 0.05)]^2 \times 12} = 2.96\,\text{kip/ft}$$

$$P_2 = w_2 L(1 - a - c) = 40\,\text{kip}$$

9.4 Continuity and Prestressing

In general, **secondary moments** are the result of **restrained deformation** and occur only in statically indeterminate structures. The cause of the deformation is immaterial to the development of secondary moments. Differential shrinkage of concrete, temperature gradients over a cross section, and deformation due to prestressing will all result in secondary reactions and moments in indeterminate structures. For instance, when the topping on a roof element is heated by the sun at a faster rate than the member itself, this differential thermal expansion causes curvature in the

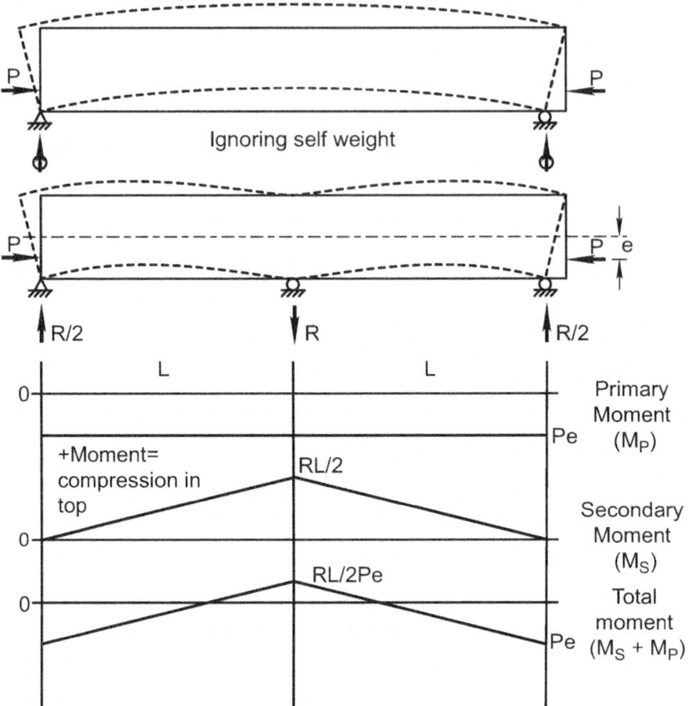

Fig. 9.9 Effect of deformations due to prestressing in statically determinate and statically indeterminate structures

element as discussed in Chap. 8. If the element is restrained or is statically indeterminate, then forces will be generated between the supports and the structure resulting in secondary moments. As secondary moments are presented here, they only occur in statically indeterminate structures and are the result of restrained deformation. Although considered a force, axial prestressing force does not cause reactions vertically because the force is self-equilibrating. Prestressing, however, results in shortening and curvature of the element, Fig. 9.9. If the shortening or deflection is restrained then reactions are generated.

Figure 9.9 illustrates how the support conditions can generate forces when prestressing is applied. In Fig. 9.9a an eccentric prestressing force is applied to a simply supported beam. If self-weight is ignored, and the supports allow axial shortening, then there the reactions are zero. The beam shortens due to the axial load effect and cambers due to the eccentricity of the prestressing. The curvature deformation imposed by the prestressing force causes an upward deflection or camber. The beam carries a constant moment which is a function of the prestress force and eccentricity, which is known as the **primary moment** (M_P). Although external in nature, the prestressing force applied to the beam must necessarily be self-equilibrating so that there is no external reaction generated since it is applied to the beam at the anchorage at both ends with equal and opposite forces.

If a redundant support B is added, however, then the beam is no longer free to camber upward and a downward force is generated at the middle support. Ignoring all superimposed loads and assuming that the spans are equal, reactions at the two outside supports would be required to maintain equilibrium, which must sum to the magnitude of the reaction at the middle support. The moments are generated by these reactions are known as **secondary moments** (M_S). Because secondary moments are generated by reactions, **secondary moment diagrams are always linear between supports**.

In statically indeterminate structures, the primary and secondary moments must exist simultaneously as the **total moment $M_T = (M_P + M_S)$**, since one is the result of the other. We are considering them separately to better understand their effect and to aid in calculating the resulting effect. This same total moment can be calculated by conducting a structural analysis of the structure with the equivalent load from the prestressing applied. The moments that result from this analysis can be combined with the primary moments to determine the secondary moments:

$$M_S = M_T - M_P \tag{9.7}$$

Example 9.2: Determine Secondary Moments
For the beam shown in Fig. 9.10, using equivalent loads, determine the secondary moments. The tendon profile is composed to two parabolically draped tendons

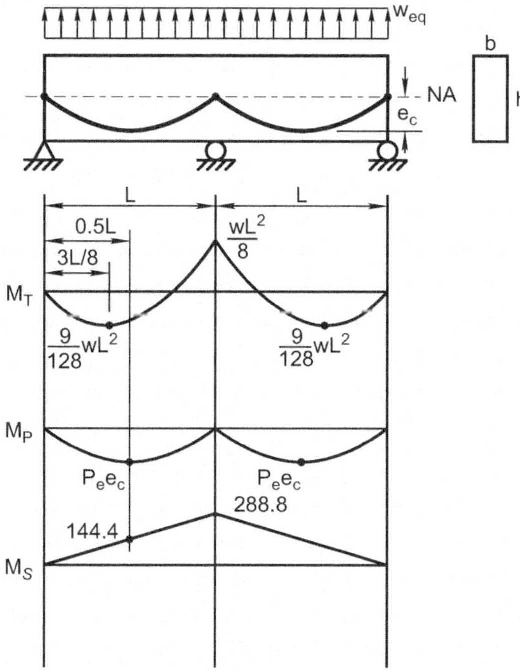

Fig. 9.10 Tendon profile and secondary moments

where the eccentricity over the support is zero. Given $L = 45$ ft, $b = 14$ in., $h = 28$ in., $e_c = 11$ in., $P_e = 315$ kip.

Solution: The equivalent load from the prestressing tendons in

$$w_{eq} = \frac{8P_e e_c}{L^2} = \frac{8 \times 315 \times 11}{(45)^2 \times 12} = 1.141\,\text{kip/ft}$$

The key moments on the total moment diagram are

$$\frac{w_{eq}L^2}{8} = \frac{1.141 \times (45)^2}{8} = 288.8\,\text{kip} \cdot \text{ft}$$

$$\frac{9w_{eq}L^2}{128} = \frac{9 \times 1.141 \times (45)^2}{128} = -162.4\,\text{kip} \cdot \text{ft}$$

The peak primary moment is

$$P_e e_c = 315 \times 11/12 = -288.8\,\text{kip} \cdot \text{ft}$$

At the middle support, the secondary moment is

$$M_S = M_T - M_P = 288.8 - (0) = 288.8\,\text{kip} \cdot \text{ft}$$

At midspan, the secondary moment is the algebraic sum of the midspan. The total moment at midspan is

$$M_T = \frac{w_{eq}L^2}{16} = \frac{1.141 \cdot (45)^2}{16} = -144.4\,\text{kip} \cdot \text{ft}$$

The secondary moment at midspan is then

$$M_S = M_T - M_P = -144.4 - (-288.8) = 144.4\,\text{kip} \cdot \text{ft}$$

which is half of the secondary moment at the support indicating that the secondary moment diagram is linear.

As demonstrated, the use of equivalent loads to analyze the effects of prestressing on an indeterminate member provides a powerful tool. Indeed, equipped with only the depth and span of the member, and without the need for a detailed structural analysis, the engineer can select the prestressing force and profile necessary to **balance** selected loads. The equivalent load is expressed as the ratio or percentage of the self-weight that is balanced. For slabs, it is customary to balance between 60 and 80% of the self-weight. For beams, this is usually increased to between 80 and 110% (Aalami and Jurgens 2003). Self-weight is used because the load is a fixed and known quantity that will be sustained to counteract the prestressing force

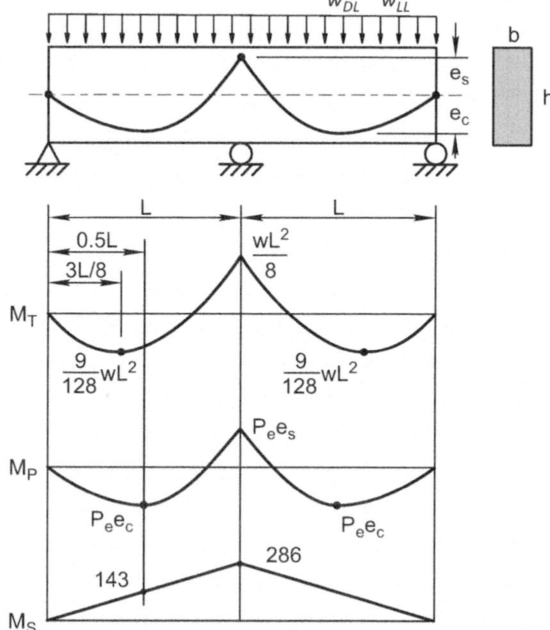

Fig. 9.11 Tendon profile, primary and secondary moments

through the life of the structure. The force used to balance the concrete is typically the effective prestressing force, meaning that the amount of prestressing required is increased to account for partial prestress losses. In general, the live loads specified in building codes are larger than the actual day to day loading. If prestressing is used to balance this portion of the load, then there are times when part of the load is not in place. As such, there is danger that the prestressing could result in excessive camber as well as time-dependent shortening due to axial creep. Axial shortening may lead to restraint cracking at connections between the flexural element and columns or shear walls.

Example 9.3: Load Balancing and Service Stresses

Given the two-span beam shown in Fig. 9.11, determine the percentage of dead load balanced by the prestressing and determine if the beam is Class U, T, or C. Superimposed DL = 1.5 klf, LL = 0.7 klf, $L = 45$ ft, $b = 14$ in., $h = 28$ in., $e_c = e_s = 11$ in., $P_e = 315$ kip, $f'_c = 5000\,\mathrm{psi}$.

Solution: Since this problem is analysis rather than design, the prestress force and eccentricity are fixed. Initially the equivalent load equation can be used to determine the equivalent upward load imposed by the tendon. This is then compared to the magnitude of the applied dead load for a balanced percentage. This member is indeterminate, so a structural analysis is needed to determine the stresses so that the beam can be classified. In practice, a commercial software package would be used to conduct this analysis. For the purposes of a simplified demonstration, indeterminate beam equations are used to determine moments. Because the member is prismatic, there is no direct effect on stresses as a result of cracking since the reduction in stiffness in each span is comparable.

Equivalent load carried by the prestressing force is

$$w_{eq} = \frac{8P_e(e_c + 0.5e_s)}{L^2} = \frac{8 \cdot 315 \cdot (11 + 0.5 \cdot 11)}{(45)^2 \cdot 12} = 1.71\,\text{kip/ft}$$

Self-weight is

$$w_{sw} = 14 \cdot 28 \cdot 0.150/144 = 0.408\,\text{kip/ft}$$

$$\%\text{balanced} = \frac{w_{eq}}{w_{sw} + w_{DL}} = \frac{1.71}{0.408 + 1.5} = 89.6\%$$

When 89.6% of the dead load is applied, this load exactly equilibrates the equivalent load due to prestressing, resulting in a net axial compressive stress over the length of the beam of

$$\frac{P}{A} = \frac{315}{14 \cdot 28} = 803.6\,\text{psi}$$

The remainder of the load will result in flexural stresses that will combine with this net axial stress. Although not explicitly stated in the ACI Building Code, the implication is that the class of the beam should be based on the location that produces the maximum tensile stress under service load conditions, which is at the middle support for this problem. The maximum unbalanced moment at the middle support is

$$\begin{aligned}
M_{unb} &= \frac{(w_{sw} + w_{DL} + w_{LL} - w_{eq})L^2}{8} \\
&= \frac{(0.408 + 1.5 + 0.7 - 1.71) \cdot (45)^2}{8} \\
&= 227.3\,\text{kip} \cdot \text{ft}
\end{aligned}$$

The moment of inertia is

$$\frac{bh^3}{12} = \frac{14 \cdot (28)^3}{12} = 25,610\,\text{in.}^4$$

Net tensile stress on the top of the section over the middle support is

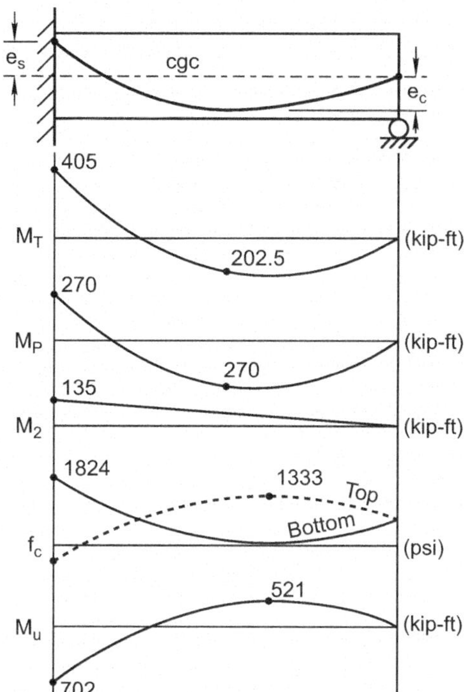

Fig. 9.12 Moment diagrams for idealized tendon in Example 9.4

$$f_t = \frac{P}{A} - \frac{M_{unbal}0.5h}{I} = \frac{315}{14 \cdot 28}1000 - \frac{227.3 \cdot 0.5 \cdot 28}{25,610}12,000 = -687\,\text{psi}$$

Limits for Class U and Class T are

$$7.5\sqrt{f'_c} = 7.5\sqrt{5000} = 530\,\text{psi}$$

$$12\sqrt{f'_c} = 12\sqrt{5000} = 848\,\text{psi}$$

The beam is Class T.

Example 9.4: Effect of Tendon Profile on Analysis

Using the results from Example 9.1 determine the secondary moments caused by the three-tendon profile. Using the service loads given, determine the concrete stresses under the full service load along with its classification (U, T, or C) and the factored moment diagram given $w_{DL} = 2.2$ klf, $w_{LL} = 2.7$ klf, $b_w = 12$ in., $h = 32$ in., $P_e = 270$ kip.

Solution: The solution for the idealized tendon is easily obtained from the above formulas. First, the total moment diagram is calculated based on the application of the equivalent load in the upward direction, Fig. 9.12. Moment diagrams are plotted on the compression face. Stresses are (+) compression and (−) tension. Important points on the total moment diagram are

$$\frac{w_{eq}L^2}{8} = \frac{3.6 \cdot (30)^2}{8} = 405 \, \text{kip} \cdot \text{ft}$$

$$\frac{w_{eq}L^2}{16} = \frac{3.6 \times (30)^2}{16} = -202.5 \, \text{kip} \cdot \text{ft}$$

at the support and midspan. The same locations on the primary moment diagram give these moments

$$P_e e_c = P_e e_s = 270 \times 12/12 = -270 \, \text{kip} \cdot \text{ft}$$

Secondary moment diagram is the total moment minus the primary moment

$$M_2 = M_T - M_P$$

Secondary moments at support and midspan are

$$M_{2,\text{support}} = 405 - 270 = 135$$

$$M_{2,\text{midspan}} = -202.5 - (-270) = 67.5$$

Midspan moment is half of the support moment, confirming that the secondary moment diagram is linear. Stresses can be calculated based on the total moment and the service moment. Calculate the full service moment at the support including self-weight.

$$w_{sw} = \frac{12 \times 32}{144} \times 0.15 = 0.4 \, \text{kip/ft}$$

Total service load is

$$w_{\text{service}} = 0.4 + 2.2 + 2.7 = 5.3$$

Total service moment at support and midspan are

$$\frac{w_{\text{service}}L^2}{8} = \frac{5.3 \times (30)^2}{8} = 596.3 \, \text{kip} \cdot \text{ft}$$

$$\frac{w_{\text{service}}L^2}{16} = \frac{5.3 \times (30)^2}{16} = -298.1 \, \text{kip} \cdot \text{ft}$$

Section properties are

$$A = 12 \cdot 32 = 384 \, \text{in.}^2$$

$$S_t = S_b = \frac{12 \times 32^2}{6} = 2048 \, \text{in.}^3$$

Concrete stresses at support are

$$f_{\text{top}} = \frac{P_e}{A} - \frac{M_{\text{service}} - M_{\text{bal}}}{S_t}$$

$$= \frac{270}{384} 1000 - \frac{596.3 - 405}{2048} 12,000$$

$$= -418 \, \text{psi}$$

Tensile stress limit for Class U is

$$7.5\sqrt{f'_c} = 7.5\sqrt{5000} = 530 \, \text{psi}$$

and for Class T is

$$12\sqrt{f'_c} = 12\sqrt{5000} = 848 \, \text{psi}$$

Since the maximum tensile stress on the top of the section at the support is less than 530 psi, the beam is Class U.

A full analysis of this indeterminate structure including the incorporation of secondary moments for service and strength design checks is completed using equivalent load procedures and common beam formulas.

Now examine the effect of trying to simulate an actual and more physically natural tendon profile using equivalent load method, Fig. 9.13. One caveat is that the loading becomes more complex and is not likely to be found in beam formula reference books. Consequently, a structural analysis program is used to conduct the analysis to obtain the total moment diagram. Using the equivalent loading diagram shown in Example 9.1 for the real tendon profile, the total moment (M_T) diagram was output and is plotted in the figure below. The moments at the support and midspan are included. The moment diagram assumes the shape of the tendon profile. The primary moment (M_P) diagram, which can be easily generated in Excel or Mathcad, is also shown. The secondary moment (M_S) diagram is then the algebraic difference between the total and primary moment diagrams. As is expected, the diagram is linear, even given the number of reversed parabolic segments of the two moment diagrams.

Concrete stresses at the fixed support are

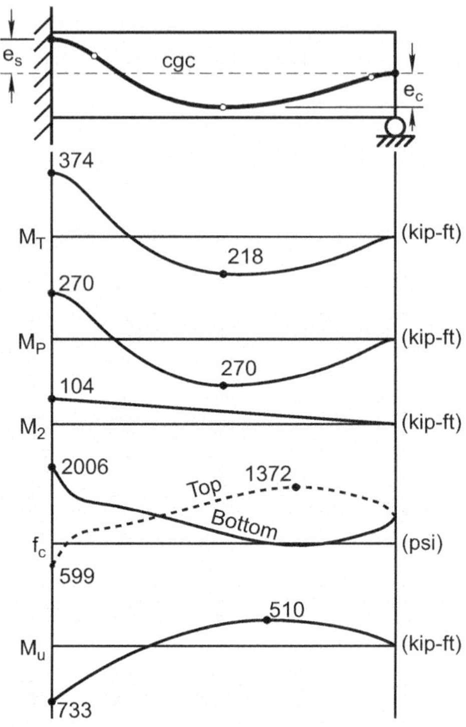

Fig. 9.13 Moment diagrams for realistic tendon in Example 9.4

$$f_{\text{top}} = \frac{P_e}{A} - \frac{M_{\text{service}} - M_{\text{bal}}}{S_t} = \frac{270}{384}1000 - \frac{596.3 - 374}{2048}12{,}000 = -599\,\text{psi}$$

$$f_{\text{bottom}} = \frac{P_e}{A} + \frac{M_{\text{service}} - M_{\text{bal}}}{S_t} = \frac{270}{384}1000 + \frac{596.3 - 374}{2048}12{,}000 = 2006\,\text{psi}$$

Because the maximum tensile stress on the top of the section at the support is greater than 530 psi, the beam is Class T. The analysis for concrete stress and factored moments assuming gross section properties is still valid; cracked section analysis must be used for determining deflections. Factored moment is calculated considering secondary moments from the ACI Building Code. Factored moment at the fixed support is

$$M_{u,\text{support}} = 1.0M_2 + 1.2M_{\text{DL}} + 1.6M_{\text{LL}}$$

$$= 1.0M_2 + 1.2\frac{w_{\text{DL}}L^2}{8} + 1.6\frac{w_{\text{LL}}L^2}{8}$$

$$= 1.0(104) + 1.2\frac{-(0.4 + 2.2) \times (30)^2}{8} + 1.6\frac{-2.7 \times (30)^2}{8}$$

$$= -733\,\text{kip} \cdot \text{ft}$$

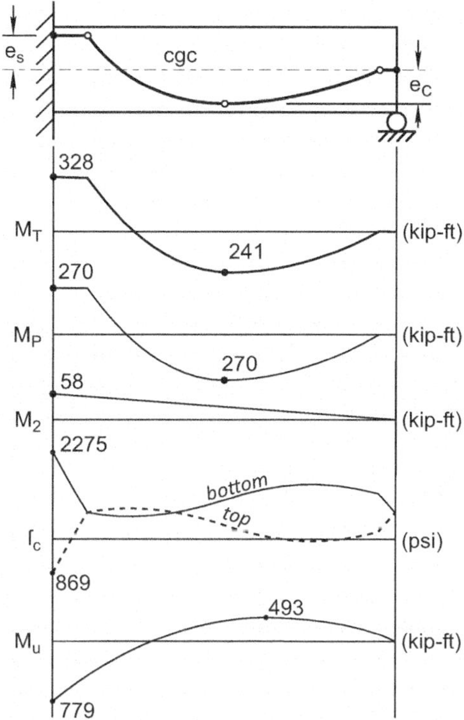

Fig. 9.14 Moment diagrams for slab tendon in Example 9.4

In the final portion of this example, the effect of using a flattened portion of tendon profile near the support is examined to avoid or reduce reinforcement congestion, Fig. 9.14. This type of layout is typically used in thin slab structures where insufficient vertical space is available around the columns to maintain a short-radius parabolic shape. Similar to the real tendon analysis, the equivalent loads from the slab tendon profile were analyzed using structural analysis software to determine the total moment diagram (M_T) as shown in the figure. Again, the moment diagram due to the equivalent load follows the unique shape of the tendon profile. The primary and secondary moments are determined as previously discussed.

Concrete stresses at the fixed support are

$$f_{top} = \frac{P_e}{A} - \frac{M_{service} - M_{bal}}{S_t} = \frac{270}{384}1000 - \frac{596.3 - 328}{2048}12,000 = -869\,\text{psi}$$

$$f_{bottom} = \frac{P_e}{A} + \frac{M_{service} - M_{bal}}{S_t} = \frac{270}{384}1000 + \frac{596.3 - 328}{2048}12,000 = 2275\,\text{psi}$$

Since the maximum tensile stress on the top of the section at the support is greater than 848 psi, the beam should be classified as Class C. Considering that the stress in the beam beyond the classification limit is small, selection of the beam as Class T would be reasonable. Factored moment is calculated considering secondary moments from the ACI Building Code. Factored moment at the fixed support is

$$M_{u,\,\text{support}} = 1.0M_2 + 1.2M_{\text{DL}} + 1.6M_{\text{LL}}$$

$$= 1.0M_2 + 1.2\frac{w_{\text{DL}}L^2}{8} + 1.6\frac{w_{\text{LL}}L^2}{8}$$

$$= 1.0(58) + 1.2\frac{-(0.4+2.2)\times(30)^2}{8} + 1.6\frac{-2.7\times(30)^2}{8}$$

$$= -895\,\text{kip}\cdot\text{ft}$$

Modern structural analysis software capable of analyzing continuous post-tensioned structures typically is capable of accommodating fairly complex tendon profiles geometries including those given in this example. Consequently, for most production work, the engineer conducts analyses using the actual tendon profile.

Several key fundamentals have been highlighted in these examples. First, no matter how complex the tendon geometry, the secondary moment diagrams will be linear because they are caused by the restraining forces generated at the supports.

Second, the simplified procedures presented here provide a way to ensure that the engineer can adequately interpret the output of the software for the larger and more complex analyses that are conducted in practice. Indeed, when comparing the results that would be used in design such as the tensile stress and factored moment, there are small differences. For instance, the maximum concrete compressive stress in the real tendon is only about 10% greater than that of the idealized tendon. There is even less difference (<5%) in the maximum factored moment. Concrete tensile stress, however, is over 40% greater in the real tendon than that of the idealized tendon. The moment gradient in the real tendon is quite steep, which leaves very little length of the beam over the cracking stress.

Example 9.5: Factored Moment

For the two-span beam from the previous problem determine the maximum positive and negative factored moments. Consider pattern loading. Use the following secondary moment diagram, Fig. 9.15.

Solution: Because there are no columns connected to the member and because cracking affects both spans similarly, the factored moments can be determined from the elastic beam formulas used for the previous example. Positive moments cause compression in the top of the section.

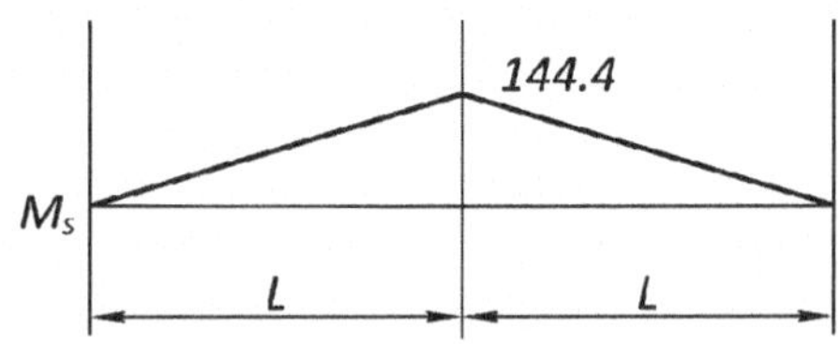

Fig. 9.15 Secondary moment diagram

Is pattern loading required in the analysis?

$$0.75 w_{\mathrm{DL}} = 0.75 \times (0.408 + 1.5) = 1.43 \, \mathrm{kip/ft}$$

which is greater than the live load. Factored moments, therefore, can be determined using the full live load on all spans. Service dead and live load moments at the support are:

$$M_{\mathrm{DL}} = \frac{(w_{\mathrm{sw}} + w_{\mathrm{DL}})L^2}{8} = \frac{(0.408 + 1.5)(45)^2}{8} = 483.0 \, \mathrm{kip \cdot ft}$$

$$M_{\mathrm{LL}} = \frac{w_{\mathrm{LL}}L^2}{8} = \frac{0.7(45)^2}{8} = 177.2 \, \mathrm{kip \cdot ft}$$

and at midspan are:

$$M_{\mathrm{DL}} = \frac{(w_{\mathrm{sw}} + w_{\mathrm{DL}})L^2}{16} = \frac{(0.408 + 1.5)(45)^2}{16} = 241.5 \, \mathrm{kip \cdot ft}$$

$$M_{\mathrm{LL}} = \frac{w_{\mathrm{LL}}L^2}{16} = \frac{0.7(45)^2}{16} = 88.6 \, \mathrm{kip \cdot ft}$$

and at $3L/8$ from end support are:

$$M_{\mathrm{DL}} = \frac{9(w_{\mathrm{sw}} + w_{\mathrm{DL}})L^2}{128} = \frac{9 \times (0.408 + 1.5) \times (45)^2}{128} = 271.7 \, \mathrm{kip \cdot ft}$$

$$M_{\mathrm{LL}} = \frac{9 w_{\mathrm{LL}}L^2}{128} = \frac{9 \times 0.7 \times (45)^2}{128} = 99.7 \, \mathrm{kip \cdot ft}$$

Maximum factored moment at support is

$$M_{\mathrm{u}} = 1.2 M_{\mathrm{DL}} + 1.6 M_{\mathrm{LL}} + 1.0 M_2$$
$$= 1.2(-483.0\,) + 1.6(-177.2) + 1.0(144.4)$$
$$= -718.7 \, \mathrm{kip \cdot ft}$$

Maximum factored moment at midspan is

$$M_{\mathrm{u}} = 1.2 M_{\mathrm{DL}} + 1.6 M_{\mathrm{LL}} + 1.0 M_2$$
$$= 1.2(241.5) + 1.6(88.6) + 1.0(0.5 \times 144.4)$$
$$= 503.8 \, \mathrm{kip \cdot ft}$$

Maximum factored moment at $3L/8$ from end support is

$$M_{\mathrm{u}} = 1.2M_{\mathrm{DL}} + 1.6M_{\mathrm{LL}} + 1.0M_2$$

$$= 1.2(271.7) + 1.6(99.7) + 1.0(3/8 \times 144.4)$$

$$= 539.7\,\mathrm{kip} \cdot \mathrm{ft}\text{—}\textbf{controls positive moment}$$

The moment at $3L/8$ from end support controls the positive flexural strength requirement. To complete the design check, the design flexural strength (ϕM_{n}) at each of the sections should be calculated as described in Chap. 5 and compared to the factored moments calculated here.

Very little testing has been done on measuring secondary stresses and as a consequence the application of the secondary moments to the overall factored moment still raises questions. The ACI Building Code requires that they be considered in the factored moments with load factor of 1.0, whether they increase the overall design moment. While it is certainly prudent to incorporate the secondary moments where they increase the design moments, one could make a case to ignore their effect when they reduce the overall design moments.

9.5 Moment Redistribution

One other advantage that continuous structures offer is the improved strength that redundancy provides. In a statically determine beam, for instance, when the flexural strength of the member is reached at midspan, a plastic hinge forms, which results in instability and the formation of a collapse mechanism. When the structure is statically indeterminate, the formation of a single plastic hinge does not result in a collapse mechanism. Consequently, load must be added until there are a sufficient number of hinges to form a collapse mechanism. The use of continuous systems, therefore, offers an added level of safety since the member is designed using elastic methods.

While the added safety is a benefit, the ACI Building Code allows the engineer to take economic advantage of this feature by allowing limited redistribution of moment in redundant systems that have sufficient ductility to accommodate the hinge. Figure 9.16 illustrates the basis for allowing redistribution of moment when considering nominal flexural strength of an indeterminate system. The beam in the figure would typically be analyzed and designed based on the elastic moments shown in the figure. If design strengths at midspan and supports were exactly equal to the elastic moment diagram, then all three hinges would form simultaneously and result in a collapse mechanism. In practical terms, though, the moment strength at a section is based on several different loading cases and depend on the live load pattern used. Consequently, the negative moment region may have a higher strength than is needed for the elastic moment. If so, then the midspan would yield first forming a hinge there. As the load increased, the moment at the supports would

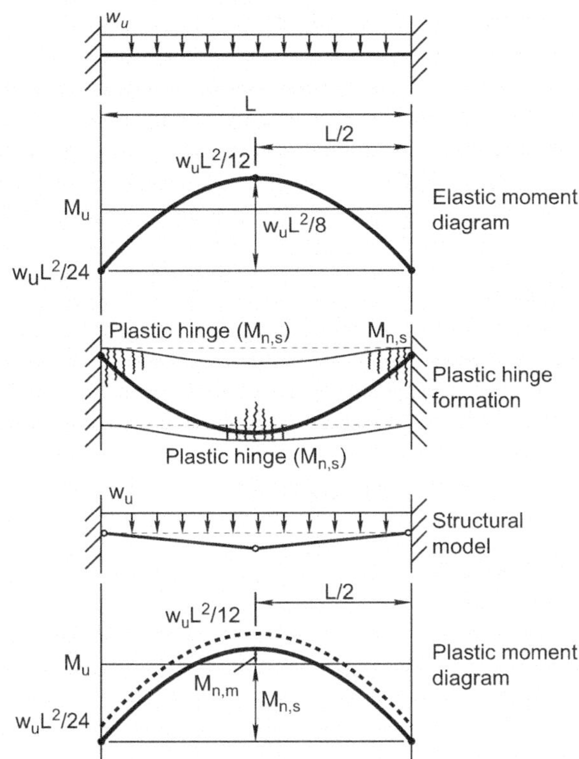

Fig. 9.16 Moment redistribution of schematic secondary moment diagrams for indeterminate beams with varying tendon configurations

increase until hinges formed at the supports. Assuming the section has adequate ductility, up to 20% redistribution of moment is allowed, which provides a method for strategic distribution of the reinforcement among the peak moment locations. Before the designer can take advantage of redistribution, however, the ACI Building Code requires that the following conditions and restrictions be satisfied:

1. Redistribution is not permitted in members that were designed using either simplified design methods or inelastic second-order analysis.
2. Members must be continuous.
3. $\varepsilon_t \geq 0.0075$ at the section where the moment is reduced.
4. For prestressed members, secondary moments must be considered.
5. Redistribution may not exceed the lesser of $1000\varepsilon_t$ percent or 20%.
6. Static equilibrium must be maintained in the span where the moment is reduced. Calculate shear and support reactions based on the static equilibrium condition.

Bondy (2003) provides a detailed discussion along with examples.

9.6 Design Approach

Table 9.3 shows suggested span/depth ratios that can be used as an initial starting point for design. Average precompression

The examples shown in this book are necessarily limited in scope to ensure that the fundamental principles are not obscured by too much detail. In practice, the engineer is required to conduct multiple analyses to account for the construction stages and the timing of post-tensioning operation relative to the construction of the building. For instance, stresses due to shoring removal, topping placement, or other temporary construction loads may have to be checked.

Example 9.6: One-Way Slab Design
Given the structural layout shown in Fig. 9.17, design the one-way slab and the interior one-way supporting girder. Assume residential occupancy.
  Solution:
  Supporting information:
Loads:
Live load = 40 psf. Assume all private rooms.
Partition allowance = 15 psf.
Superimposed dead load = 15 psf
Materials:
Concrete: $f'_c = 6000\,\text{ksi}$
Prestressing steel: ASTM A416 ½ in. unbonded single-strand tendons Grade 270 $f_{pu} = 270\,\text{ksi}$. Use an estimated effective prestress after all losses of $f_{se} = 175\,\text{ksi}$, which is an effective prestress force of

$$P_{se} = f_{se}A_{ps} = 175 \times 0.153 = 26.8\,\text{kip}$$

Mild steel: ASTM A615 Grade 60 $f_y = 60\,\text{ksi}$
Fire rating: Only structural considerations are addressed in this problem. Slab thickness and clear cover of tendons and reinforcement should be checked to ensure compliance with applicable building code fire protection requirements.

Table 9.3 Span/depth ratio from post-tensioning manual (from PTI TAB.1-06 2006)

Floor system	Span/depth ratio
One-way slabs	48
Two-way slabs	45
Two-way slab with drop panel (minimum drop panel at least $L/6$ each way)	50
Two-way slab with two-way beams	55
Two-way waffle slab (5 ft × 5 ft grid)	35
Beams, $b \sim h/3$	20
Beams, $b \sim 3\,h$	30
One-way joists	40

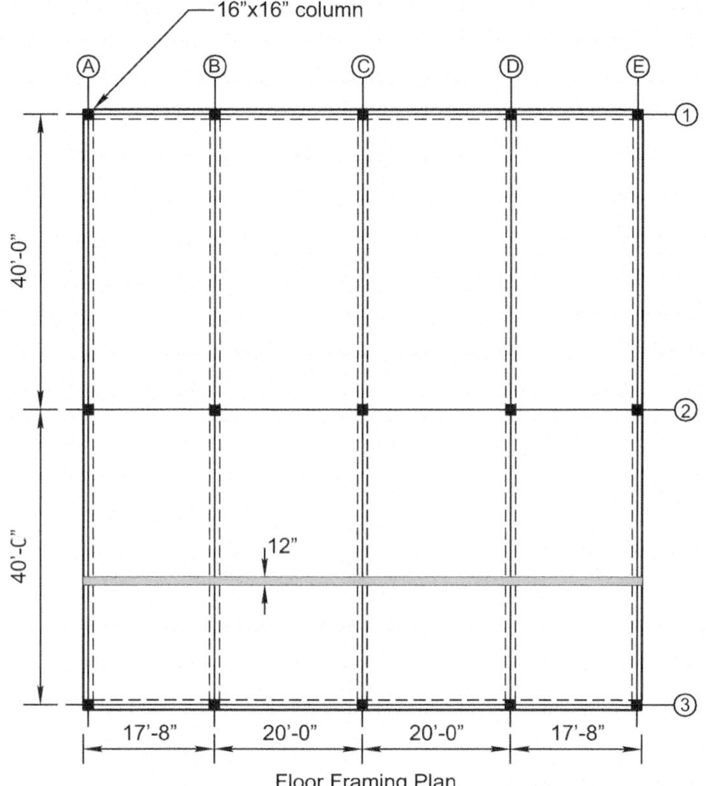

Fig. 9.17 Plan view of building

Determine preliminary slab thickness. Using Table 9.3, the suggested span-to-depth ratio for one-way slabs considering the interior span gives a slab thickness of

$$h_{\text{slab}} = \frac{L}{48} = \frac{12 \times 20}{48} = 5\,\text{in.}$$

Live load is

$$w_{\text{LL}} = 15 + 40 = 55\,\text{psf}$$

Dead load is self-weight and superimposed dead load

$$w_{\text{DL}} = \frac{5}{12} \times 150 + 15 = 62.5 + 15 = 77.5\ \text{psf}$$

Although ACI318-14 does not explicitly require a minimum precompression for one-way slabs, use Section 8.2.3 of ACI 318-14 provisions for two-way slabs. Tendon spacing necessary to provide this minimum required compression of 125 psi is

$$s_{\max} = \frac{P_e}{125h} = \frac{26.8 \times 1000}{125 \times 5} = 42.9\,\text{in.}$$

Use load balancing to layout the tendon profile and determine required spacing. Balance 70% of the self-weight gives a required equivalent load of

$$w_b = 0.70 \times 62.5 = 43.75\,\text{psf}$$

Tendon eccentricity is based on the minimum clear cover requirements from Table 20.6.1.3.2 of ACI 318-14. Maximum eccentricity both above and below the NA is

$$e = \frac{5}{2} - 0.75 - 0.25 = 1.5\,\text{in.}$$

The required prestress force per foot of slab for the exterior span is

$$P_e = \frac{w_b L^2}{8e_{\text{ext}}} = \frac{43.75 \times 17^2}{8 \times 2.25}\frac{12}{1000} = 8.43\,\text{kip/ft}$$

and for the interior span is

$$P_e = \frac{w_b L^2}{8e_{\text{int}}} = \frac{43.75 \times 20^2}{8 \times 3}\frac{12}{1000} = 8.75\,\text{kip/ft}$$

Interior spans have slightly more prestress required and will govern the tendon spacing. Place the tendons at

$$s_{\text{tendon}} = \frac{P_e}{p_e} = \frac{26.8 \times 12}{8.75} = 36.7\,\text{in.}$$

which is less than $s_{\max}$. If desired, the eccentricity in the exterior span could be reduced slightly to reduce the balance load so that it is equivalent to the balanced load for the interior spans. Use an even spacing of 36 in., which gives a precompression of

$$p_e = \frac{P_e}{s_{\text{tendon}}} = \frac{26.8}{3} = 8.93\,\text{kip/ft}$$

$$f_{pe} = \frac{p_e}{h} = \frac{8.93 \times 1000}{5 \times 12} = 149\,\text{psi}$$

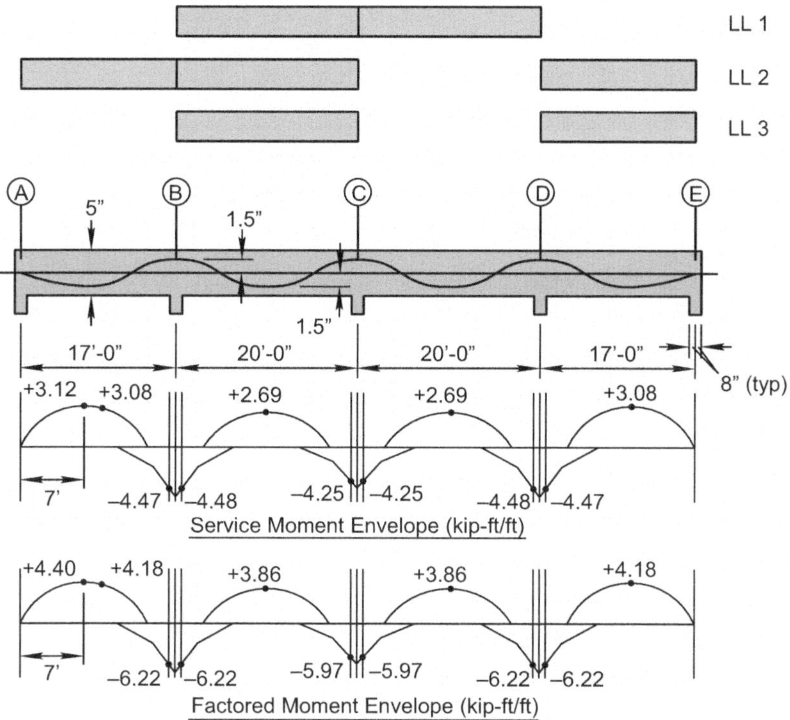

Fig. 9.18 Loading and moment diagrams

The tendon layout is shown in Fig. 9.18. In this problem, a realistic tendon layout will be used to conduct the structural analysis. Conduct structural analysis of the system for both service and strength conditions using the pattern live loading shown in the figure. LL1 produces maximum negative moment at support C. LL2 along with its mirror produces the maximum negative moment at support B and D. LL3 along with its mirror produces maximum positive moment in span BC/CD and AB/DE, respectively. The supporting beam's torsional stiffness was ignored rendering roller supports. The slab was analyzed using the center-to-center spans. Moments and shears were extracted from the analysis at the face of the supports.

Total moments were calculated using a realistic tendon layout as shown in the figure below. The tendon was subdivided into segments with each segment imposing a unique equivalent load over its length. The loads used in the analysis are

$$w_{\text{eq.1}} = \frac{2p_{\text{e}}e_{\text{c}}}{[(1-\alpha)L]^2} = \frac{2 \times 8.93 \times 1.50 \times 1000}{[(1-0.5)17]^2 \cdot 12} = 30.9\,\text{plf}$$

$$w_{\text{eq.2}} = \frac{2p_{\text{e}}(e_{\text{c}}+e_{\text{s}}-h_3)}{[(\alpha-\beta)L]^2} = \frac{2 \times 8.93 \times (1.50+1.50-0.6) \times 1000}{[(0.5-0.1)17]^2 \cdot 12} = 77.2\,\text{plf}$$

$$w_{\text{eq.3}} = \frac{2p_{\text{e}}h_3}{(\beta L)^2} = \frac{2 \times 8.93 \times 0.6 \times 1000}{(0.1 \times 17)^2 \cdot 12} = 309\,\text{plf}$$

$$w_{\text{eq.4}} = \frac{2 \times 8.93 \times (1.50+1.50-0.6) \times 1000}{[(0.5-0.1)20]^2 \cdot 12} = 55.8\,\text{plf}$$

$$w_{\text{eq.5}} = \frac{2 \times 8.93 \times 0.6 \times 1000}{(0.1 \times 20)^2 \cdot 12} = 223\,\text{plf}$$

The total moment diagram from the analysis is plotted in Fig. 9.19 along with the primary moment diagram obtained from the product of the effective prestress force and the tendon eccentricity. These are combined to produce the secondary moment diagram also shown in the plot below. For example, the secondary moment at midspan of AB is

$$M_{\text{S}} = M_{\text{T}} - M_{\text{P}} = -0.872 - (-1.12) = +0.25\,\text{kip} \cdot \text{ft/ft}$$

Net concrete stresses are calculated from the unbalanced moments, which are the full service load moments minus the total moments. Section modulus as a unit width of the slab is

$$S_{\text{slab}} = \frac{bh^2}{6} = \frac{12 \times (5)^2}{6} = 50\,\text{in.}^3/\text{ft}$$

For example, net tension stress at the outside face of the first interior support is

$$f_{\text{top}} = \frac{m_{\text{ext}}}{S_{\text{slab}}} + f_{\text{pe}} = \frac{(-4.47+1.51) \times 12,000}{50} + 149 = -533\,\text{psi}$$

which are located in the top of the slab. For the positive moment in the outside span (AB), the stresses were calculated at the location of maximum moment, which is located at $7.0'$ from A and not at midspan. A summary of stresses are shown in Table 9.4 Beam tensile stresses (psi).

Comparing these tensile stresses to the limit for Class U members of

$$7.5\sqrt{f_{\text{c}}'} = 7.5\sqrt{6000} = 581\,\text{psi}$$

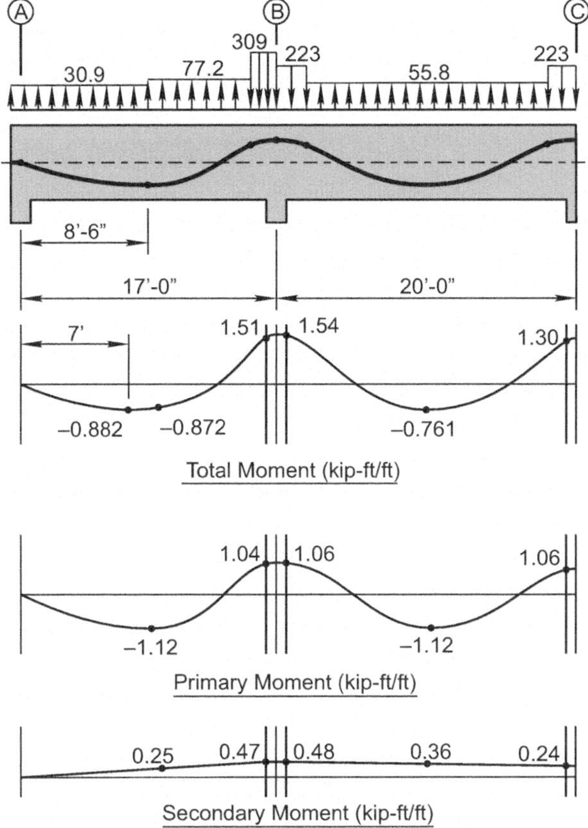

Fig. 9.19 Beam dimensions, loading and moment diagrams

Table 9.4 Beam tensile stresses (psi)

		Support		
Span		Outside	Middle	Inside
Exterior		0	388	561
Interior		557	314	559

makes this a Class U member. Stresses and deflections may then be calculated using gross section properties.

Determine flexural strength and add mild reinforcement, if necessary, to provide adequate strength. Section 7.6.2.3 of ACI 318-14 requires minimum bonded steel reinforcement in slabs with unbonded tendons. Minimum area of bonded reinforcement is

$$A_{s,\min} = 0.004 A_{cs} = 0.004 \times \frac{5}{2} \times 12 = 0.12 \text{ in.}^2/\text{ft}$$

Use #4 bars at a spacing of $12 \times \frac{0.20}{0.12} = 20\,\text{in}$.

Flexural strength is calculated using the empirical formula for f_{ps} as demonstrated in Chap. 5. The span-to-depth ratio for the shorter span is based on the clear span (l_n):

$$\frac{l_n}{h} = \frac{17 - 1.33}{5} \times 12 = 37.6$$

which is greater than 35. Calculate the parameters for the equation to determine the stress in the tendon at nominal strength. Reduced d_p to account for the tendon drop at the face of the support.

$$\rho_p = \frac{A_{ps}}{bd_p} = \frac{0.153}{36 \times 3.9} = 0.00109$$

$$f_{ps} = f_{ps} + 10\,\text{ksi} + \frac{f'_c}{300\rho_p} = 175 + 10 + \frac{6000}{300 \times 0.00109} \times \frac{1}{1000} = 203\,\text{ksi}$$

$$f_{ps} = f_{ps} + 30\,\text{ksi} = 175 + 30 = 205\,\text{ksi}$$

$$f_{ps} = f_{py} = 0.9f_{pu} = 0.9 \times 270 = 243\,\text{ksi}$$

203 ksi controls

Depth of the stress block is

$$a = \frac{f_{ps}A_{ps}}{0.85f'_c b} = \frac{203 \times 0.153}{0.85 \times 6000 \times 36} \times 1000 = 0.169\,\text{in}.$$

$$\varepsilon_t = 0.003\left(\frac{\beta_1 d_p}{a} - 1\right) = 0.003\left(\frac{0.75 \times 3.9}{0.169} - 1\right) = 0.0489$$

which is greater than 0.005, so the phi-factor is set to 0.9. For a 36-in. tendon spacing the unit moment strength is

$$\phi M_n = \phi A_{ps} f_{ps}\left(d_p - \frac{a}{2}\right)$$

$$= 0.9 \times 203 \times 0.153 \times \left(3.9 - \frac{0.169}{2}\right) \times \frac{1}{12 \times 3}$$

$$= 2.96\,\text{kip} \cdot \text{ft/ft}$$

Flexural strength is inadequate. Try minimum reinforcement, No. 4 at 20 in. spacing

$$a = \frac{f_{ps}A_{ps} + A_s f_y}{0.85 f'_c b} = \frac{203 \times 0.153 \times {}^{1}/_{3} + 60 \times 0.20 \times {}^{12}/_{20}}{0.85 \times 6000 \times 12} \times 1000 = 0.287\,\text{in.}$$

$$\varepsilon_t = 0.003 \left(\frac{\beta_1 d_p}{a} - 1 \right) = 0.003 \left(\frac{0.75 \times 3.9}{0.287} - 1 \right)$$
$$= 0.0276$$

Phi-factor is still 0.9

$$\phi M_n = 0.9 \left[203 \times 0.153 \times {}^{1}\!/_{3} \left(3.9 - \frac{0.287}{2} \right) + 60 \times 0.20 \times {}^{12}\!/_{20} \left(4 - \frac{0.287}{2} \right) \right] \times \frac{1}{12}$$
$$= 5.00\,\text{kip} \cdot \text{ft/ft}$$

Does not satisfy negative bending requirements, but does satisfy all positive bending requirements. Try No. 4 at 12 in. spacing

$$a = \frac{203 \times 0.153 \times {}^{1}/_{3} + 60 \times 0.20 \times {}^{12}/_{12}}{0.85 \times 6000 \times 12} \times 1000 = 0.366\,\text{in}$$

$$\varepsilon_t = 0.003 \left(\frac{\beta_1 d_p}{a} - 1 \right) = 0.003 \left(\frac{0.75 \times 3.9}{0.425} - 1 \right) = 0.0210$$

Phi-factor is still 0.9

$$\phi M_n = 0.9 \left[203 \times 0.153 \times \frac{1}{3} \left(3.9 - \frac{0.366}{2} \right) + 60 \times 0.20 \times \frac{12}{14} \left(4 - \frac{0.366}{2} \right) \right] \frac{1}{12}$$
$$= 6.33\,\text{kip} \cdot \text{ft/ft}$$

Ok. No. 4 at 12 in. spacing for negative reinforcement is adequate.

Shear strength does not typically control one-way slab design assuming that typical proportions and material properties are used. Check shear strength at the location of maximum shear based on the shear envelopes developed with the pattern loading analysis, Table 9.5.

Section 7.4.3.2 of ACI 318-14 allows the critical shear to be checked for the shear at $h/2$ from the face of the support as long as the joint is in compression. Because shear is not likely to control and for simplicity check the shear strength at the face of the support and ignore the contribution of prestressing to shear strength by using ACI 318-14 equation 22.5.5.1. Section 22.5.2.1 of ACI 318-14 allows the effective

Table 9.5 Factored shear at face of supports (kip/ft)

Span	Support	
	Outside	Inside
Exterior	1.14	1.86
Interior	1.86	1.80

Table 9.6 Deflections (in.)

Span	Unbalanced dead load		Live load	
	Deflection (in.)	Span ratio	Deflection (in.)	Span ratio
Exterior	0.0067	$\frac{l}{30,500}$	0.026	$\frac{l}{7800}$
Interior	0.0038	$\frac{l}{63,000}$	0.031	$\frac{l}{7700}$

depth used to calculate the shear to be no less than 0.8 $h = 4$ in. The resulting nominal design shear strength is then

$$\phi V_{\mathrm{n}} = \phi 2\sqrt{f_{\mathrm{c}}'}b_{\mathrm{w}}\mathrm{d} = 0.75\left[2\sqrt{6000\,\mathrm{psi}} \times 12 \times 4\right]\frac{1}{1000} = 5.5\,\mathrm{kip/ft}$$

which is well above any of the factored shear values at critical sections.

The primary advantage of using prestressed concrete is to maintain gross section properties under service loads, which reduces deflections considerably. If properly designed and adequately proportioned, rarely will deflections control the design of one-way slabs. Ignoring the torsional stiffness of the supporting beam, the resulting deflections are shown in Table 9.6.

Table 24.2.2 of ACI 318-14 limits live load deflections to $l/360$ for floor systems which is well above the maximum live load deflection for this problem. Long-term deflections are limited to $l/240$. Assuming a long-term multiplier of 3.0, the combined time-dependent and intermittent deflections is

$$0.0067 \times 3.0 + 0.026 = 0.16\,\mathrm{in.}$$

or $l/1270$, which is still well below the allowable.

Secondary reinforcement must be provided perpendicular to the slab span to ensure control of shrinkage and temperature effects and to provide transverse distribution of concentrated loads. Section 24.4.4.1 of ACI 318-14 allows prestressed reinforcement to be used for this reinforcement as long as the average compressive stress is at least 100 psi on the gross concrete area, which includes the beam and slab section halfway to the adjacent beams. If prestressed shrinkage and temperature reinforcement is used, then Section 7.6.4.2 of ACI 318-14 requires that at least one tendon be placed between beams. In addition, Section 7.7.6.3.1 requires that these tendons be spaced at 6 ft. Also, if the tendons are placed farther apart that 4.5 ft, then deformed bars are required at the edge of the slab where these tendons are anchored. The reason for this requirement is that widely spaced tendons result in nonuniform compressive stresses at the slab edge. The deformed bars help control the potential cracking that may occur. First, check the precompression on the gross section for the beam on column line C. The gross section area of beam and slab is

$$A_g = 20 \times 12 \times 5 + 16 \times (36 - 5) = 1696\,\text{in.}^2$$

Using the beam tendon force from the next example problem the precompression is

$$f_c = \frac{374.9 \times 1000}{1696} = 221\,\text{psi}$$

which is greater than 100 psi, and thus maximum tendon spacing controls. The tendons should be spaced at 6 ft and placed no more than 6 ft away from the side face of a beam. Three tendons equally spaced between beams provides a spacing of $20 \times 12/4 = 60\,\text{in.} < 72\,\text{in.}$ which is ok, but would require that deformed bars be placed at the slab edge. One additional tendon in each bay will reduce the spacing to below 4.5 ft and avoid having to place the deformed bars. Provide four tendons in each bay. Shrinkage and temperature tendons should be placed at the mid thickness of the slab to avoid unwanted flexural tensile stresses. ACI Building Code commentary states that the tendon location may be adjusted vertically to assist in supporting primary tendons as long as the tendons remain in the middle third of the slab.

Primary deformed bars can be terminated near the inflections points similar to that of reinforced concrete slabs. Section 9.7.4.4 of ACI 318-14 provides specific required lengths of deformed bars in each region for post-tensioned slabs. In the positive moment region the bar length is

$$\frac{l_n}{3} = \frac{20 \times 12}{3} = 80\,\text{in.}$$

And in the negative moment area is

$$b_w + 2\frac{l_n}{6} = 16 + \frac{20 \times 12}{3} = 96\,\text{in.}$$

The final tendon and reinforcement layout is seen in Fig. 9.20.

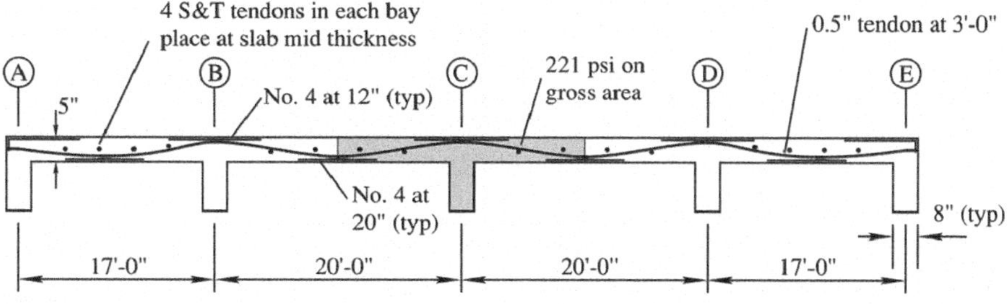

Fig. 9.20 Final tendon layout

Example 9.7: One-Way Beam Design

Using the information given in the previous example design the one-way beam along column line B. Assume a 12-ft story height.

Determine preliminary beam depth assuming a width of 16 in. to match the column. Using Table 9.3, the suggested span-to-depth ratio for beams gives a depth of

$$h_{\text{beam}} = \frac{l}{20} = \frac{12 \times 60}{20} = 36\,\text{in.}$$

Try 36 in. deep beam. Tributary width is 20/2 on one side and 17 on the other. Live load is then

$$w_{\text{LL}} = 55 \times \frac{(20 + 17)}{2} = 1.02\,\text{klf}$$

Dead load is composed of superimposed dead load, slab self-weight, and beam self-weight

$$w_{\text{DL}} = 0.0775 \times 18.5 + 0.15 \times \frac{16}{12} \times \frac{36 - 5}{12} = 1.95\,\text{klf}$$

The effective flange width used to determine section properties is based on ACI Building Code provisions and is the smallest of

$$b_{\text{e}} = b_{\text{w}} + 16h_{\text{f}} = 16 + 16 \times 5 = 96\,\text{in.}$$

$$b_{\text{e}} = s_{\text{beams}} = 18.5 \times 12 = 222\,\text{in.}$$

$$b_{\text{e}} = 0.25L = 0.25 \times 60 \times 12 = 180\,\text{in.}$$

where 96 in. controls. Section properties of the beam are:

$$I_{\text{g}} = 119,750\,\text{in.}^4$$

$$A = 976\,\text{in.}^2$$

$$y_{\text{t}} = 11.65\,\text{in.}$$

$$y_{\text{b}} = 24.35\,\text{in.}$$

Tendon profile is shown in the figure below. Tendon eccentricity is based on a cover to CGS of 4 in. This will be checked at final design or during shop drawing phase. Cover should also allow space for mild steel reinforcement cage including longitudinal bars for supplemental flexural strength and stirrups for shear strength. Eccentricity at midspan is

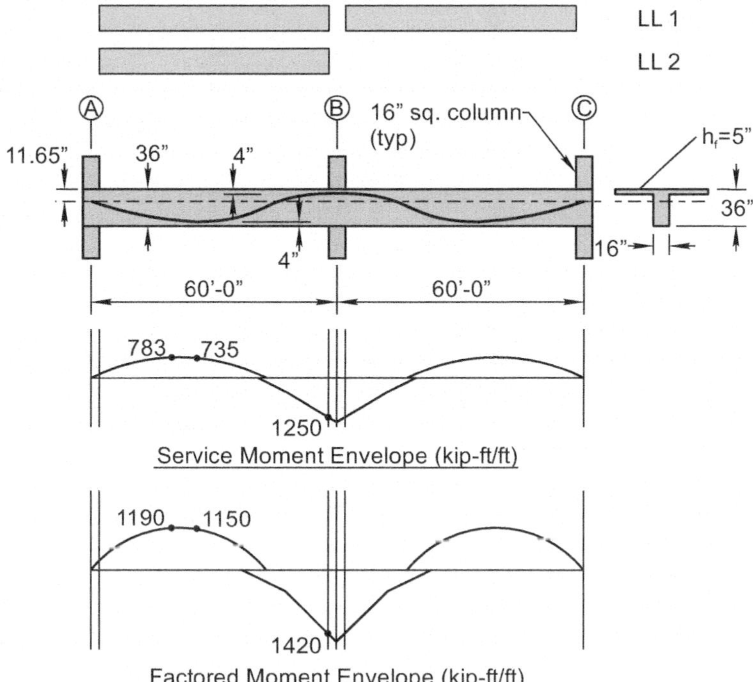

Fig. 9.21 Beam loads and moments

$$e_c = 11.65 - 4 = 7.65\,\text{in.}$$

and eccentricity at the middle support is

$$e_s = 24.35 - 4 = 20.35\,\text{in.}$$

Preliminary design iterations indicated that 14 0.5-in. diameter prestressing strands were necessary to meet serviceability requirements. The effective prestressing force is then

$$P_e = 175 \times 0.153 \times 14 = 374.9\,\text{kip}$$

The balanced load using this tendon force and geometry is

$$w_b = \frac{8P_e e}{L^2} = \frac{8 \times 374.9 \times \left(20.35 + {}^{7.65}\!/_2\right)}{(60)^2 \times 12} = 1.68\,\text{klf}$$

The tendon layout is shown in Fig. 9.21. In this problem, an idealized tendon layout will be used to conduct the structural analysis. Conduct structural analysis of the system for both service and strength conditions using the pattern live loading

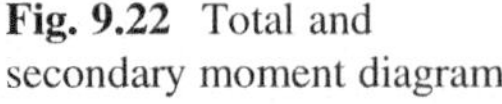

Fig. 9.22 Total and secondary moment diagram

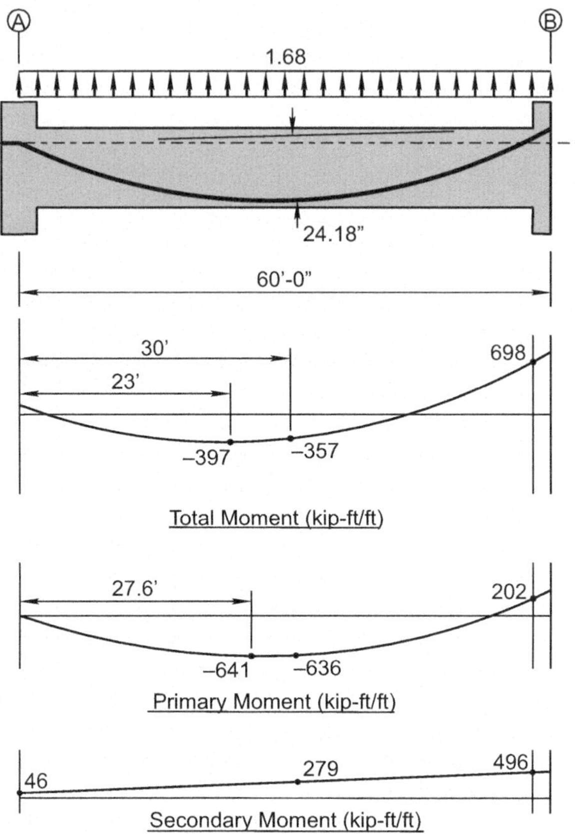

shown in the figure. LL1 produces maximum negative moment at support 2. LL2 produces maximum positive moment in span 1-2. The columns above and below the floor were included in the design. They were assumed to be fixed at the far ends. The beam was analyzed with spans that are center-to-center of columns. Moments and shears were extracted from the analysis at the face of the columns.

Total moments were calculated using an idealized tendon geometry. The tendon was subdivided into segments with each segment imposing a unique equivalent load over its length and are the loads used in the analysis.

The total moment diagram from the analysis is plotted in Fig. 9.22 along with the primary moment diagram obtained from the product of the effective prestress force and the tendon eccentricity. These are combined to produce the secondary moment diagram also shown in the plot below.

Net concrete stresses at the top of the section at the interior column and at the bottom of the section at midspan are calculated from the unbalanced moments, which are the full service load moments minus the total moments.

For this example, the concrete tensile stress is maximum near the midspan. Effective precompression of concrete is

$$f_{pe} = \frac{P_e}{A} = \frac{428.4 \times 1000}{976} = 439\,psi$$

This gives a net tension in the bottom of the section of

$$f_{bottom} = -\frac{(M_{DL+LL} + M_{balanced})y_b}{I_g} + f_{pe}$$

$$= -\frac{(783 - 397) \times 24.35 \times 12,000}{119,750} + 384$$

$$= -558\,psi$$

This stress was located at approximately 25 ft from the support. Tension stress in the top of the section at the middle support is

$$f_{top} = \frac{(M_{DL+LL} + M_{balanced})y_t}{I_g} + f_{pe}$$

$$= \frac{(-1250 + 698) \times 11.65 \times 12,000}{119,750} + 384$$

$$= -261\,psi$$

Compare these tensile stresses to the limit for Class U members of

$$7.5\sqrt{f_c'} = 7.5\sqrt{6000} = 581\,psi$$

makes this a Class U member. Stresses and deflections may then be calculated using gross section properties.

Determine flexural strength and add mild reinforcement, if necessary, to provide adequate strength at the locations of maximum positive and negative factored moment as shown in the figure above. Section 9.6.2.3 of ACI 318-14 requires minimum bonded steel reinforcement in beams with unbonded tendons. Minimum area of bonded top reinforcement is from the area above the neutral axis, including the flange

$$A_{s,min} = 0.004A_{ct} = 0.004 \times [96 \times 5 + 16 \times (11.65 - 5)] = 2.35\,in.^2$$

and area of bottom reinforcement required is the area below the neutral axis

$$A_{s,min} = 0.004A_{ct} = 0.004 \times 16 \times 24.35 = 1.56\,in.^2$$

Check positive reinforcement for flexural strength. The span-to-depth ratio is:

$$\frac{l_n}{h} = \frac{60 - 1.33}{36} \times 12 = 19.6$$

which is less than 35. Calculate the parameters for the equation to determine the stress in the tendon at nominal strength:

$$\rho_p = \frac{A_{ps}}{bd_p} = \frac{14 \times 0.153}{96 \times (36 - 4)} = 0.000697$$

$$f_{ps} = f_{ps} + 10\,\text{ksi} + \frac{f'_c}{100\rho_p} = 175 + 10 + \frac{6000}{100 \times 0.000697} \times \frac{1}{1000} = 271\,\text{ksi}$$

$$f_{ps} = f_{ps} + 60\,\text{ksi} = 175 + 60 = 235\,\text{ksi}$$

$$f_{ps} = f_{py} = 0.9 f_{pu} = 0.9 \times 270 = 243\,\text{ksi}$$

$235\,\text{ksi controls}$

Depth of the stress block is

$$a = \frac{f_{ps}A_{ps} + A_s f_y}{0.85 f'_c b} = \frac{235 \times 14 \times 0.153 + 60 \times 1.56}{0.85 \times 6000 \times 96} \times 1000 = 1.22\,\text{in.}$$

$$\varepsilon_t = 0.003\left(\frac{\beta_1 d_p}{a} - 1\right) = 0.003\left(\frac{0.75 \times 32}{1.22} - 1\right) = 0.0560$$

Phi-factor is 0.9

$$\phi M_n = 0.9\left[235 \times 14 \times 0.153\left(32 - \frac{1.22}{2}\right) + 60 \times 1.56\left(32 - \frac{1.22}{2}\right)\right]\frac{1}{12}$$

$$= 1405\,\text{kip} \cdot \text{ft} \geq M_u = 1190\,\text{kip} \cdot \text{ft}$$

OK. Consider negative moment strength. Recalculate tendon stress with the

$$\rho_p = \frac{A_{ps}}{bd_p} = \frac{14 \times 0.153}{16 \times (36 - 4)} = 0.00418$$

$$f_{ps} = f_{ps} + 10\,\text{ksi} + \frac{f'_c}{100\rho_p} = 175 + 10 + \frac{6000}{100 \times 0.00418} \times \frac{1}{1000} = 199\,\text{ksi}$$

$$f_{ps} = f_{ps} + 60\,\text{ksi} = 175 + 60 = 235\,\text{ksi}$$

$$f_{ps} = f_{py} = 0.9 f_{pu} = 0.9 \times 270 = 243\,\text{ksi}$$

$235\,\text{ksi controls}$

$$a = \frac{f_{ps}A_{ps} + A_s f_y}{0.85 f'_c b} = \frac{235 \times 14 \times 0.153 + 60 \times 2.35}{0.85 \times 6000 \times 16} \times 1000 = 7.90 \, \text{in.}$$

$$\varepsilon_t = 0.003 \left(\frac{\beta_1 d_p}{a} - 1 \right) = 0.003 \left(\frac{0.75 \times 32}{7.90} - 1 \right) = 0.00611$$

Phi-factor is still 0.9

$$\phi M_n = 0.9 \left[235 \times 14 \times 0.153 \left(32 - \frac{7.90}{2} \right) + 60 \times 2.35 \left(32 - \frac{7.90}{2} \right) \right] \frac{1}{12}$$

$$= 1355 \, \text{kip ft} \le M_u = 1420 \, \text{kip} \cdot \text{ft}$$

Does not meet strength requirement. Add mild steel reinforcement to satisfy strength requirements.

From analysis the maximum factored shear occurs at the face of the interior support. Use factored shear at $h/2$ from the support to check the shear design.

$$V_u = 138 \, \text{kip}$$

To demonstrate the shear check, use the simplified equation from the ACI Building Code for concrete contribution to shear.

$$V_c = 2\sqrt{f'_c} b_w d = 2\sqrt{6000} \times 16 \times 32 = 79.3 \, \text{kip}$$

Required contribution of stirrups to the shear strength is then

$$V_{s,\text{reqd}} = \frac{V_u}{\phi} - V_c = \frac{138}{0.75} - 79.3 = 104.7 \, \text{kip}$$

If #3 stirrups are used, then the required spacing is

$$s_{\text{reqd}} = \frac{A_{st} f_y d}{V_{s,\text{reqd}}} = \frac{0.11 \times 2 \times 60 \times 32}{104.7} = 4.03 \, \text{in.}$$

Use 4 in. spacing for stirrups near the support. Refining the calculation of the concrete contribution would increase the required spacing somewhat. Spacing may also be increased away from the support where shear is less. Minimum shear reinforcement and minimum spacing requirements covered in Chap. 8 must also be checked.

Deflections from the analysis are presented here. The gross moment of inertia of the columns in the analysis is multiplied by 70% to account for cracking as recommended in the ACI Building Code. Live load deflections near midspan are

$$\Delta = 0.12\,\text{in.}$$

$$\sim \frac{l}{6000}$$

Table 24.2.2 of ACI 318-14 limits live load deflections to $l/360$ for floor systems, which is well above the maximum live load deflection for this problem.

Long-term deflections are limited to $l/240$. Assuming a long-term multiplier of 3.0, the combined time-dependent and intermittent deflections is

$$\Delta_{\text{LT}} = (0.141 - 0.121) \times 3.0 + 0.12$$

$$= 0.18\,\text{in.}$$

$$\sim \frac{l}{4000}$$

which is still well below the allowable.

One-way slab and beam design problem demonstrates the process used along with the usual results that are obtained from a typical layout. In general, as with these problems, flexural stresses at key points in the exterior span will usually control the design as long as the exterior to interior span ratio is close to 1.0. If not, then critical sections may occur at other locations. Iterations are generally needed by adjusting section geometry, prestressing force, and tendon eccentricity to attain flexural stresses that ensure that the member is classified as Class U. Little iteration should then be required for the remainder of the design.

References

Aalami, B. O., & Jurgens, J. D. (2003). Guidelines for the design of post-tensioned floors. *Concrete International, 25*(3), 77–83.

Bondy, K. B. (2003, Jan). Moment redistribution: Principles and practice using ACI 318-02. *PTI Journal, 1*(1), 3–21.

PTI TAB.1-06. (2006) *Post-tensioning manual* (6th ed., 354 pp). Farmington Hills, MI: Post-Tensioning Institute.

Chapter 10
Composite Beams

10.1 Introduction

Composite flexural members are composed of two separate concrete placements that are connected to behave as a single beam element; the separate elements can be either precast or cast-in-place. The most common form is a precast element topped with cast-in-place concrete and include decks on precast bridge beams, or toppings on precast double T or hollowcore, Fig. 10.1. Prestressed composite girders used for bridge structures allow the girders to be placed and then the deck construction continues with the area below generally open to traffic. Double-T and hollowcore beams include topping for finished floors, for additional strength, or to create a floor diaphragm, Fig. 10.2. Section properties for composite members produced with precast elements are provided in the PCI Design Handbook (2017). Stresses in prestressed girders used in bridges are given in the *AASHTO Bridge Design Specification* (2017).

Composite topping adds to the loading stages for prestressed beams and for the beam to perform satisfactorily, each load stage is checked. The load stages are:

1. Initial prestress P_i plus self-weight of the beam.
2. Effective prestress P_e plus non-composite and composite dead loads.
3. Effective prestress plus all dead load and live load.
4. Factored loads.

Stage 2 requires consideration of the sequence of loading. Loads added prior to the composite topping acquiring the full design strength are applied to the precast section. Loads applied following the hardening of the composite topping act on the composite section. To carry the composite topping on the composite section, temporary intermediate shoring is required to support the added weight until the topping has gained sufficient strength to carry the load. At such a time, removal of the shoring adds the weight of the topping on the composite section.

© Springer Nature Switzerland AG 2019
C. W. Dolan, H. R. Hamilton, *Prestressed Concrete*,
https://doi.org/10.1007/978-3-319-97882-6_10

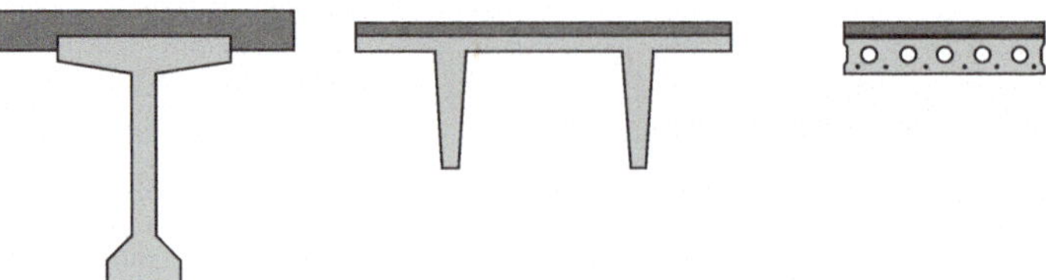

Fig. 10.1 Composite beams

Fig. 10.2 Composite floor
being finished

Stage 3 places the superimposed loads on the composite section. For buildings superimposed dead loads may include flooring, ceiling fixtures, suspended utilities, and partitions. For bridge structures superimposed dead loads may include surface treatments, guard rails, utilities, and sidewalks. Live loads are added last and are considered short duration loadings.

Stresses for load stages 1–3 are calculated based on elastic behavior and require transformation of the elastic properties if the concrete strength of the topping is different than the precast beam. Factored loads are calculated on a strength basis. The strains at factored loads are sufficiently high, that any strain discontinuities at the interface are ignored in the strength calculations. Three facets of composite beam behavior and design are addressed; service level stresses, nominal strength of the composite beam, and interface shear between the topping and the girder.

10.2   Service Level Stresses

Flexural stress distributions at Load Stages 1 and 2 for a simple span beam are illustrated in Fig. 10.3. The beam is first subjected to initial prestress and its self-weight. Over time losses occur. For simplicity, the losses are assumed to occur prior to casting the deck slab. The deck slab is cast without shoring for the precast beam so the weight of the deck slab is carried completely by the precast beam. Service level

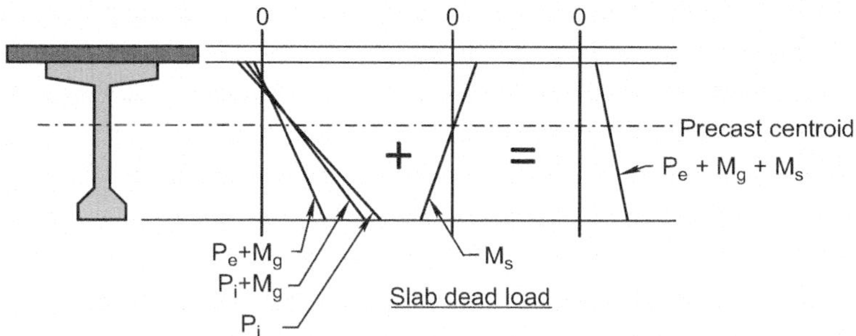

Fig. 10.3 Stresses on precast beam

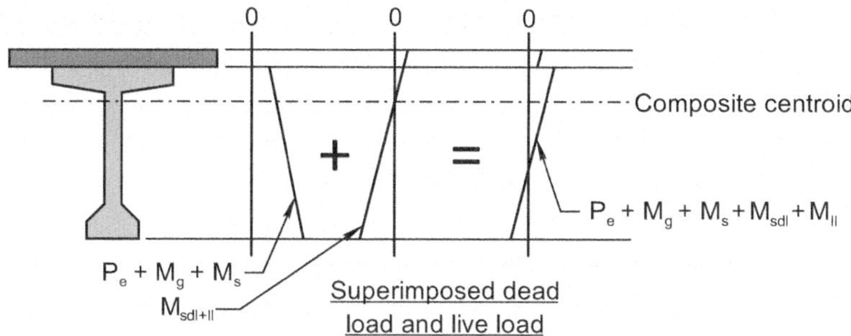

Fig. 10.4 Stresses on composite beam

stresses for these stages are calculated using the procedures discussed in Chap. 5 and the properties of the precast beam.

After the deck slab hardens, several behavior changes occur that affect the stress calculations. The centroid of the composite beam moves upward. Superimposed dead load and live load are carried by the composite section using the composite section properties, Fig. 10.4. Incremental stresses due to the loads on the composite section are added to the stresses already present in the precast beam. The effect of the change in properties is evident in the applied moment stresses in Fig. 10.4. The bending moment results in no change in stress at the new neutral axis and a reduction in stress at the original neutral axis of the precast beam. One additional stress calculation is to check stresses at the composite interface. Figure 10.4 illustrates the possible stress checks at the beam composite interface and the bottom of the composite slab, shown by the dashed line.

The section modulus S_c at the interface between the precast beam and the composite beam is calculated as

$$S_c = \frac{I_c}{y_c} \tag{10.1}$$

where I_c is the composite moment of inertia and y_c is the distance from the neutral axis to the composite interface. Using the section modulus allows calculation of stresses at the composite interface, which is located at the top of the precast beam and the bottom of the deck slab;

$$f_c = \frac{M_{sdl} + M_{ll}}{S_c} \tag{10.2}$$

Equation 10.2 may be separated into multiple steps to evaluate the addition of various elements of superimposed dead load and live loads.

Calculation of the composite beam properties takes into consideration the differences in strength between the precast beam and the composite topping. The differences in strength and stiffness require that an equivalent flange width must be used for the cast-in-place slab or topping. The **equivalent flange width** is based on the difference in material properties between the beam and the cast-in-place concrete and is separate from the effective flange width based on ACI Building Code requirements. From consideration of plane sections remaining plane, the strain at the composite interface must be the same for both the precast beam and the composite topping. Therefore, knowing the strain is the stress divided by the modulus of elasticity:

$$\frac{f_c}{E_c} = \frac{f_{cp}}{E_{cp}} \tag{10.3}$$

where E_c and E_{cp} are the elastic modulus for the precast concrete and the concrete topping, and f_c and f_{cp} are the stresses in the precast beam and composite topping, respectively. Rearranging terms gives

$$f_{cp} = \frac{E_{cp}}{E_c} f_c = n f_c \tag{10.4}$$

where n is the modular ratio. The equivalent section provides the same compression resistance, or

$$dC = f_c b \, dy \; \textbf{Giving } \mathbf{C} = f_{cp} b_{eq} y \tag{10.5}$$

where b is the effective flange width based on Code limits or experimental validation and b_{eq} is the equivalent width, Fig. 10.5. Using nf_c for f_{cp} and cancelling terms gives

$$b_{eq} = \frac{b}{n} \tag{10.6}$$

The equivalent width is used in the calculation of the composite section properties.

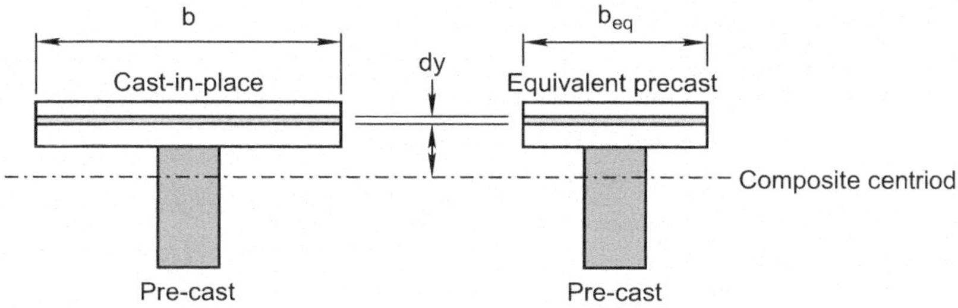

Fig. 10.5 Actual and equivalent composite sections

Table 10.1 Precast and composite beam properties

		Precast		Composite	
A_g	$=$	401	in.^{2}	558	in.^{2}
I_g	$=$	20,985	in.^{4}	27,982	in.^{4}
y_b	$=$	17.15	in.	19.36	in.
y_t	$=$	6.85	in.	6.64	in.
S_b	$=$	1224	in.^{3}	1446	in.^{3}
S_t	$=$	3064	in.^{3}	4212	in.^{3}
S_c	$=$			6026	in.^{3}

Example 10.1: Calculate Composite Stresses

Calculate the stresses in the 8 ft wide 24 in. deep simple span DT beam 48 ft long. The double-T beam concrete strength is 6000 psi and the 2-in. thick topping strength is 4000 psi. The beam is subjected to 10 psf superimposed dead load and 40 psf live load.

Solution: While the ACI Building Code contains provisions for effective flange width, most precast is exempt based on experience. Therefore, only the topping needs adjustment for equivalent width. The modular ratio of the beam to topping is

$$n = \frac{E_{c,\text{top}}}{E_c} = \sqrt{\frac{f'_{c,\text{top}}}{f'_c}} = \sqrt{\frac{4000}{6000}} = 0.82$$

resulting in an equivalent width of 78.4 in. Using the equivalent width, the precast beam and composite beam properties are given in Table 10.1:

A trial design using the PCI Design Handbook selects 12-½ in. diameter strands with a constant eccentricity of 7 in. The initial prestress force is 367 kips and an effective prestress after losses is 312 kips. The service moments in the beam are

M_g	$=$	1437.7	in-kip
M_{slab}	$=$	691.2	in-kip
M_{sdl}	$=$	276.5	in-kip
M_{ll}	$=$	1105.9	in-kip

Table 10.2 Stresses in precast beam

	Initial stresses		Stresses on precast beam Final Stresses			Top	Bottom
	Top	Bottom					
	(psi)	(psi)				(psi)	(psi)
P_i/A_g	916	916	P_e/A_g			778	778
$P_i ey/I_g$	−839	2101	$P_e ey/I_g$			−713	1786
$M_g y/I_g$	469	−1175	$M_g y/I_g$			469	−1175
Sub total	546	1841	Sub total			534	1389
			$M_{slab} y/I_g$			226	−565
			Subtotal			760	824
			$M_l y/I_g$			0	0
			Total			760	824

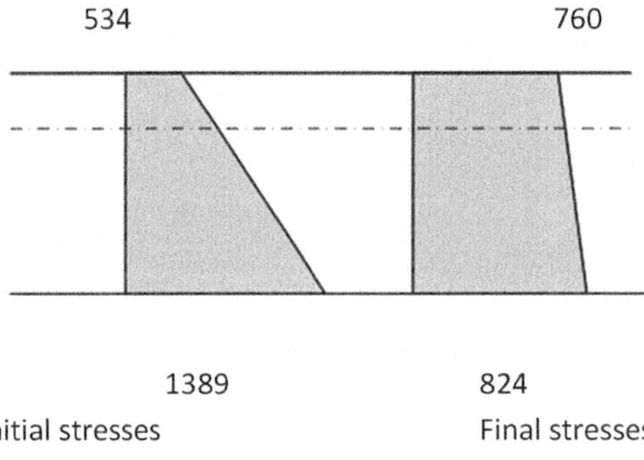

The moments for the girder and the slab are applied to the precast beam resulting in the stresses given in Table 10.2.

The superimposed dead load and the live load are applied to the composite section resulting in the stresses given in Table 10.3. Stresses from the previous loading are reproduced for completeness.

A check establishes that all stresses are within the Code limits.

Table 10.3 Stresses in composite beam

Stresses

Final Stresses on composite
beam

	Top (psi)	Interface (psi)	Bottom (psi)
P_e/A_g		778	778
$P_e ey/I_g$		−713	1786
$M_g y/I_g$		469	−1175
$M_{slab} y/I_g$		226	−565
Subtotal		760	824
$M_{sdl+ll} y/I_g$	328	229	−956
Total	328	990	−132

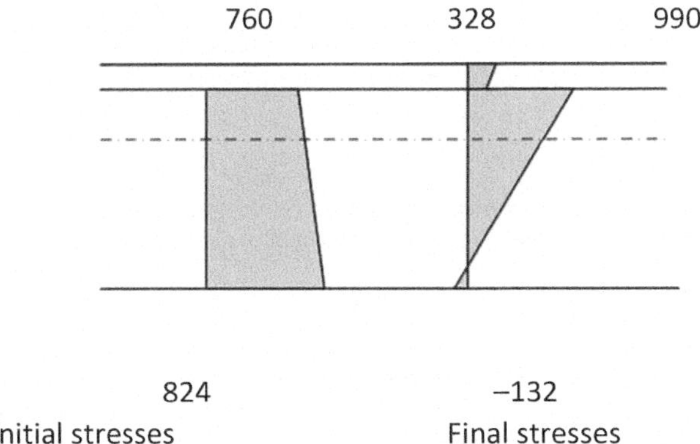

10.3 Nominal Flexural Strength

Nominal flexural strength is dependent on the horizontal shear strength across the
interface between the two components. If the interface is properly prepared, then the
entire cross section is considered effective for calculation of the nominal strength of
the section. These requirements are presented in Sect. 10.4.

At the large compressive strains associated with nominal strength, the difference
in elastic moduli between the precast and cast-in-place concrete are inconsequential
and transformed sections are not required. The full effective flange width of the
flange is used. The difference in the stress-strain behavior, if different concrete

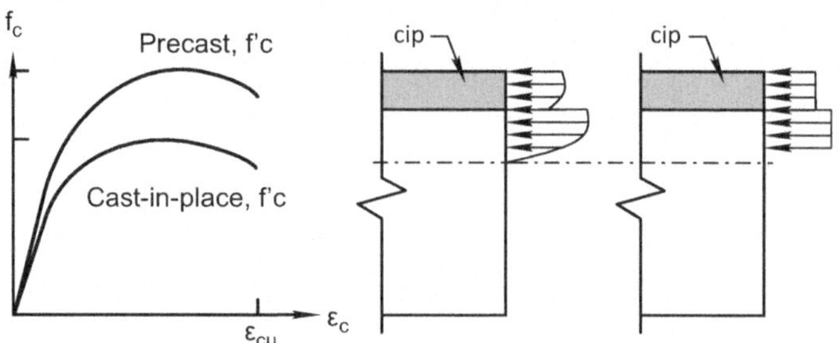

Fig. 10.6 Concrete compressive stress distribution

strengths are present, does create a stress incompatibility at the interface as illustrated in Fig. 10.6.

Accounting for the abrupt stress change complicates the determination of nominal moment strength; however, such a correction is often unnecessary. In most cases, the flange is wide and the compression block shallow, lying completely within the cast-in-place topping, so no change is stress calculation is needed. Second, a conservative value of the nominal moment is possible using the lower topping concrete compression strength, which is allowed in Section 22.3.3.4 of ACI 318-14. Third, tests have shown that the use of the equivalent rectangular stress block is adequate if the rectangular stress block is in the flange even when the neutral axis is below the flange (Dolan 1967).

Example 10.2: Confirm Nominal Flexural Strength

Confirm the nominal flexural strength of the beam in Example 10.1 is adequate. Use the ACI equation (Eq. 5.7) to estimate the nominal prestress using low relaxation strand.

Solution: The eccentricity of the 12 strands is 7 in. from the precast neutral axis. This gives a d_p of the distance to the top of the beam, plus the eccentricity, plus the thickness of the topping or $d_p = 6.85$ in. $+ 7$ in. $+ 2$ in. $= 15.85$ in. For low relaxation strand $\gamma = 0.28$ and for 4000 psi concrete $\beta_1 = 0.85$. Calculation of the prestress reinforcement ratio is based on the full flange width and that the compression block remains in the flange. Thus

$$\rho = \frac{A_p}{b_w d_p} = \frac{12 \cdot 0.153}{96 \cdot 15.85} = 0.00121$$

The nominal stress in the strand is

$$f_{ps} = f_{pu}\left[1 - \frac{\gamma_p}{\beta_1}\left(\rho_p \frac{f_{pu}}{f'_c}\right)\right] = 270\left[1 - \frac{0.28}{0.85}\left(0.00121\frac{270}{4}\right)\right] = 262.8\,\text{ksi}$$

Resulting in a depth of the equivalent rectangular stress block of

$$a = \frac{A_p f_{ps}}{0.85 f'_c b} = \frac{1.836 \cdot 262.8}{0.85 \cdot 4 \cdot 96} = 1.48 \, \text{in.}$$

which is less than the depth of the composite topping. The nominal moment strength is

$$M_n = f_{ps} A_{ps} \left(d_p - \frac{a}{2} \right) = 262.8 \cdot 1.836 \left(15.85 - \frac{1.48}{2} \right) = 7290 \, \text{in-kip}$$

The factored moment on the beam is

$$\begin{aligned}
M_u &= 1.2 \left(M_g + M_{slab} + M_{sdl} \right) + 1.6 \cdot M_{ll} \\
&= 1.2(1438 + 691 + 276) + 1.6 \cdot 1106 \\
&= 5098 \, \text{in-kips}
\end{aligned}$$

Finally, noting that $c/d_p \ll 3/8$, then $\phi = 0.90$ and $\phi M_n = 0.90 \times 7290 = 6561$ in-kips, concluding that the nominal moment strength is adequate.

Comment: In this example the equivalent rectangular stress block lies within the cast-in-place topping therefore there is no concern about mixed concrete strengths. The value for ρ is based on the total flange width of 96 in. The width of the beam stem is used to calculate ρ for shear; however, the full width is needed for flexure. If just the width of the stem was used in these calculations, the calculated nominal moment strength would suggest that the beam has insufficient capacity.

This example uses the actual structural depth for calculation of the nominal moment strength. The ACI Building Code allows the use of the actual depth or 0.8 h for shear calculations but would not be applicable in this situation

10.4 Horizontal Shear

As a beam deforms due to applied load, shear accumulates in both the horizontal and vertical directions. Consider the stresses on a small element in a flexural member, Fig. 10.7. The element size is selected to provide a horizontal cut above the neutral axis of the member. Flexural stresses are applied to both ends of the element as

$$f_A = \frac{M_A y}{I} \tag{10.7}$$

and

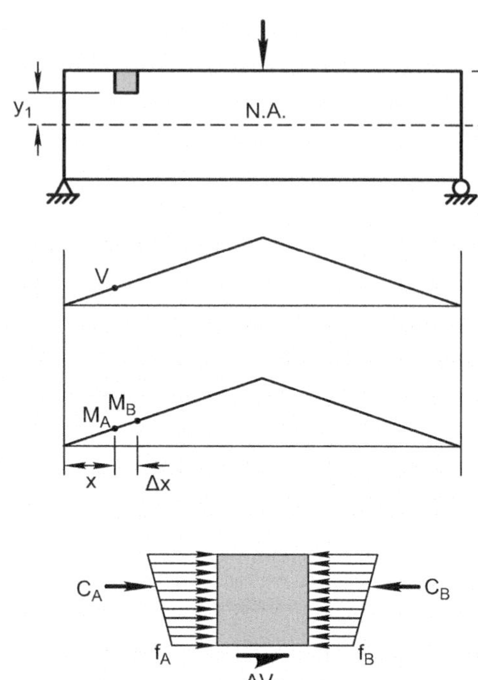

Fig. 10.7 Differential element between segments A and B

$$f_{\mathrm{B}} = \frac{M_{\mathrm{B}}y}{I} \tag{10.8}$$

The resultants of the stresses integrated over the height of the element are slightly different due to the moment gradient. This difference is equilibrated by the horizontal shear force ΔV_{H}

$$\Delta V_{\mathrm{H}} = \int_{y_1}^{c} \frac{M_{\mathrm{B}}y}{I}\,dy - \int_{y_1}^{c} \frac{M_{\mathrm{A}}y}{I}\,dy \tag{10.9}$$

which reduces to

$$\Delta V_{\mathrm{H}} = \frac{M_{\mathrm{B}} - M_{\mathrm{A}}}{I} \int_{y_1}^{c} y\,dy \tag{10.10}$$

The first moment of area of the portion of the section above the cut is defined as

$$Q = \int_{y_1}^{c} y\,dy \tag{10.11}$$

Divide both sides of equation 10.10 by dx to give infinitesimal form of the equation

$$\frac{dV_{\mathrm{H}}}{dx} = \frac{dM}{dx}\frac{Q}{I} \qquad (10.12)$$

where $V = dM/dx$ and shear flow $q = dV_{\mathrm{H}}/dx$

$$q = \frac{VQ}{I} \qquad (10.13)$$

Divide the shear flow by the web width to give horizontal shear stress

$$v_{\mathrm{h}} = \frac{VQ}{Ib_{\mathrm{w}}} \qquad (10.14)$$

where

v_{h} is the horizontal shear stress
V is the vertical shear force
Q is the static moment of area above the section under investigation
I is the moment of inertia of the composite section
b_{w} is the width of the section at the point being considered

The horizontal shear stress v_{h} generates a condition where the cast-in-place topping slips relative to the precast element if the interface bond is not adequate, Fig. 10.8. If the cast-in-place topping slips relative to the web, the load is resisted by two independent elements behaving in non-composite action, Fig. 10.8. Preventing slip requires development of shear along the interface and allows the development of composite action in the two elements.

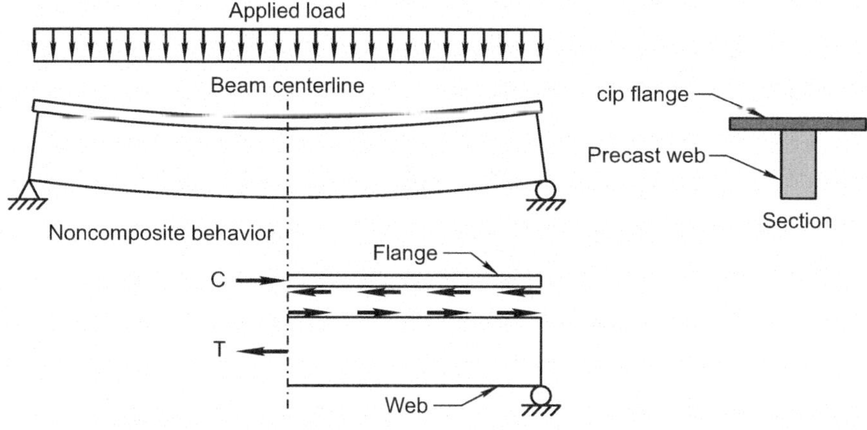

Fig. 10.8 Composite action in a T-beam

Fig. 10.9 Roughening top
surface of precast element to
ensure composite action
when deck is placed

Shear resistance is provided by the adhesion between the precast and cast-in-place concrete surfaces. The amount of resistance depends on the surface preparation (Kaar 1966; Hanson 1961). If the top surface of the precast beam is intentionally roughened, the horizontal shear capacity is greater than if the surface is troweled smooth, Fig. 10.9. Intentional roughening is commonly applied to the top of wide precast members such as hollowcore, double-T, single-T, and box sections. Two types of roughening are typical; broom finish and raked finish, the latter having a much coarser finish. These finishes provide sufficient shear capacity to develop the full nominal moment of the section.

Narrower beams required additional restraint to prevent slip. In these cases, dowel reinforcement is extended through the interface to ensure the interface remains intact. The dowels typically are extensions of the vertical shear reinforcement.

While Eq. 10.14 was used in early composite research, it does not provide an accurate assessment of the shear in a cracked section (Saemann and Washa 1964). A simpler equation for shear developed that correlates equally well with the shear behavior.

$$v_{\mathrm{hu}} = \frac{V_{\mathrm{u}}}{b_{\mathrm{v}} d_{\mathrm{p}}} \qquad (10.15)$$

where d_{p} is the depth to the centroid of the prestress reinforcement and b_{v} is the width of the cross section contact surface in horizontal shear. Section 22.5.8.3 of ACI 318-14 allows d_{p} to be 0.8 h or the centroid depth, whichever is greater, for shear calculations. Discussion of horizontal shear in Chap. 16 of the ACI Building Code address the structural depth d, so is not clear whether the 0.8 h provision applies to composite beams or to horizontal shear calculations. Thus, the equation for nominal shear strength is

Table 10.4 Nominal horizontal shear strength

Shear transfer reinforcement	Surface preparation	V_{nh} (lb)	
$A_v \geq A_{v,min}$	Concrete placed against hardened concrete intentionally roughened to a full amplitude of approximately ¼ in.	Lesser of:	$\lambda\left(260 + 0.6\frac{A_{vf}f_{yt}}{b_v s}\right)b_v d$
			$500\,b_v d$
	Concrete placed against hardened concrete not intentionally roughened	$80\,b_v d$	
Other cases	Concrete placed against hardened concrete intentionally roughened	$80\,b_v d$	

$$V_{nh} = v_{nh}b_v d \tag{10.16}$$

and

$$\phi V_{nh} \geq V_{uh} \tag{10.17}$$

and ϕ equals 0.75 from Table 21.2.1 of ACI 318-14. The values of v_{nh} are provided in Table 16.4.4.2 of ACI 318-14 and are summarized in Table 10.4. In all cases, the roughened surfaces should be clean and free of laitance.

The minimum shear reinforcement provision for horizontal shear transfer is the lesser of Eqs. 10.18 and 10.19.

$$A_{v,min} = \frac{50b_w s}{f_{yt}} \tag{10.18}$$

$$A_{v,min} = 0.75\sqrt{f'_c}\frac{b_w s}{f_{yt}} \tag{10.19}$$

where b_w is the web width of the precast beam, s is the longitudinal spacing of the stirrups along the beam and f_{yt} is the yield stress of the transverse reinforcement. Maximum stirrup spacing for horizontal shear transfer is four times the least dimension of the supported element or 24 in. Stirrups may be single bars, multiple leg stirrups or welded wire reinforcement with the reinforcement anchored on both sides of the shear plane. Normally, stirrups designed for vertical shear are extended into the cast-in-place topping to provide the horizontal shear transfer capacity.

Section 16.4.5 of ACI 318-14 provides an alternative method to provide shear reinforcement by requiring that Eq. 10.16 be met at any location along the contact surface. The simplest approach to this method is to apply the total compression or tension force to the topping and calculating the total area of reinforcement to resist that force using shear friction methods. Stirrups are then apportioned approximately in relation to the accumulated force. Thus, the horizontal force to be resisted is the lesser of

$$C = 0.85f'_c b_v h_f \tag{10.20}$$

$$T = A_p f_{ps} \tag{10.21}$$

and f'_c is the strength of the composite topping. Substituting the depth of the equivalent stress block a in lieu of h_f is allowed if that depth is entirely within the topping.

Example 10.3: Investigate the Shear Transfer
Investigate the shear transfer in Example 10.2.

Solution: The shear at the end of the beam is

$$V_u = \frac{w_u l}{2} = \frac{1347 \cdot 48}{2} = 32,200\,\text{lb}$$

giving a horizontal shear stress of

$$v_{hu} = \frac{V_u}{b_v d_p} = \frac{32,200}{96 \cdot 15.85} = 21.2\,\text{psi}$$

this is less than the 80-psi allowed so the design is adequate.

Comment: The results of this horizontal shear check are consistent with extensive industry practice. The chapter on spliced girders provides an additional solution for horizontal shear transfer when these minimums are exceeded.

10.5   Vertical Shear

In accordance with Section 22.5.4 of ACI 318-14, the vertical shear may be designed the same as a monolithic concrete beam, if the entire concrete section resists shear. The allowance is made because most of the shear is carried in the web of the beam. Therefore, in standardized elements, the composite topping has little effect on the shear area.

If the composite section extends into the web of the beam, the horizontal shear from the composite action may affect the shear at the interface and more than one concrete strength may be present. This condition requires a detailed examination of the concrete contribution and any differential shrinkage effects that may be additive to the shear created by vertical loads.

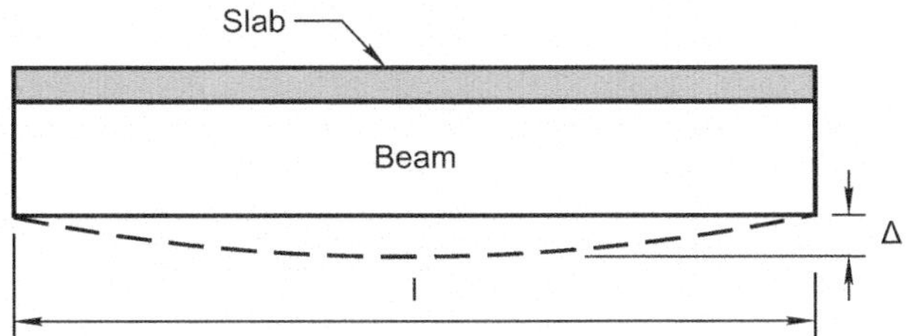

Fig. 10.10 Deflection due to differential shrinkage

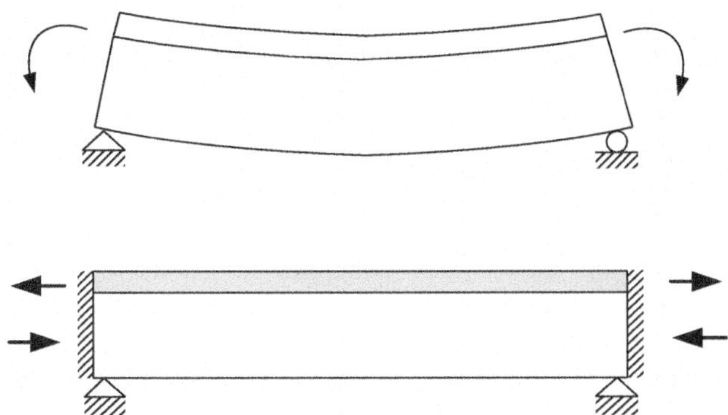

Fig. 10.11 Effect of shrinkage on simple and continuous beam

10.6 Differential Shrinkage in Composite Beams

Composite topping assists in controlling deflections in beams subjected to external loads due to the higher moment of inertia. At the same time, the addition of a composite topping increases the downward deflection due to shrinkage of the topping as it cures, Fig. 10.10.

The deflection due to differential shrinkage can be significant on simple span beams and rather modest on continuous beams, Fig. 10.11. In the case of the simple span restraining the shrinkage creates an internal moment in the beam, whereas the restraining forces for the continuous beam go into the supports or adjacent beam.

Several solutions to calculate the deflection due to shrinkage are available (Birkeland 1960; Branson 1977; Kim and Lee 1998). All are based on the principal that plane sections remain plane and that when the shrinkage strains have stabilized there is no slippage between the beam and the cast-in-place topping. Birkeland (1960) for example uses a classical mechanics of materials solution for two bonded materials subjected to differential strains. Therefore, the shortening of the topping is compensated by a fictional tension force. The equilibration of the tensile force results in a compression force in the beam and the beam shortens, Fig. 10.12.

Complicating any analysis is the time related behavior of the cast-in-place concrete. The cast-in-place concrete begins shrinkage almost as soon as it is cast.

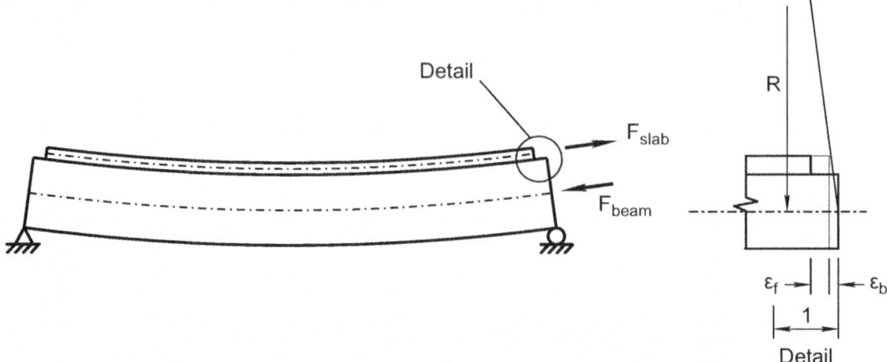

Fig. 10.12 Differential movement of a composite topping subjected to shrinkage

Concurrently, the bond to the precast concrete exceeds the shrinkage strain from the beginning. Thus, the differential shrinkage occurs with a continually varying modulus of elasticity of the cast-in-place concrete. Considering these variables, the following approach is a simplified method to obtain an approximate value for the deflection due to shrinkage. A tension stress is applied to the cast-in-place concrete equal to the shrinkage strain times an adjusted modulus of elasticity equal to one-half of the final modulus. The reduced modulus accounts for the relaxation in the concrete topping due to the composite strains and is the inverse of the long-term creep coefficient in the PCI Design Handbook.

$$f_{ct} = \varepsilon_{sh}\frac{E_{ct}}{2}$$

(10.22)

The force to equilibrate the cast-in-place concrete shrinkage is

$$T = f_{ct}b_f h_{ct}$$

(10.23)

and is applied at the centroid of the cast-in-place concrete. This tensile force is compensated by an equal and opposite compression force applied at the centroid of the precast beam. The two forces create a constant couple internal to the composite beam.

$$M_{eq} = T \cdot z$$

(10.24)

where z is the distance between the tensile and compressive centroids. The deflection is then calculated using equations for a constant moment applied to the ends of the composite beam.

$$\Delta_{\text{sh}} = \frac{M_{\text{eq}}l^2}{8E_{\text{b}}I_{\text{cb}}} \tag{10.25}$$

where the moment is applied to the composite section properties of the precast beam. In a precise mechanic of materials solution, the force would be further reduced because the extension of the cast-in-place topping would reduce due to the elongation of the beam. The use of an adjusted modulus of elasticity avoids this complication.

Example 10.4: Estimate the Deflection Due to the Topping

Estimate the deflection due to the topping in Example 10.1 assuming the total shrinkage is 0.0002.

Solution: The moduli of elasticity of the precast beam and cast-in-place topping are 4,415,000 psi and 3,604,000 psi, respectively. The tensile force to compensate the shrinkage is

$$T = \varepsilon_{\text{sh}}\frac{E_{\text{ct}}}{2}b_f h_{\text{ct}} = 0.0002 \cdot \frac{3,604,000}{2} \cdot 96 \cdot 2 = 69,200\,\text{lb}$$

The moment arm between the tension and compression force is $y_t = 1$ in. $+$ 6.85 in. $=$ 7.85 in., resulting in an equivalent moment of

$$M_{\text{eq}} = T \cdot z = 69,200 \cdot 7.85 = 543,200\,\text{in.-lb}$$

and a deflection of

$$\Delta_{\text{sh}} = \frac{M_{\text{eq}}l^2}{8E_{\text{b}}I_{\text{b}}} = \frac{543,200 \cdot 48^2 \cdot 1728}{8 \cdot 4,415,000 \cdot 27,982} = 2.19\,\text{in.}$$

Comment: This compares to the 2.4 in. camber predicted in the PCI Design Handbook for a 8 ft wide by 24 in. deep double-T beam prestressed with 12 strands. Thus, after application of the topping and shrinkage, the beam would sag of about 1 in. or 1/5670 of the span length. The above calculation is intended as a design check on deflection behavior. If a more refined value for deflection is needed, finite element analysis or methods in the cited references provide guidance.

References

AASHTO. (2017). *LRFD bridge design specification* (8th ed.). Washington, DC: American Association of State Highway and Transportation Officials (AASHTO).

Birkeland, H. W. (1960, May). Differential shrinkage in composite beams. *Journal Proceedings of ACI, 56*(5), 1123–1136.

Branson, D. E. (1977). *Deformation of concrete structures*. New York: McGraw-Hill Companies.

Dolan, C. W. (1967). *Ultimate capacity of reinforced concrete sections using a continuous stress-strain function*. MS Thesis, Cornell University, Ithaca, NY.

Hanson, J. A. (1961, July). Tensile strength and diagonal tension resistance of structural lightweight concrete. *ACI Journal Proceedings, 58*(1), 1–40.

Kaar, P. H. (1966, May). High strength bars as concrete reinforcement, part 8: Similitude in flexural cracking of T-beam flanges. *Journal, PCA Research and Development Laboratories, 8*(2), 2–12.

Kim, J. K., & Lee, C. S. (1998). Prediction of differential drying shrinkage in concrete. *Cement and Concrete Research, 28*(7), 985–994.

PCI. (2017). *PCI design handbook* (8th ed.). Chicago, IL: Precast and Prestressed Concrete Institute.

Saemann, J. C., & Washa, G. W. (1964, November). Horizontal shear connections between precast beams and cast-in-place. *ACI Journal Proceedings, 61*(11), 1383–1410.

Chapter 11
Two-Way Slabs

11.1 Introduction

Prestressed concrete slabs are used in buildings, parking garages, roofs, decks, and other similar applications. They may be either **one-way** slabs, spanning in predominately one direction, or **two-way** slabs, spanning in two orthogonal directions. One-way slabs are discussed in Example 6.2 and in Chap. 9. Two-way slabs are most efficient when the column spacing in each direction is approximately the same. Aspect ratios greater than 2:1 generate predominately one-way action and are designed as one-way slabs with attention paid to localized bending at the corners. Two-way slab systems are typically more efficient and therefore more slender structures. As such, punching shear can control the slab thickness. Design typically assumes that the slab is subjected to a uniformly distributed area loading. Slabs subjected to concentrated loadings require additional attention during the analysis and design.

11.2 Two-Way Slab Systems

Two-way slabs may be supported by beams at the slab perimeter or by columns only. Most two-way post-tensioned slabs are supported by columns only, Fig. 11.1. The slab in Fig. 11.1b is designated a **flat plate** and is distinguished by a constant thickness slab throughout. Flat plate systems carry relatively light loads and are commonly used for residential structures. Because large open spaces are not required in residential structures, the column grid can be tighter, resulting in shorter spans and thinner slabs with spans up to 25–30 ft. In many cases, the punching shear capacity of flat plate systems controls the slab thickness. Stud rails are often used to improve the punching shear strength so that thinner slabs can be used. The thinner slabs with unobstructed soffit reduce the story height, dead load, and allow for more efficient

© Springer Nature Switzerland AG 2019

C. W. Dolan, H. R. Hamilton, *Prestressed Concrete*,

https://doi.org/10.1007/978-3-319-97882-6_11

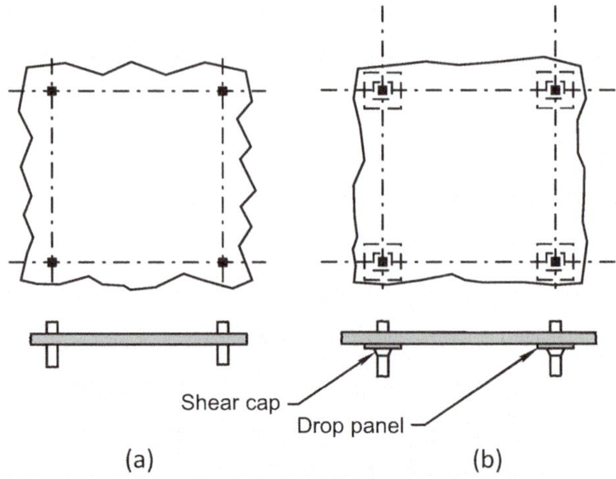

Fig. 11.1 Two-way slabs (**a**) flat plate (**b**) flat slab

Fig. 11.2 Flat slab with drop panels

use of narrower space for mechanical and electrical equipment. This all combine to provide a more efficient use of the vertical space.

A **flat slab** is similar to a flat plate but contains shear caps, drop panels, or column capitals, Fig. 11.1. Column capitals are enlarged column tops that provide additional punching shear capacity and reduce the effective span length. Historically, the tops of columns were flared, but the cost and complexity of forming led to the use of **shear caps**, which are typically box-shaped projections below the slab intended solely to increase shear resistance around the column. A **drop panel** is also a projection below the slab, but it is used to increase the effective depth of the negative reinforcement near a column in addition to increasing the slab shear strength. The ACI Building Code requires that drop panels be at least one-sixth of the span length and extend below the slab at least one-quarter of the slab thickness, Fig. 11.2. Shear caps are required to extend at least the thickness of the slab both below the slab and one thickness away from the column face. As such, drop panels tend to have larger areas than shear caps. While these elements help to extend the span, they come at the

Fig. 11.3 Waffle or grid slab (Courtesy Larry Novak, Portland Cement Association)

cost of increased reinforcement and formwork complexity. Furthermore, the projections below the slab soffit tend to increase the plenum space required for mechanical and electrical equipment. While suitable for residential construction, larger column spacing make this system more attractive for light commercial or parking structures.

For heavier loads such as heavy commercial or manufacturing or perhaps institutional such as hospitals, **grid slab** or **waffle slab** can be used, Fig. 11.3. Waffle slabs are constructed using metal, plastic, or wood pressboard inverted pans placed on the form floor. The space between the pans becomes the rib, and a thin slab is cast between the ribs. Pans may be omitted around the column top to create a flush drop panel as seen in Fig. 11.3.

11.3 Analysis and Design

The analysis of two-way slabs is performed by hand calculations or computer analysis. The ACI Building Code recognizes two methods of hand calculation, Direct Design and Equivalent Frame Method. Both methods calculate an equivalent statical moment $wl^2/8$ for the entire width of the slab then distribute that moment over the width and length of the slab. The analysis is conducted in both directions for both service and strength loadings and with the full dead and live load used in each analysis. Full loads in each direction assure that the structural compatibility of deflections at midslab are maintained.

The restrictions on Direct Design make the method unsuitable for prestressed slabs and it is specifically limited in the ACI Building Code to nonprestressed structures. The **equivalent frame** method considers the torsional stiffness of slab edges and edge beams then distributes the statical moment to adjust for the torsional edge stiffness. Equivalent frame analysis results in **column strips**, the portion of the slab one quarter of the span length either side of the column, and **middle strips**, the central one half of the slab bounded by the column strips. Equivalent frame method is prescribed in the ACI 318-14 Building Code but is likely to be removed in future versions of the code. Detailed design guidance for equivalent frame calculations is found in reinforced concrete design texts (Darwin et al. 2015; Wight 2015).

The most common method of slab analysis and design uses computer software programs. Software programs such as Autodesk, Adina, and Bentley Systems or specialty software programs such as ADAPT, Structurepoint, and other commercial programs can conduct either an equivalent frame analysis (EFA) or a finite element analysis (FEA), or both. EFA analysis uses equivalent two-way frame elements to determine moments and shears using the direct stiffness method of analysis. In implementing EFA, software programs do not always differentiate between column strips and middle strips and may distribute the moments and shears according the overall structural stiffness. FEA is a three-dimensional analysis method that uses discrete elements to determine stresses from the applied loading. Those stresses are then transformed into unit moments and shears. Depending on their features available some of the software packages will also provide post-tensioning designs. The engineer should be familiar with the underlying principles of the software programs.

Computer generated solutions have variable tendon spacing across the slab and concentrate or **band** the tendons in the highest moment regions. The ACI Building Code requires that at least one direction of a two-way slab have uniform tendon spacing. Therefore, post-tensioning tendons may only be banded in one direction. Column strips and middle strips retain their meaning in the ACI Building Code even if they are not part of the detailed analysis.

11.4   Design of Flat Slabs

The interaction between computer analysis and design complicate the presentation of design methodology. The following presentation of two-way slab design assumes the computer analysis is independent of the text. The focus is on determination of the initial slab post-tensioning and a series of checks to validate both the initial design and the check computer output. The emphasis on ACI Building Code requirements reflects the detail required to complete a design.

Requirements in the ACI Building Code impact the tendon spacing and required supplemental nonprestressed reinforcement. To be considered a prestressed two-way slab, the minimum effective prestress must be at least 125 psi. If a drop panel or shear cap is used the 125-psi minimum stress applies to the thickened portion of the slab in addition to the thinner central part of the slab. Prestressed slabs are designed as Class U members with the restriction that the maximum tensile stress is less than or equal to $6\sqrt{f'_c}$. Other stresses must be within the ACI limits given in Tables 5.1 and 5.2. End reactions introduced by the prestress force, the **secondary moments**, are included in the design. The maximum spacing of tendons or groups of tendons in at least one direction is $8h$ not to exceed 5 ft. This requirement leads to design that have **banded tendons** in one direction and uniformly spaced tendons in the

Fig. 11.4 Plan view of distributed and banded tendons

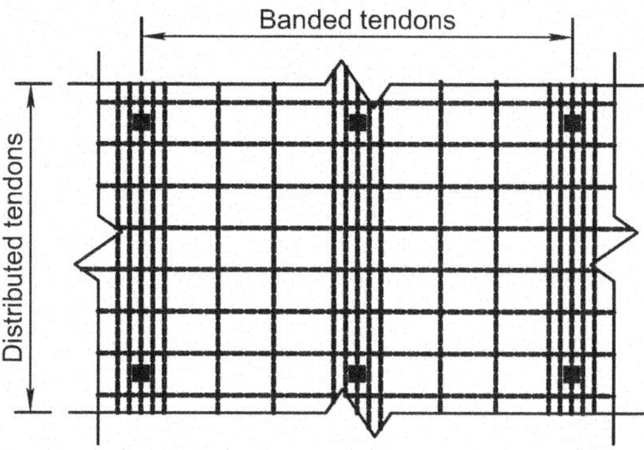

Table 11.1 Suggested span/depth ratio

One-way slabs	48
Two-way slabs	45
Two-way slab with drop panel	50
Two-way slab with two-way beams	55
Two-way waffle slab (5 ft^2 grid)	35
Beams ($b \approx \frac{h}{3}$)	20
Beams ($b \approx 3h$)	30
One-way joists	40

Adapted from PTI Manual (2006)

orthogonal direction, Fig. 11.4. Banded tendons are closely spaced tendons in regions of high moment and more widely spaced tendons in areas of lower demand. For example, tendons may be banded in the region of a drop panel to provide resistance for negative moments in that area.

11.4.1   Slab Thickness

Section 8.3 of ACI 318-14 specifies minimum thickness limits for nonprestressed concrete slab systems but not for prestressed slabs. The ACI Building Code does require deflection checks for prestressed two-way slabs. The commentary notes that the span-to-depth ratios for prestressed concrete slabs should generally not exceed 42 for floors and 48 for roofs. Using this guidance, a 30–40-ft flat slab would have thicknesses in the range of 8–10 in. The Post-Tensioning Institute Design Manual suggests the span-to-depth ratios listed in Table 11.1. Regardless of the method used to select and initial slab thickness to initiate the design, the slab thickness is checked for strength and serviceability requirements.

Deflection checks include short- and long-term deflections and should determine that camber and vibrational frequencies are not objectionable. Deflection limits for serviceability conditions are given in Chap. 8 and procedures for calculating

Table 11.2 Minimum bonded reinforcement in two-way slabs

Region	Calculated f_t after all losses	$A_{s,min}$, in.^{2}	Notes
Positive moment	$f_t \leq 2\sqrt{f'_c}$	Not required	
	$2\sqrt{f'_c} \leq f_t \leq 6\sqrt{f'_c}$	$N_c/0.5f_y$	N_c is the tensile force acting on the portion of the concrete cross section that is subjected to tensile stresses due to the combined effects of service loads and effective prestress
Negative moment at columns	$f_t \leq 6\sqrt{f'_c}$	$0.00075A_{cf}$	A_{cf} is the greater cross-sectional area of the slab-beam strip of the two orthogonal equivalent frames intersecting at a column of a two-way slab

Adapted from Section 8.6.2.3 of ACI 318-14

deflections are in Sect. 11.4.5. The ACI Building Code provides no guidance for vibration response limits.

11.4.2 Supplemental Reinforcement

Section 8.6.2 of ACI 318-14 requires supplemental nonprestressed reinforcement in two-way slabs. The reinforcement, in conjunction with the prestressing reinforcement, must be able to develop 1.2 times the moment of the slab that causes cracking, that is, $M_n \geq 1.2 M_{cr}$. M_{cr} is based on the modulus of rupture including prestressing times the section modulus $S = I_g/y$.

$$M_{cr} = \frac{(f_r + f_{pe})I_g}{y} \tag{11.1}$$

where $f_r = 7.5\lambda\sqrt{f'_c}$, and f_{pe} is the effective stress due to prestressing at the tension face of the beam.

Bonded deformed longitudinal reinforcement in addition to the bonded or unbonded prestressing reinforcement is also required in positive and negative moment regions according to Table 11.2. The nonprestressed reinforcement is based on tests of two-way slab systems (Odello and Mehta 1967; Burns and Hemakom 1977). These tests indicated minimum reinforcement ratios of 0.0075 provide sufficient ductility in the column strips. Determination of the reinforcement is based, in part, on equivalent frame models and is conservative for rectangular panels. The ACI Building Code restricts the specified yield stress of the bonded reinforcement to 60 ksi to limit cracking even if higher strength reinforcement is used.

For slabs with bonded tendons, $A_{s,min}$ may be reduced by the area of the bonded prestressed reinforcement located within the area used to determine N_c for positive

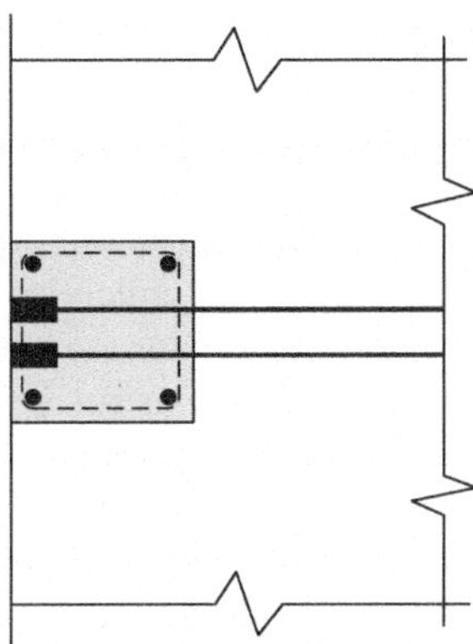

Fig. 11.5 Plan view of anchorage of structural integrity tendons

moment or within the width of the slab between lines that are 1.5 *h* outside opposite faces of the column support for negative moment. Bonded reinforcement required in negative moment regions is placed at the top of the slab. At least four deformed bars, deformed wires or bonded strands are provided in each direction within the lines 1.5 *h* outside the column face at a maximum spacing of 12 in.

11.4.3   Structural Integrity

Structural integrity provisions are intended to prevent disproportionate or progressive collapse of a structure. Two-way slab provisions are given in Section 8.7.4.2 of ACI 318-14 and are summarized as follows. At least two 1/2 in. diameter or larger strands are required in each direction at columns and they must pass through the region bounded by the longitudinal reinforcement of the column, sometimes referred to as the column core. Tendons are anchored within the region bounded by the longitudinal reinforcement of the column and the anchorage is located beyond the column centroid and away from the anchored span, Fig. 11.5. Outside of the column and shear cap faces, the two structural integrity tendons pass under any orthogonal tendons in adjacent spans.

If the tendons do not satisfy the above requirements, minimum bonded bottom deformed reinforcement is required in each direction to be the greater of the amount in Eq. 11.2.

$$A_s = \frac{4.5\sqrt{f_c'}\,b_w d}{f_y}$$

$$A_s = \frac{300 b_w d}{f_y}$$

$$(11.2)$$

where b_w is the width of the column face through which the reinforcement passes. The bottom deformed reinforcement must pass within the region bounded by the longitudinal reinforcement of the column and be anchored at exterior supports to develop f_y beyond the column or shear cap face.

The two tendons through the column core are met for most designs. However, the placement of the integrity tendons beneath the transverse tendons may not occur if the banded tendons are also intended to be the integrity tendons because the distributed tendons run below the banded tendon in the maximum negative moment region. In such cases, the minimum reinforcement indicated in Eq. 11.2 may be less than that suggested in ACI 352.1-R (2012). ACI 352.1 recommends a minimum area of reinforcement in accordance with Eq. 11.3 be placed continuously in the bottom of the slab along the column line. This amount of reinforcement is sufficient to develop catenary action to support the slab.

$$A_s = \frac{0.5 w_u l_1 l_2}{\phi f_y} \tag{11.3}$$

where w_u is the factored load, l_1 and l_2 are the length of the slab in each direction. The value for f_y is the grade of the reinforcement or the yield stress of the post-tensioning tendon. Alternatively, two tendons used exclusively for integrity reinforcement may be added below the distributed tendons to meet the structural integrity provisions.

11.4.4   Moment Transfer at Columns

Moments transferred to the column at the slab-to-column interface M_{sc} result from the bending restraint of the column and any lateral forces the slab–column joint resists. Design of the slab–column interface addresses this moment transfer through direct bending $\gamma_f M_{sc}$ and moment transfer by shear $\gamma_v M_{sc}$. Slab moments transferred to the column by shear generate torsional stresses at the column edge and the column face in addition to the contribution of vertical shear, Fig. 11.6.

The assumed fraction of the factored slab moment resisted by the column and transferred by flexure $\gamma_f M_{sc}$ may be calculated by Eq. 11.4

$$\gamma_f = \frac{1}{1 + \left(\frac{2}{3}\right)\sqrt{\frac{b_1}{b_2}}} \tag{11.4}$$

where b_1 and b_2 are the length of the critical slab sections in the direction of the span and perpendicular to the span, Fig. 11.7. The values of c_1 and c_2 are the column dimensions in the corresponding directions. The fraction of the moment transferred in shear and applied at the centroid of the critical section is then given in Eq. 11.5.

Fig. 11.6 Slab shears resulting from moment transfer at column

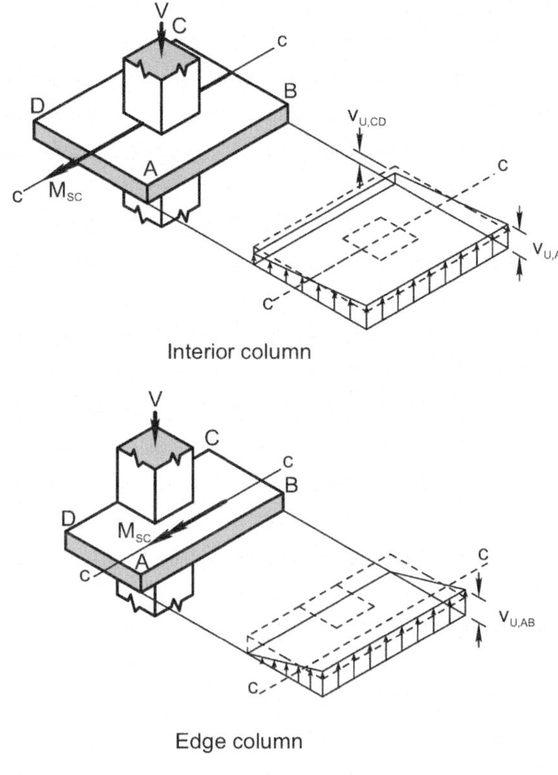

$$\gamma_v = 1 - \gamma_f \tag{11.5}$$

Research by Hanson and Hanson (1968) found that where moment is transferred between a column and a slab, 60% of the moment should be considered transferred by flexure across the perimeter of the critical section defined Fig. 11.7, and 40% by eccentricity of the shear about the centroid of the critical section. For rectangular columns, the portion of the moment transferred by flexure increases as the width of the face of the critical section resisting the moment increases.

The factored shear stress $v_{u,AB}$ resulting from $\gamma_v M_{sc}$ is assumed to vary linearly about the centroid of the critical section c-c and is calculated along the face of the critical section AB by Eq. 11.6.

$$v_{u,AB} = v_{ug} + \frac{\gamma_v M_{sc} b_2}{J_c} \tag{11.6}$$

where v_{ug} is the uniform shear due to gravity load and J_c is the polar moment of inertia of the critical section of the slab around the centroid of the perimeter line, designated by line c-c in Fig. 11.6. For an interior column, J_c may be calculated as:

Fig. 11.7 Geometry of critical section

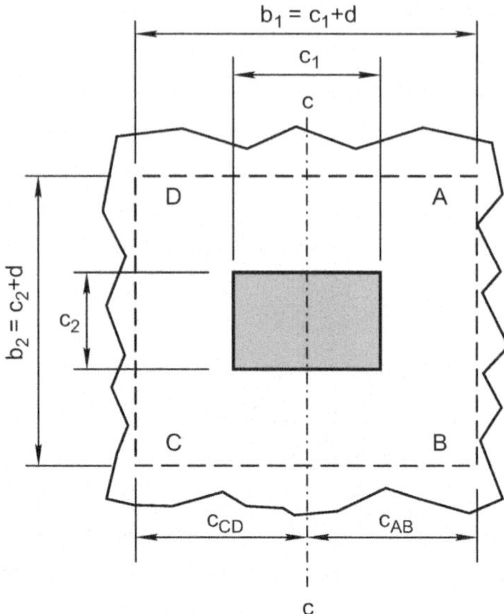

Interior column

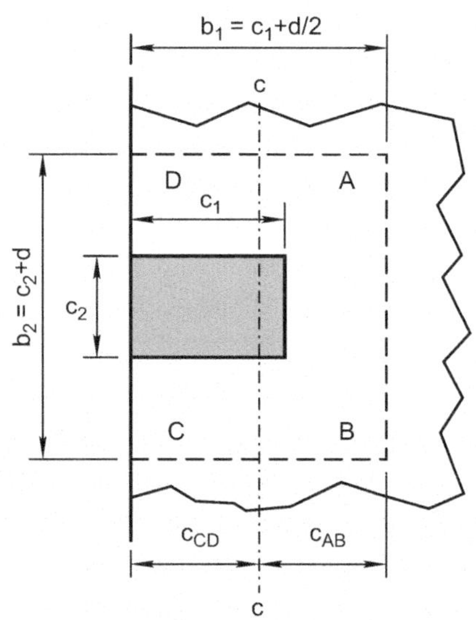

Exterior column

$$J_c = \frac{d(c_1 + d)^3}{6} + \frac{(c_1 + d)d^3}{6} + \frac{d(c_2 + d)(c_1 + d)^2}{2} \tag{11.7}$$

and similar equations can be developed for J_c at other locations. For the edge column, the centroid moves toward the interior because the perimeter is no longer symmetrical. The shear along the face of the critical section CD is given by Eq. 11.8, and is zero at the free edge of an exterior column. Terms in Eq. 11.8 are defined above.

$$v_{u,CD} = v_{ug} - \frac{\gamma_v M_{sc} b_2}{J_c} \tag{11.8}$$

The fraction of M_{sc} not transferred by eccentricity of the shear is transferred by flexure. A conservative method assigns the fraction transferred by flexure over an effective slab width of the width of the column or column capital plus $1.5\,h$ either side of the column. Often, column strip reinforcement is concentrated near the column to accommodate M_{sc}. Available test data from Hanson and Hanson (1968) suggests that this reinforcement does not increase shear strength but may be desirable to increase the stiffness of the slab–column junction. Collaborating test data indicate that the moment transfer strength of a prestressed slab-to-column connection can be calculated using these procedures (Hawkins 1981).

Where headed shear stud reinforcement is used, the critical section beyond the shear reinforcement generally has a polygonal shape, Fig. 11.7. Equations for calculating shear stresses on such sections are given in ACI 421.1 (2008). Research indicates that orthogonal layout of the studs is satisfactory; however, higher capacity may be possible using radial layouts (Dam et al. 2017).

Most of the data in Hanson and Hanson (1968) were obtained from tests of square columns. Limited information is available for round columns. Information indicates that round columns can be approximated as square columns of the same area. Tests indicate that some flexibility in distribution of M_{sc} transferred by shear and flexure at both exterior and interior columns is possible providing there is sufficient shear capacity. Interior, exterior, and corner columns refer to slab–column connections for which the critical perimeter for rectangular columns has four, three, or two sides, respectively.

For M_{sc} resisted about an axis parallel to the edge of exterior columns, the portion of moment transferred by eccentricity of shear $\gamma_v M_{sc}$ may be reduced, provided that the factored shear at the column, excluding the shear produced by moment transfer, does not exceed 75% of the shear strength ϕv_c for edge columns, or 50% for corner columns. Tests indicate that there is no significant interaction between shear and M_{sc} at the exterior column in such cases (Moehle 1988; ACI 352.1R 2012). To maintain equilibrium, as $\gamma_v M_{sc}$ is decreased, $\gamma_f M_{sc}$ is increased.

Evaluation of tests of interior columns indicates that some flexibility in distributing M_{sc} transferred by shear and flexure is possible, but more limited than for exterior columns. For interior columns, M_{sc} transferred by flexure is permitted to be increased up to 25%, provided that the factored shear, excluding the shear caused by the moment transfer, at the interior columns does not exceed 40% of the shear strength ϕv_c.

If the factored shear for a slab–column connection is large, the slab–column joint cannot always develop all of the reinforcement provided in the effective width. The modifications for interior slab–column connections in this provision are permitted only where the reinforcement required to develop $\gamma_f\,M_{sc}$ within the effective width has a net tensile strain ε_t not less than 0.010. The use of Eq. 11.4 without the modification permitted in this provision generally indicates overstress conditions on the joint. This provision is intended to improve ductile behavior of the slab–column joint. If reversal of moments occurs at opposite faces of an interior column, both top and bottom reinforcement should be concentrated within the effective width. A ratio of top-to-bottom reinforcement of approximately 2 has been observed to be appropriate.

11.4.5   Deflections

Flat plates are relatively thin structures and, while prestressing may meet strength and cracking requirements, slabs may be inadequate for service conditions due to excessive deflections or camber. Therefore, calculation of the deflection in a two-way slab is required by the ACI Building Code. Calculation of the slab center deflection is a two-step process. First the midspan deflection along a column line is calculated δ_{cy}. Then the deflection at midspan between two adjacent column lines is calculated δ_{my} and added to δ_{cy} to obtain the total deflection at the middle of the slab, Fig. 11.8. The same middle deflection should be obtained whether the X or the

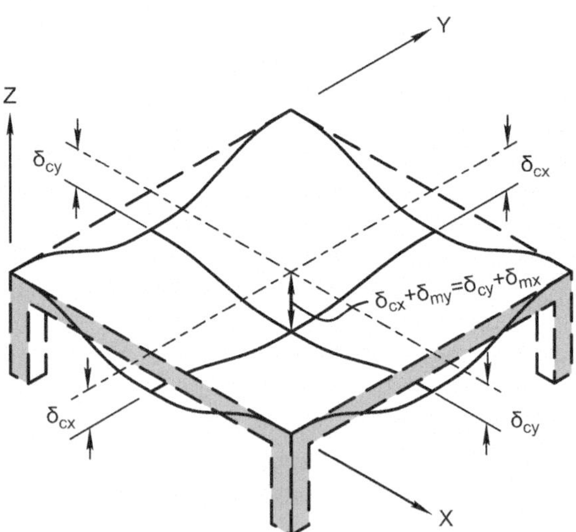

Fig. 11.8 Deformations in a two-way slab

Y direction is initially selected. If the slab has edge beams, the effect of the beams is included in the calculation.

Computer analyses provide the mid-slab deflection. If the calculation is done manually, it is helpful to use column and middle strips. For simple spans, 55–60% of the moment is assigned to the column strip M_{cs} and for continuous slabs 65–75% of the moment is assigned to the column strip. The remaining moment is applied in the transverse middle strip M_{ms}.

Begin by assuming the two-way slab is fixed at the ends. Correction for less than full moment restraint is addressed later. A reference deflection δ_r for a fixed-fixed member is than calculated as:

$$\delta_r = \frac{wl^4}{384\, E_c I_s} \tag{11.9}$$

where w is the service load per foot on the full width of the slab, E_c is the modulus of elasticity of the concrete, and I_s is the moment of inertia of the full slab width including any edge beams. Reference deflections are calculated in the X and the Y direction as δ_{rx} and δ_{ry}. Two-way slabs are designed for a maximum tensile stress of $6\sqrt{f'_c}$, which is less than the cracking stress. Consequently, gross moments of inertia are used for deflection calculations. Individual deflections are then obtained by multiplying the ratio of the total slab stiffness to the strip stiffness and by the ratio of the moments each segment carries. Thus, the deflection along the column strip in the Y direction is:

$$\delta_{cy} = \frac{M_{cs}}{M_s} \cdot \frac{E_c I_s}{E_c I_{cs}} \delta_{ry} \tag{11.10}$$

and for the middle strip the deflection δ_{my} is:

$$\delta_{mx} = \frac{M_{ms}}{M_s} \cdot \frac{E_c I_s}{E_c I_{ms}} \delta_{rx} \tag{11.11}$$

giving a total deflection at the middle of the slab of

$$\delta_m = \delta_{cy} + \delta_{mx} = \delta_{cx} + \delta_{my} \tag{11.12}$$

For a flat slab with no edge girders and 60% of the moment to the column strip, the total deflection at midslab becomes

$$\delta_m = 0.6 \cdot 0.5\delta_{ry} + 0.4 \cdot 0.5\delta_{rx} = 0.3\delta_{ry} + 0.2\delta_{ry} \tag{11.13}$$

The initial assumption of the slab being fixed along the edges is reasonable for a structure with all spans equally loaded and approximately equal dimensions and for slabs when the ratio of live load to dead load is small. If this condition is not met, then the deflection due to rotation at the slab edge is added to the gravity deflection. Nilson (1987) approached this by noting that the slab edge rotation is calculated using the rotation of the equivalent column:

$$\theta = \frac{M_{\text{net}}}{K_{\text{ec}}} \tag{11.14}$$

where θ is the angle change at the edge of the slab in radians,

M_{net} is the difference in moment between the slab edges, and
K_{ec} is the stiffness of the equivalent column from Appendix C.

Once the rotation is known, the midspan deflection is found by noting that the midspan deflection for a beam with a rotation at one end and one end fixed is approximately:

$$\delta_\theta = \frac{\theta l}{8} \tag{11.15}$$

Adding subscripts 1 and 2 to denote the rotation at each end of the span and noting that δ_θ is calculated for both the X and the Y direction respectfully. The deflections along the column strip including end rotations are:

$$\delta_{cy} = \frac{M_{\text{cs}}}{M_{\text{s}}} \cdot \frac{E_C I_{\text{s}}}{E_c I_{\text{CS}}} \delta_{\text{ry}} + \delta_{\theta 1} + \delta_{\theta 2} \tag{11.16}$$

And in the middle strip are:

$$\delta_{mx} = \frac{M_{\text{ms}}}{M_{\text{s}}} \cdot \frac{E_C I_{\text{s}}}{E_c I_{\text{mS}}} \delta_{\text{rx}} + \delta_{\theta 1} + \delta_{\theta 2} \tag{11.17}$$

Finally, the total mid-slab deflection is given by Eq. 11.12 using the values from Eq. 11.16 or Eq. 11.17. The example in Sect. 11.6 illustrates these principles.

11.4.6　Corner Slab Restraint

Exterior corners of two-way slabs are prone to lift upward due to loads on interior panels. The corner column, edge walls, or edge beams restrain this movement leading to the possibility of negative moment cracking at the corner. Reinforcement at the corners of the slab is designed for the factored moment in the corner and detailed according to Fig. 11.9.

Fig. 11.9 Corner
reinforcing details

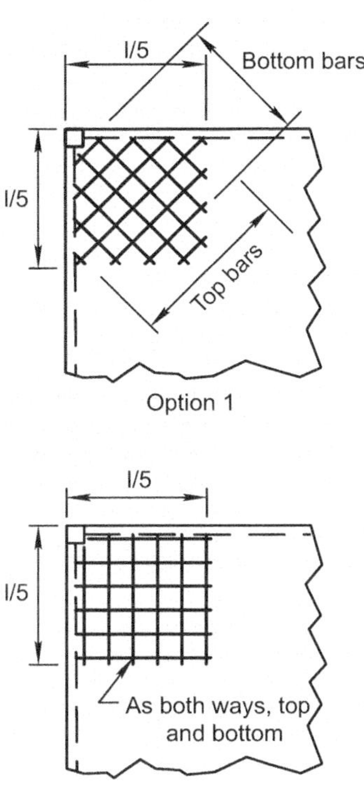

11.4.7   Openings in Slabs

The ACI Building Code permits openings of any size in the slab providing the
analysis indicates that the strength and serviceability criteria are satisfied. As an
alternate to analysis, Section 8.5.4 of ACI 318-14 provides conditions where openings are permitted. All permitted openings come with the condition that any reinforcement cut by the opening be replaced around the perimeter of the opening. If
post-tensioning tendons must be relocated around an opening, the lateral effects of
the post-tensioning realignment must be considered. Supplemental reinforcement to
prevent slab splitting due to separating tendons is one solution, Fig. 11.10. Assuming
an equivalent horizontal load due to tendon curvature provides a basis for selecting
the reinforcement. Sizing the reinforcement using $2/3 f_y$ as a design strength assists
in crack control.

Allowable opening conditions include:

1. Openings of any size are permitted in the area common to middle strips.
2. At intersecting column strips, openings of no more than one eighth of the width of
   the column strip are allowed.

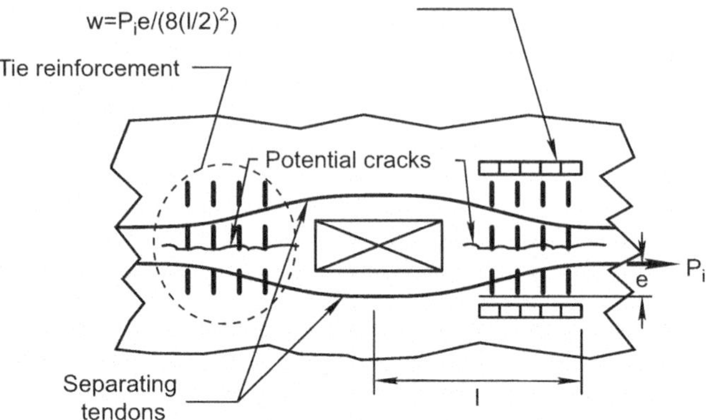

Fig. 11.10 Splitting forces from realigned tendons and tie reinforcement

3. At the intersection of a column strip and a middle strip, openings no more than
   one fourth of the either strip are allowed.
4. Openings in column strips within a distance 10 h of a reaction area or concentrated load require reduction of the perimeter shear area, according to Fig. 11.13,
   when calculating the shear capacity.

Openings in slabs near columns or edges reduce the shear capacity of the slab–
column connection and are discussed in Sect. 11.4.

Example 11.1: Reinforcement Around Openings
A 5 ft by 10 ft long opening in an 8 in. thick slab requires that a single monostrand
tendon with 30 kips initial prestress is to be offset by 5 ft in a 10-ft curve.

Solution: Half of the total offset occurs in the portion of the slab creating tension.
Thus, $e = 5$ ft and $l = 5$ ft. The uniform splitting load is then

$$w = \frac{Pe}{8(l/2)^2} = \frac{30 \cdot 2.5}{8(5/2)^2} = 1.5\,\text{kip/ft}$$

Using Grade 60 reinforcement, the required area of reinforcement is

$$A_s = \frac{w}{2/3 f_y} = \frac{1.5}{2/3 \cdot 60} = 0.04\,\text{in.}^2/\text{ft}$$

Over a 5-ft length, 0.19 in.^{2} of reinforcement is needed. As a practical design one
No. 4 would be placed 2 in. from the face of the opening and 2.5 ft from the face of
the opening at each end of the opening.

Comment: The splitting stress in the slab is w/slab area = 1.5 kips/ft/ (8 in. × 12 in.) = 16 ksi. The slab is unlikely to crack in this condition, even if the concrete is at an early age. The two bars would prevent any crack that might form from "unstitching" behind the opening. In this case the jacking load is used as this is the largest force that may cause splitting.

11.5 Two-Way Slab Shear Design

Shear design for two-way slabs involves gravity loads and bending effects discussed in Sect. 11.4.4. Gravity loads are carried by **punching shear** and the slab fails around the perimeter of the column, Fig. 11.11. The perimeter of the failure plane is located a distance $d/2$ from the face of the column resulting in a total failure perimeter of $b_o = 4(b + d)$ on an interior column, where b is the distance across the face of the column and d is the average distance to the effective depth of the reinforcement in both directions.

The factored concrete shear at the column or under a concentrated load is then

$$v_u = \frac{V_u}{b_o d} \tag{11.18}$$

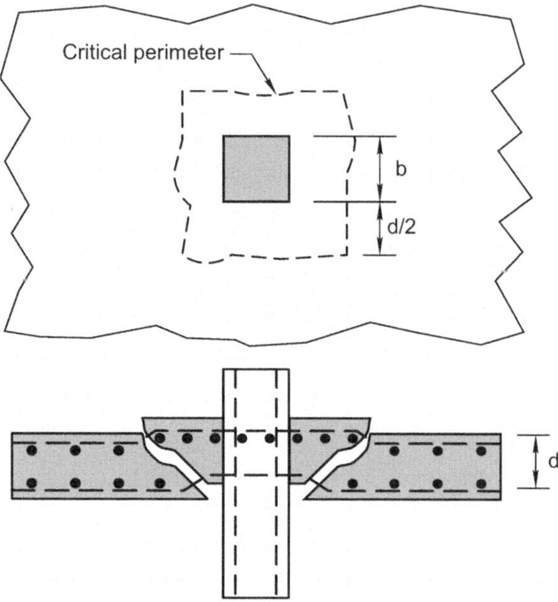

Fig. 11.11 Column punching shear failure

11.5.1   *Allowable Shear Stresses*

Reinforcement through the joint and two directional confinement of the concrete provides higher shear resistance than one-way slabs or beams. A value of

$$v_c = 4\lambda\sqrt{f'_c} \qquad\qquad (11.19)$$

not to exceed 100 psi is permitted for square and round columns. Shear at a round column is calculated using an equivalent square column with the same gross area as the round column.

Tests reported by Joint ACI-ASCE Committee 426 (1974) indicate that the value of $4\lambda\sqrt{f'_c}$ is unconservative when β, the ratio of the longer face of the column divided by the shorter face of the column, is larger than 2.0. In such cases, the actual shear stress on the critical section at punching shear failure varies from a maximum of approximately $4\lambda\sqrt{f'_c}$ around the corners of the column or loaded area, down to 2λ $\sqrt{f'_c}$ or less along the long sides. Therefore, the value of v_c diminishes for rectangular columns and approaches the value for one-way shear as the aspect ratio of the column increases, Table 11.3.

β is the ratio of the long side to short side of the column, concentrated load or reaction area and α_s has a value of 40 for interior columns 30 for edge columns and 20 for corner columns.

Other tests (Vanderbilt 1972) indicate that v_c decreases as the ratio b_o/d increases or edge confinement around the column perimeter decreases. Expressions in Table 11.3 modifying the factor of 4 were developed to account for these two effects. For shapes other than rectangular, β is taken to be the ratio of the longest overall dimension of the effective loaded area to the largest overall perpendicular dimension of the effective loaded area. For example, in an L shaped column the effective loaded area is that area totally enclosing the actual loaded area of the L-shaped reaction area and has a minimum perimeter, Fig. 11.12.

If a shear cap or drop panel is used the concrete shear capacity around the column is given by Eq. 11.19. At the edge of the shear cap or drop panel, the shear capacity reduces to $2\lambda\sqrt{f'_c}$ to reflect the lack of confinement.

Table 11.3 v_c for two-way shear

	v_c
Least of	$4\lambda\sqrt{f'_c}$
	$\left(2+\dfrac{4}{\beta}\right)\lambda\sqrt{f'_c}$
	$\left(2+\dfrac{\alpha_s d}{b_o}\right)\lambda\sqrt{f'_c}$

Fig. 11.12 Value for β for a non-rectangular loading

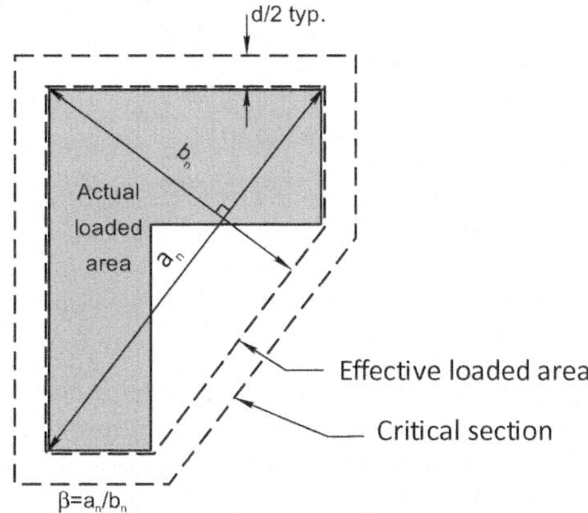

Prestressed concrete slabs having the bonded reinforcement described in Sect. 11.4.2, no portion of the column closer than four times the slab thickness from an edge, and the 125-psi minimum prestress, are allowed a higher concrete shear strength. Section 22.6.5.5 of ACI 318-14 allows calculation of the shear stress per Eq. 11.20.

$$v_c = 3.5\lambda\sqrt{f'_c} + 0.3f_{pc} + \frac{V_p}{b_o d}$$

$$v_c = \left(1.5 + \frac{\alpha_s}{b_o}\right)\lambda\sqrt{f'_c} + 0.3f_{pc} + \frac{V_p}{b_o d} \tag{11.20}$$

where α_s is defined in Table 11.3 and f_{pc} is the average of the prestress in both directions, not to exceed 500 psi. V_p is the vertical component of all effective prestress forces crossing the critical section. The ACI Building Code limits $\sqrt{f'_c}$ to 70 psi.

Example 11.2: Design the Slab for Punching Shear

A flat plate floor of normalweight concrete has thickness $h = 11$ in. and is supported by 18 in. square columns spaced 28 ft on centers each way. The floor carries a total factored load of $q = 235$ psf. Check the adequacy of the slab in resisting punching shear at a typical interior column. An average effective depth $d = 9$-1/2 in. may be used. Material properties are $f_y = 60{,}000$ psi, $f'_c = 4000$ psi, and $\lambda = 1.0$.

Solution: The first critical section for punching shear is a distance $d/2 = 4$-3/4 in. from the column face, providing a shear perimeter $b_o = 4\,(18 + 9.5) = 110$ in. Based on the tributary area of loaded floor, the factored shear is

$$V_u = q \cdot \left(l^2 - (b+d)^2\right) = 235 \cdot \left(28^2 - 2.29^2\right) = 183\,\text{kip}$$

and the design strength of the slab is

$$\phi V_c = \phi 4\lambda\sqrt{f'_c}b_o d = 0.75 \cdot 4 \cdot 1.0 \cdot \sqrt{4000} \cdot 110 \cdot 9.5 = 198\,\text{kip}$$

confirming that shear capacity is adequate and use of the expanded shear equations are not required.

Calculation of the shear perimeter around a column is affected by the location of openings in the slab. The ACI Building Code requires the perimeter be reduced by an amount dependent on the placement of the opening. Openings more than 4 slab thicknesses away from the support need not be considered. Otherwise, the perimeter is reduced by the amount indicated in Fig. 11.13.

In-slab duct systems can route HVAC systems in the plane of the floor system Fig. 11.14. There are no Code provisions addressing the use of these systems or their effect on flexure, shear, and deflections. The ACI Building Code indicates that their presence should be accounted for in design. For the purposes of shear design, ducts within 4 h of the column, should be considered like openings in the slab indicated in Fig. 11.13. Similarly, if at any point, the equivalent rectangular stress block, exceeds the duct cover, a thicker slab may be required.

Fig. 11.13 Reduction of shear perimeter due to openings

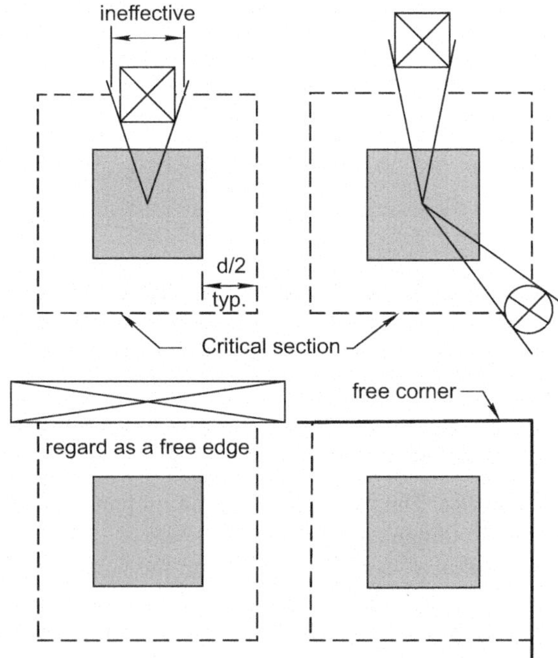

Fig. 11.14 In-slab HVAC duct system (Courtesy of Eccoduct, www.eccoduct. com)

11.5.2 Headed Shear Stud Systems

If the allowable shear stress values in Example 10.3 are exceeded, additional shear strengthening is required. Drop panels and shear caps are often solutions. Shear reinforcement is an option that maintains the slab thickness. The ACI Building Code recognizes several shear reinforcement alternatives; single or double leg stirrups, shearheads, and headed shear stud strip assemblies. Stirrups and shearheads are described in reinforced concrete textbooks and are seldom used in prestressed flat slabs. Headed shear studs are the most common reinforcement for prestressed two-way slab systems.

The basic principle of shear strengthening is to extend the shear reinforced area until the shear stress is reduced to an acceptable level. This principle applies to shear caps as well as shear reinforcement. For all shear strengthening options, the maximum shear stress at the perimeter of the strengthened zone is $2\lambda\sqrt{f'_c}$.

The slab–column joint region is usually congested, with top and bottom slab reinforcement running in two perpendicular directions, with vertical bars in the column, and possibly column ties. Congestion becomes critical when the slab has openings at or near the column faces. Shear stud reinforcing strips, sometimes called stud rails, as shown in Fig. 11.15 are widely used to alleviate this congestion and provide shear strength (Ghali 1989; Elgabry and Ghali 1990).

Multiple strips are arranged in two perpendicular directions for square and rectangular columns or sometimes in radial directions for circular columns. They are secured in position in the forms before the top and bottom flexural reinforcement and tendons are placed. The steel strip rests on bar chairs to maintain the needed concrete cover below the steel and is held in position by nails through holes in the strip. For design purposes, an individual stud is the equivalent of one vertical leg of one stirrup. Design procedures recommended by Ghali are:

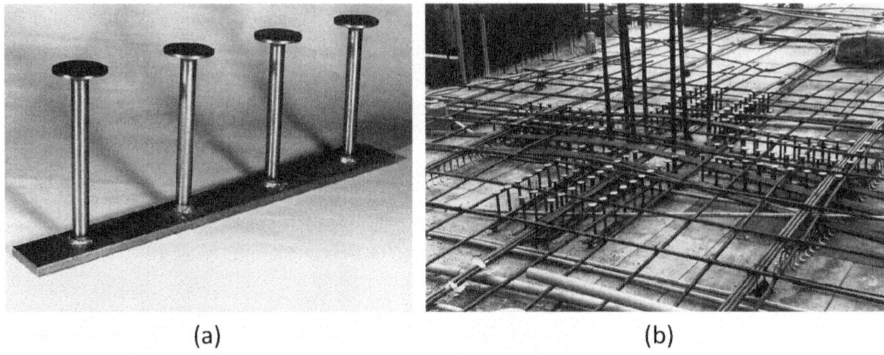

(a)   (b)

Fig. 11.15 Headed shear studs. (**a**) Studs and support bar. (**b**) Shear stud installation. *Courtesy of Amin Ghali and Walter H. Dilger*

1. The upper limit for the nominal shear stress at $d > 2$ from the column face is increased to $8\sqrt{f'_c}b_o d$.
2. The allowable stud spacing is increased to between $2d > 3$ in. and $3d > 4$ in., depending on the maximum nominal shear stress at factored loads.
3. Within the shear-reinforced zone, the contribution of the concrete is increased in addition to the above; Ghali has recommended the following details:

   (a) Top anchors are in the form of circular or square plates, the areas of which are at least ten times the area of the stem—as is required for "headed shear studs."
   (b) When the top anchor plates and the bottom strips are of uniform thickness, the thickness should be greater than or equal to one-half the stud diameter.
   (c) If the top anchor plate is tapered, the thickness at the connection with the stem should be greater than or equal to the stud diameter.
   (d) The width of the bottom strip should be greater than or equal to 2.5 times the stud diameter.
   (e) Bottom anchor strips should be aligned with the column faces of square or rectangular columns.
   (f) In the direction parallel to the column face, the distance between anchor strips should not exceed twice the effective depth of the slab.
   (g) The minimum concrete cover above and below the stud strips should be as normally specified for slab bars, and the cover should not exceed the minimum plus the bar diameter of the flexural reinforcement.

Further recommendations are found in Hawkins et al. (1989) and ACI 352-02 (2002) pertaining to the use of shear stud reinforcement at exterior and corner columns, where special problems always exist because of lack of symmetry, reduced perimeter of the critical section, and relatively large unbalanced moments.

Example 11.3: Design Slab Shear Reinforcement
A flat plate floor of normalweight concrete has thickness $h = 11$ in. and is supported by 18 in. square columns spaced 28 ft on centers each way. The floor carries a total

factored load of $q = 300$ psf. Design the slab to resist punching shear at a typical interior column. An average effective depth $d = 9\text{-}1/2$ in. may be used. Material properties are $f_y = 51{,}000$ psi, for headed shear studs, $f'_c = 4000$ psi, and $\lambda = 1.0$.

Solution: A check of the expanded concrete shear capacity (Eq. 11.20) using the minimum prestress of 125 psi giving a shear capacity of 280,150 lb, in which case shear reinforcement is not required. The following calculations use the lower nominal shear strength to illustrate the design of a stud rail system.

The first critical section for punching shear is a distance $d/2 = 4\text{--}3/4$ in. from the column face, providing a shear perimeter $b_o = 4 \times (18 + 9.5) = 110$ in. Based on the tributary area of loaded floor, the factored shear is

$$V_u = q \cdot \left(l^2 - (b+d)^2\right) = 300 \cdot \left(28^2 - 2.29^2\right) = 233\,\text{kip}$$

and the design strength of the slab is

$$\phi V_c = \phi 4\lambda\sqrt{f'_c}\,b_o d = 0.75 \cdot 4 \cdot 1.0 \cdot \sqrt{4000} \cdot 110 \cdot 9.75 = 205\,\text{kip}$$

$\phi V_c < V_u$ indicating that shear reinforcement is required.

Try two stud rails on each column face spaced 14 in. apart, which is less than twice the slab depth. At the perimeter of the headed shear studs the allowable concrete shear stress is $2\phi\sqrt{f'_c}$ leading to a trial length of the stud rail of 32 in. resulting in a perimeter $p_o = 264$ in. and an enclosed area $A_o = 34.2$ ft^2, see Fig. 11.16.

Fig. 11.16 Stud rail pattern

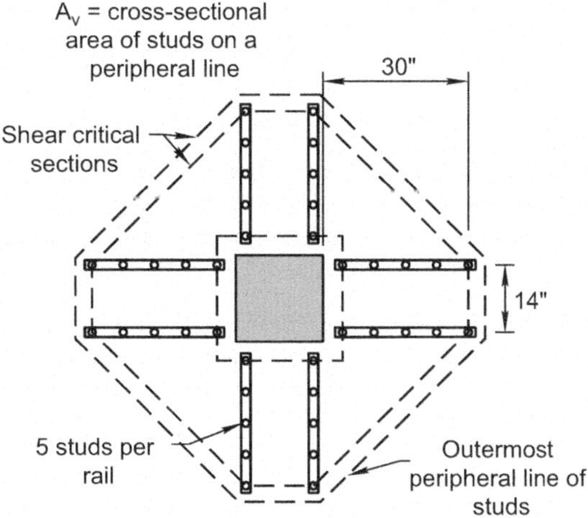

The factored shear is

$$V_u = q \cdot \left(l^2 - A_o\right) = 300 \cdot \left(28^2 - 34.4\right) = 224\,\text{kip}$$

The concrete shear capacity is then:

$$\phi V_c = \phi 2\lambda\sqrt{f_c'}\,p_o d = 0.75 \cdot 2 \cdot 1.0 \cdot \sqrt{4000} \cdot 264 \cdot 9.5 = 244\,\text{kip}$$

confirming that the selected perimeter is satisfactory. The concrete shear capacity at the column is reduced to $2\lambda\sqrt{f_c'}$ when shear reinforcement is used, thus the required strength of the shear reinforcement is:

$$V_s = V_u - 2\phi\lambda\sqrt{f_c'}\,b_o d = 198,300 - 2 \cdot 0.75 \cdot 1.0 \cdot \sqrt{4000} \cdot 110 \cdot 9.5$$
$$= 130\,\text{kip}$$

The stud spacing must be less than $d/2$ so select 4.0 in. and use the yield strength of headed studs of 51 ksi. The thickness of the slab is 11 in. so using 3/4 in. clear top and bottom give $d_s = 9.5$ in. The required area of stud is then

$$A_v = \frac{(V_u - \phi V_c)s}{\phi f_{ys} d} = \frac{130,900 \cdot 4.0}{0.75 \cdot 51,000 \cdot 9.5} = 1.66\,\text{in.}^2$$

with two rows of studs, the critical perimeter line crosses 8 studs and the required stud diameter is 1/2 in. The base plate is 1 in. wide and 1/2 in. thick.

11.6 Two-Way Slab Flexural Design Example

As indicated in the slab analysis design of two-way slabs is an iteration between the indeterminate analysis and the selection of post-tensioning. Example 11.4 is in two parts. The first part deals with the selection of the post-tensioning. The second part uses the selected post-tensioning and approximate analysis to check the selection. The approximate analysis is presented to indicate at least one set of hand calculations that may be useful for independent validation the computer analysis.

Example 11.4: Design Slab Post-Tensioning
Design the post-tensioning for the transverse strip of the building in Fig. 11.17 and perform an approximate check on the solution validity. Given properties are $f_c' = 4000\,\text{psi}$, $f_y = 60,000$ psi and ½ in. diameter low relaxation strand with $f_{pu} = 270$ ksi. Assume that the strand is stressed to $0.70\,f_{pu}$ and has losses of 15 ksi giving an effective prestress force of (189 ksi $-$ 15 ksi losses) $\times$ 0.153 in.^{2} = 26.6 kips/strand. The floor carries a superimposed deal load of 10 psf and an

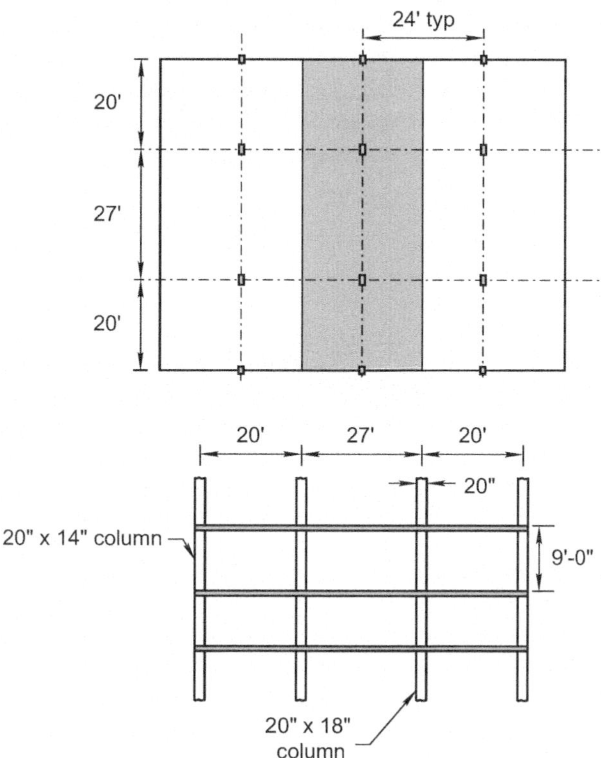

Fig. 11.17   Building plan and elevation

effective live load on the building is 30 psf, which includes consideration of live load reduction factors.

Solution—Design of post-tensioning: Assume a span-to-depth ratio of 45. The resulting slab thickness is 7.2 in., which is rounded up to 7.5 in. The slab loads are then:

$w_{slab} = 7.5$ in./12 in./ft $\times$ 150 pcf $= 93.75$ psf
$w_{sdl} = \quad 10$ psf
$w_{ll} = \quad 30$ psf
giving a total service load $\mathbf{w_{service} = 133.75\ psf}$

The factored load is $\mathbf{w_u} = 1.2(93.75 + 10) + 1.6(30) = \mathbf{172.5\ psf.}$

The post-tensioning design begins by selecting the tendon spacing to provide the minimum 125 psi compression required by code in the section:

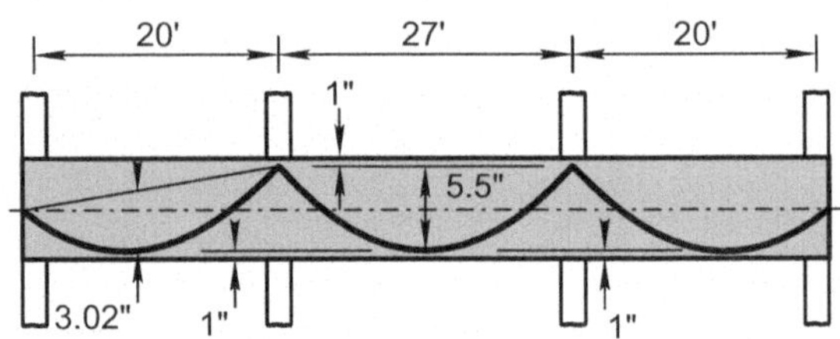

Fig. 11.18 Tendon profile

$$s = \frac{26.6\,\text{kip/strand}}{125\,\text{psi} \cdot 12\frac{\text{in.}}{\text{ft}} \times h} = 2.36\,\text{ft}$$

This spacing results in 10 strands in the 24 ft width of the section. The amount of load to be balanced is 56 psf or about 60% of the slab dead load. To better balance the slab self-weight, 16 tendons at 1.5 ft are selected. This provides an average effective prestress force of 17.5 kips/ft or 197 psi, above the 125-psi minimum required.

In the center panel, the maximum drape in the strand is the slab thickness less ¾ in. cover top and bottom plus ¼ in. to the center of the strand, see Fig. 11.18. Thus, the drape is 7.5 in. − 2 in. = 5.5 in. The effective balanced load is then:

$$w_{\text{bal}} = \frac{8P_e e_{\text{mid}}}{l^2} = \frac{8 \cdot 17.5 \cdot 5.5}{27^2} = 89.3\,\text{psf}$$

This is about 85% of the slab dead load and superimposed dead load. This level of prestress suggests that long term deformations are largely compensated by the post-tensioning. With considering the 15% losses, the initial post-tension closely balances the slab dead load.

To maintain the same load balance in the shorter end span, the tendon eccentricity is adjusted

$$e_{\text{end}} = \frac{w_{\text{bal}}l^2}{8P_e} = \frac{89.3 \cdot 20^2}{8 \cdot 17.75} = 3.02\,\text{in.}$$

which is less than the 4.125 in. available. Therefore, the design proceeds with a balanced load of 89.3 psf.

At this point the design becomes analysis and stress check of the indeterminate structure. The analysis used a long-term service load of $w = 93.75 + 10 - 89.3 = 14.5$ psf and full dead plus live load $w = 93.75 + 10 - 89.3 + 30 = 44.5$ psf. Because such an analysis would be calculated using computer software, it is not provided here.

Check the slab design: An equivalent frame analysis of the middle span results in a long-term unfactored dead load moment of 1.17 ft-kips/ft and a short-term unfactored moment with dead plus live load of 3.60-ft kips/ft. The corresponding sectional stresses are for sustained dead load, $f_{\text{top}} = 72$ psi, $f_{\text{bottom}} = 332$ psi and with live load, $f_{top} = -187$ psi, $f_{bottom} = 581$ psi. The ACI Building Code limit in tension is $6\sqrt{f_c'} = 379$ psi, therefore the design is satisfactory.

A check of the nominal strength is conducted at the maximum moment at the first interior column. The factored moment from the analysis, including consideration of pattern loading, is 15.7 kip-ft/ft. The nominal resistance for the stand is the effective stress plus 10,000 psi. Thus, the strand stress at nominal capacity is $f_{ps} = 174 + 10 = 184$ ksi. The nominal bending capacity is then:

$$a = \frac{f_{ps}A_p}{0.85f_c's} = \frac{184 \cdot 0.153}{0.85 \cdot 4 \cdot 18} = 0.46\,\text{in.}$$

and

$$\phi M_n = \phi A_p f_{ps}\left(d - \frac{a}{2}\right) = 0.9 \cdot 0.153 \cdot 184 \cdot \left(6.5 - \frac{0.46}{2}\right) = 113.2\,\text{ft-kip/ft}$$

This is slightly less than required. Therefore, supplemental reinforcement must be provided. The amount of reinforcement is approximately

$$A_s = \frac{M_u - M_n}{\phi f_y\left(d - \frac{a}{2}\right)} = \frac{15.7 - 13.2}{0.9 \cdot 60 \cdot \left(6.5 - \frac{0.46}{2}\right)} = 0.074\,\frac{\text{in.}^2}{\text{ft}}$$

Table 11.2 requires $0.00075\,A_{cf}$ negative moment. Because the slab has the same section and length in both directions, the amount of reinforcement for a 1-ft strip is conservatively determined using the gross area. Thus:

$$A_s = 0.00075 \cdot b \cdot h = 0.00075 \cdot 12 \cdot 7.5 = 0.0675\,\text{in.}^2/\text{ft}$$

Select No. 4 bars at 18 in. giving 0.13 in.^{2}/ft, which is more than adequate to provide the required excess reinforcement. This value is used to check the computer output values.

The positive moment is taken as approximately $wl^2/16$ or 2.03 in-kip/ft giving an approximate stress distribution shown in Fig. 11.19.

Requirements from Table 11.2 indicate that no supplemental reinforcement is required. In this case, the tensile force is the resultant of the 19 psi across the width of the section by 0.35 in. deep.

Solution check punching shear: The factored shear at the column is w_u time the tributary area, so for an interior column

$$V_u = 172.5 \times 28 \times 20 = 93.5\,\text{kips}$$

and the nominal resistance for a perimeter around an 18 in. by 20 in. column with $b_o = 102$ in. is

Fig. 11.19 Stress distribution

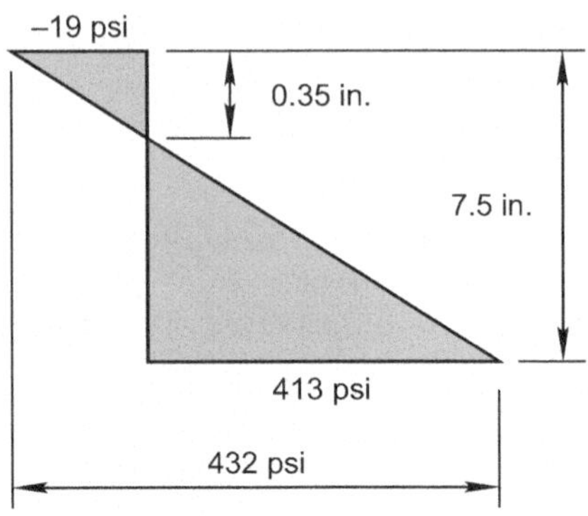

$$V_\text{c} = 4\phi\sqrt{f'_\text{c}}\,b_\text{o}d = 4 \times 0.75 \times \sqrt{4000} \times 102 \times 6.5 = 125.8\,\text{kips}$$

Therefore, shear capacity is adequate and no shear reinforcement is required.

Solution Check deflections: For checking purposes assume 60% of the moment goes to the column strip and 40% goes to the middle strip. The deflection is checked against the total service load less the balance load across the 24 ft width of the strip or 1068 kip/ft. A reference deflection is calculated for the total slab width. Thus:

$I_\text{slab} = \frac{24\,\text{ft}\cdot h^3}{12} = 10,125\,\text{in.}^4$ and the column strip and middle strip have I values one half of the total strip. The reference deflection is then

$$\delta_\text{ref} = \frac{wl^4}{384E_\text{c}I_\text{slab}} = 1068 \cdot \frac{27^4 1728}{384 \cdot 3,600,000 \cdot 10,125} = 0.07\,\text{in.}$$

The deflection in the y direction is then $\delta_y = 0.6 \cdot \delta_\text{ref}\,\frac{I_\text{slab}}{I_\text{col}} = 0.02\,\text{in.}$ and then in the x direction $\delta_x = 0.4 \cdot \delta_\text{ref}\,\frac{I_\text{slab}}{I_\text{mid}} = 0.01\,\text{in.}$ giving a total deflection of 0.03 in. or $l/$ 8200. This implies that the computer-generated deflection should be small, a result of the load balancing, and should meet all code requirements.

Comment: This example is designed to allow the engineer to develop the load balancing required for a detailed analysis. It is not complete as a detailed analysis is not included. The above calculations suggest that a small amount of negative reinforcement is required but that the reinforcement is within the bound of the required supplemental reinforcement required by the ACI Building Code and deflections are not a controlling issue.

References

ACI 352.1R. (2012). Guide for design of slab-column connections in monolithic concrete structures. Farmington Hills, MI: ACI. 28 p.

ACI 352-02. (2002). *Recommendations for design of slab-column connections in monolithic reinforced concrete structures*. Farmington Hills, MI: ACI. 38 p.

ACI 421.1. (2008). *Guide to shear reinforcement for slabs*. Farmington Hills, MI: ACI. 15 p.

Burns, N. H., & Hemakom, R. (1977, June). Test of scale model post-tensioned flat plate. *Proceedings, ASCE, 103*(ST6), 1237–1255.

Dam, T. X., Wight, J. K., & Parra, G. M. (2017, January/February). Behavior of monotonically loaded slab-column connections reinforced with shear studs. *ACI Structural Journal*, 221–232.

Darwin, D., Dolan, C. W., & Nilson, A. H. (2015). *Design of concrete structures* (786 pp). New York: McGraw Hill Education.

Elgabry, A. A., & Ghali, A. (1990). Design of stud-shear reinforcement for slabs. *ACI Structural Journal, 87*(3), 350–361.

Ghali, A. (1989). An efficient solution to punching of slabs. *Concrete International, 11*(6), 50–54.

Hanson, N. W., & Hanson, J. M. (1968, Jan). Shear and moment transfer between concrete slabs and columns. *Journal, PCA Research and Development Laboratories, 10*(1), 2–16.

Hawkins, N. M. (1981, Jan–Feb). Lateral load resistance of unbonded post-tensioned flat plate construction. *PCI Journal, 26*(1), 94–116.

Hawkins, N. M., Bao, A., & Yamazaki, J. (1989). Moment transfer from concrete slabs to columns. *ACI Structural Journal, 86*(6), 705–716.

Moehle, J. P. (1988, Jan–Feb). Strength of slab-column edge connections. *ACI Structural Journal, 85*(1), pp. 89–98.

Odello, R. J., & Mehta, B. M. (1967). *Behavior of a continuous prestressed concrete slab with drop panels*. Report, Division of Structural Engineering and Structural Mechanics, University of California, Berkeley, Berkeley, CA.

PTI. (2006). *Post-tensioning manual* (6th ed., p. 354). Farmington Hills, MI: Post-tensioning Institute.

Wight, J. K. (2015). *Reinforced concrete: Mechanics and design* (7th ed.). New York: Person.

Nilson, A. H. (1987). *Design of prestressed concrete* (2nd ed.). New York: J. Wylie.

Vanderbilt, M. D. (1972). Shear Strength of Continuous Plates, *J. Struct. Div*, ASCE 98(ST5), p. 961–973

Chapter 12
Axially Loaded Members

12.1 Introduction

The primary motivation in prestressing concrete is to offset tensile stresses that develop in flexural members. Consequently, using prestressed concrete in members that are largely loaded in axial compression or loaded in tension might appear contradictory. Furthermore, concrete's low tensile strength makes its use as a tension member immediately doubtful. Nonetheless, there are cases where axially loaded prestressed concrete contributes to the engineer's design options. Applications include ring girders for shells and foundations, circular tanks, columns, and piles.

12.2 Tension Members

Prestressed tension members offer several advantages to solve design concerns. Consider a tension ring girder used to restrain a shell roof, the walls of a circular tank, or a tie for a tied arch structure, Fig. 12.1. In these cases, the concrete is precompressed so that entire section remains in compression under full service loads. Structures such as water tanks that are expected to remain watertight are designed for a net compression stress of 300–400 psi after losses to ensure the tank wall does not leak, especially at cold joints (ACI 350 2006; ACI 372 2013). A criterion of 350 psi residual compression was used on the Riverton Heights Reservoir tank described in Sect. 2.6.3.

Design of axially loaded tension members is based on service level stresses. A residual compressive stress or allowable tensile stress f_{res} is selected. The applied service tensile demand T_s is calculated and the effective prestressing force determined:

© Springer Nature Switzerland AG 2019

C. W. Dolan, H. R. Hamilton, *Prestressed Concrete*,

https://doi.org/10.1007/978-3-319-97882-6_12

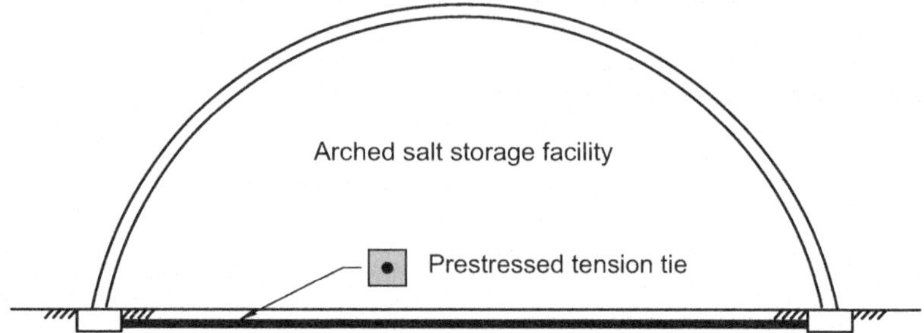

Fig. 12.1 Tension tie for arched storage facility

$$\frac{P_e}{A_g} - \frac{T_s}{A_g} \geq f_{res} \tag{12.1}$$

The effective prestress force is used to ensure that the residual compressive stress is retained. Prestressing reinforcement or nonprestressed reinforcement is used to obtain the required nominal strength.

Prestressed tension elements have advantages over using a stressed steel tie. Because the tie is pretensioned, the tie supports the structure during initial construction. The overall axial stiffness of the prestressed concrete tie is greater than the steel alone. The additional stiffness reduces the tie deformation during construction and in-service. The concrete provides environmental protection for the steel. Lastly, tension stiffening between any cracks that may occur in a bonded prestressed member reduces deflections compared to a bare steel tie.

Example 12.1: Design of a Building Tension Tie

Design of a tension tie for the building shown in Fig. 12.1. The analysis determines that the lateral service load on each frame requiring a tie is a tension due to dead load of 65 kips and tension due to live load of 20 kips resulting in a total service load of 85 kip and a factored load of 1.2D + 1.6L or 110 kip. The maximum allowable service deflection of the tie is 3/8 in. The concrete properties are $f'_c = 5000\,\text{psi}$, $f_t = 424$ psi, $E_c = 4031$ ksi, and the creep coefficient $C_c = 2.2$. The strand is ½ in. diameter 270 ksi low relaxation steel with $E_p = 28{,}500$ ksi and $f_y = 246$ ksi. The tie is post-tensioned using a 1.5 in. diameter duct that is grouted after tensioning. The structure holds salt, so the design requires a permanent residual compression of at least 100 psi to prevent cracking. The initial prestress is 200 ksi, and the effective prestress is 175 ksi.

Determine: the required post-tensioning, the tie section, the deformation under service load, and check the design strength.

Solution: Estimate the required tie area based on the total deflection

$$A_c = \frac{F_s l}{\Delta_{max} E_c} = \frac{85 \cdot 75 \cdot 12}{0.375 \cdot 4,031,000} = 50.6\,\text{in.}$$

Try an 8-in.^{2} tie having a gross area of 64 in.^{2} and a net area of $A_{net} = 8^2 - 1.5^2 \pi/4 = 62.2$ in.^{2} The number of required strands to maintain 100 psi compression are

$$n = \frac{F_s + A_{net} \cdot f_{res}}{A_p \cdot f_{pe}} = \frac{85 + 50.6 \cdot 0.100}{0.153 \cdot 175} = 3.4\,\text{strands}$$

And for strength limiting the stress to the strand yield so the tie does not expand unduly

$$n = \frac{F_u}{\phi \cdot A_p \cdot f_y} = \frac{110}{0.9 \cdot 0.153 \cdot 246} = 3.2\,\text{strands}$$

Select four strands and the total $A_p = 0.612$ in.^{2} giving an initial prestress force of 122.4 kips and an effective force after losses of 107 kips. Finally check stresses and deflections.

The initial stress in the tie is

$$f_i = \frac{P_i}{A_{net}} = \frac{122.4 \cdot 1000}{62.2} = 1967\,\text{psi}$$

The long-term stress on the tie is based on the initial prestress acting on the net area and the long-term dead load acting on the gross area after grouting.

$$f_t = \frac{P_e}{A_{net}} - \frac{F_d}{A} = \frac{107 \cdot 1000}{62.2} - \frac{65 \cdot 1000}{64} = 705\,\text{psi}$$

and the stress in the tie due to combined prestress, dead load, and live load is

$$f_1 = \frac{P_e}{A_{net}} - \frac{F_s}{A} = \frac{107 \cdot 1000}{62.2} - \frac{85 \cdot 1000}{64} = 393\,\text{psi}$$

thereby meeting the 100-psi minimum requirement. The initial deflection of the tie due to prestress is

$$\delta = \frac{P_i l}{A_{net} E_c} = \frac{122.4 \cdot 1000 \cdot 75 \cdot 12}{62.2 \cdot 4,031,000} = 0.439\,\text{in.}$$

This exceeds the 3/8 in. limit; however, this deflection occurs prior to construction of the superstructure. Critical to the structural behavior is the deflection occurring after the dead load is placed. Both the dead load and the prestress act on the gross section and both are subject to long-term creep. Thus, the long-term deflection is a shortening of

$$\delta = \frac{C_{\mathrm{c}}(P_{\mathrm{e}} - F_{\mathrm{d}})l}{AE_{\mathrm{c}}} = \frac{2.2 \cdot (107 - 65) \cdot 1000 \cdot 75 \cdot 12}{64 \cdot 4,031,000} = 0.17\,\mathrm{in.}$$

The deflection due to live load is

$$\delta = \frac{F_{\mathrm{l}}l}{AE_{\mathrm{c}}} = \frac{20 \cdot 1000 \cdot 75 \cdot 12}{64 \cdot 4,031,000} = 0.07\,\mathrm{in.}$$

Giving a total deflection of 0.24 in. within the 3/8 in. limit. Finally, the stress in the tie under the factored load is

$$f_{\mathrm{u}} = \frac{P_{\mathrm{e}}}{A_{\mathrm{net}}} - \frac{F_{\mathrm{u}}}{A} = \frac{107 \cdot 1000}{62.2} - \frac{110 \cdot 1000}{64} = 2.2\,\mathrm{psi}$$

Indicating that the tie does not crack under the factored load.

Comment: The elongation of the tendon due to stressing is 6.3 in. Once stressed, the tie deformations are small enough that serviceability issues due to the tie contracting and expanding are minimal. The stress under factored load is still in compression, indicating that cracking remains unlikely.

12.3 Compression Members

Prestressed concrete has been used for columns, frames, and other compression members (Zia and Moreadith 1966; Zia and Guillermo 1967; Lin and Itaya 1957; Itaya 1965). Prestressed concrete is used to reduce cracking in the columns and thereby ensure that the full sectional bucking strength is available. This is a benefit for both columns and prestressed frames.

Prestressing forces that are aligned with the compression member axis do not affect the bucking behavior of the column. Figure 12.2 indicates two prestressing configurations. The sketch to the left in Fig. 12.2 illustrates a column with the pretensioned strands bonded to the concrete. Any curvature of the column carries the strand with it and the strands immediately provide a counteracting effect. That is,

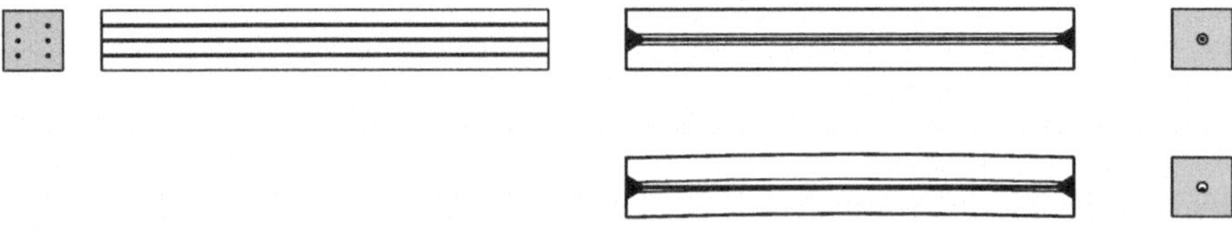

Fig. 12.2 Tendon and concrete in pretensioned and post-tensioned configurations

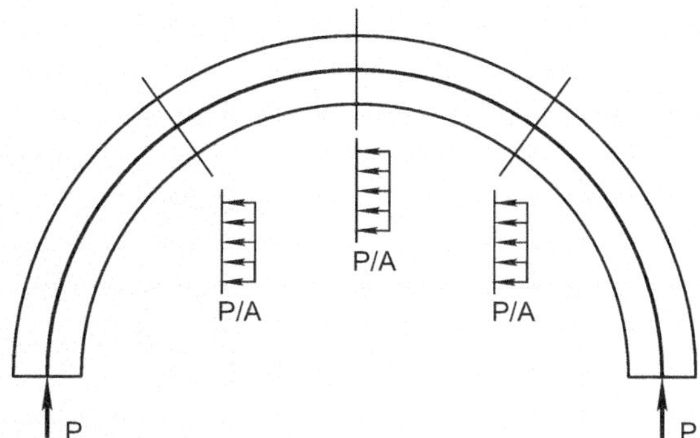

Fig. 12.3 Concentric tendons generate uniform stresses

the P_e/A stress condition remains unchanged along the length. A tendon in a hollow duct is shown to the right in Fig. 12.2. The column deflects laterally until the tendon contacts the side of the duct, then the tendon provides a lateral compensating force.

A free body diagram of a tendon located on the centerline of a compression member indicates that, regardless of the external geometry, the prestressing force remains concentric, Fig. 12.3. In this example, a uniform compression P/A results and the curved member is subjected to only axial shortening.

12.4 Piles

Prestressed concrete piles are commonly used as prestressed concrete compression members (Gerwick 1968). Prestressing provides three important advantages for the pile. First, the prestressing compensates for bending forces resulting during handling and transportation to the construction site, Fig. 12.4. Second, the prestress provides resistance for the pile from cracking during driving. Third, the uncracked prestressed concrete pile provides protection for the prestressed steel, especially in marine environments.

Precast piles range in shape from a 12-in. square section up to 66 in. diameter hollow circular sections, Fig. 12.5. Actual pile dimensions and prestress levels vary by manufacturer and need to be verified for a specific application. In addition to the sections in Fig. 12.5, precast concrete sheet piles are available. The pile length is dependent on the construction site soils conditions. Precast pretensioned piles over 160 ft long have been used. These long piles are designed to be lifted and supported at multiple points along their length to avoid cracking during handling and transportation. Piles are typically pretensioned to 700 psi after losses. The PCI Design Handbook (2017) provides service axial strength of a pile N as:

$$N = A_\mathrm{g}\left(0.33f'_\mathrm{c} - 0.27f_\mathrm{pc}\right) \tag{12.2}$$

where f_pc is the effective compressive stress in the concrete.

Fig. 12.4 Prestressed concrete piles (**a**) pile being lifted from form, (**b**) pile being driven

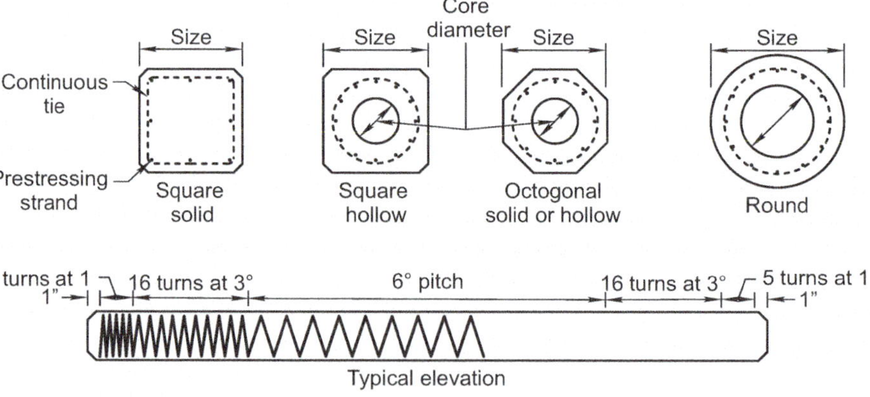

Fig. 12.5 Typical pile sections. From PCI Design Handbook (2017)

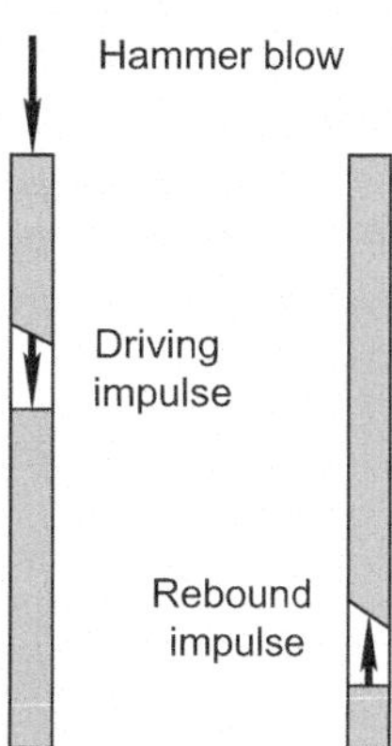

Fig. 12.6 Impulse loading from pile driving

The pile driving hammer generates a compressive impulse stress that travels down the pile. That stress rebounds off the pile tip to create a tensile impulse return stress, Fig. 12.6. The rebound wave is a function of the mass ratio of the hammer and the pile, the input energy, for example velocity or drop, the stiffness of the cushion block, the stiffness of the pile and the driving resistance of the soil (Anderson and Moustafa 1970, 1971). The **Cushion block** or **packing** is typically a stack of semi-compressible material. Green hemlock and Douglas fir are preferred, and plywood or oak is often too stiff. The packing serves several functions including reduction of the peak impact stress and protection of the pile end from spalling. The packing becomes compressed during the pile driving operation and is replaced regularly. Overly stiff packing leads to transverse cracking of the pile. The 700-psi effective prestress is satisfactory to keep the rebound stress from cracking the pile in a well-monitored pile driving operation.

12.4.1 Pile Termination

Piles are fabricated to the specified design length. The pile driving process requires an unobstructed pile end. The end detail results in square pile ends or ends detailed with ducts lapping the strand. Reinforcement is grouted into the ducts to create a connection to the pile cap. The detail is effective if the pile is driven to its design elevation or splicing the pile if additional length is required, Fig. 12.7. An alternative pile splice casts a steel plate into the end of the pile then field welds the splice plates (Bruce Jr. and Hebert 1974). In the situation where the pile is not driven to its full length, the top of the pile is cut off, the exposed strands splayed open, and the strand is used as reinforcement to tie the pile to the pile cap.

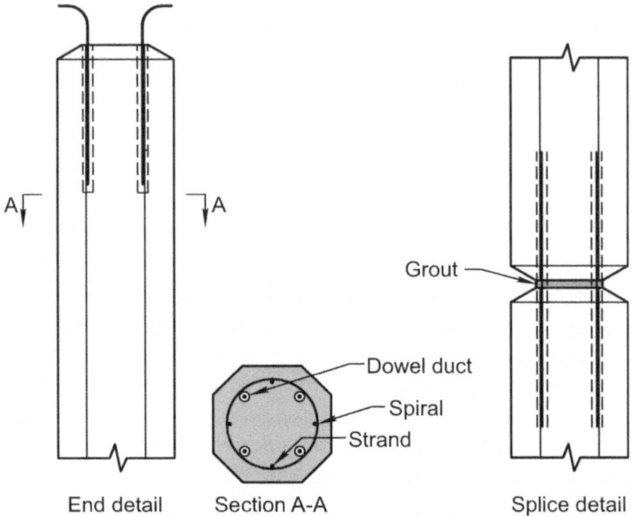

Fig. 12.7 Pile end and splice detail

12.4.2 *Nominal Strength of Piles*

The nominal strength of a pile is established using an interaction diagram like that used for columns. The primary differences are that the reinforcement has a tensile preload so it does not act in compression and the strains are superimposed on the effective strain from the prestress ε_{pe}, Fig. 12.8. By including the effective prestress strain in the calculations, the maximum usable concrete compressive strain is reached earlier than in a reinforced column.

The nominal axial load and moment strength are established by selecting several locations for the apparent neutral axis location c' and calculating the resulting true neutral axis c, the compression block, and the stress in the prestressed reinforcement. Constructing the interaction diagram allows for inclusion of nonprestressed reinforcement if present.

The change in stress in the strand is measured from the apparent neutral axis c'; the location on the section where the strain gradient crosses the effective prestress in the section. The change in strain at the compression face $\Delta\varepsilon'$ is 0.003 less the effective precompression of the concrete. Thus, at the depth of any strand at location i, the stress is:

$$f_i = f_{pe} - \Delta\varepsilon' \frac{c' - d_i}{c'} E_p \qquad (12.3)$$

If the quantity $c' - d_i$ is negative, the strand stress increases. The true neutral axis is located at

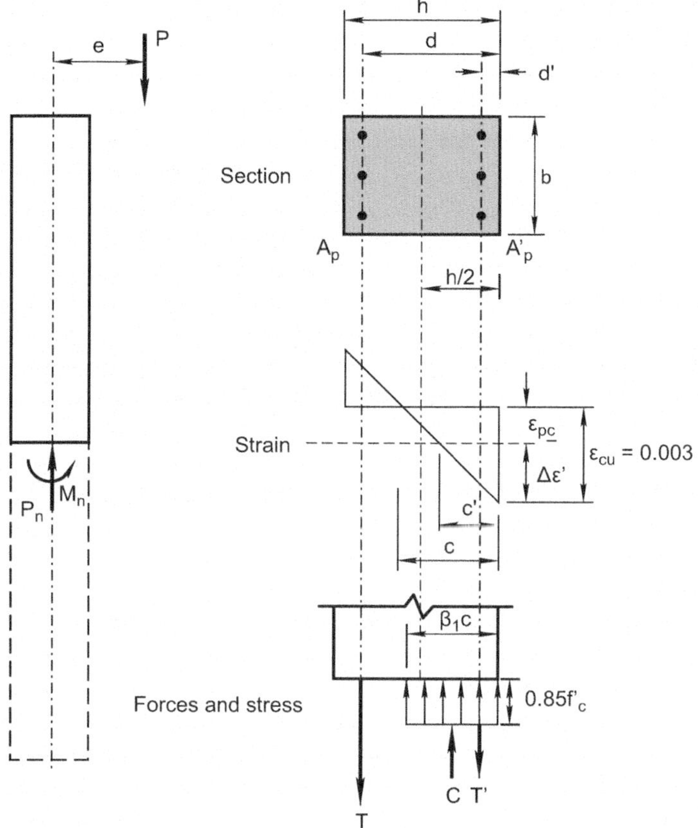

Fig. 12.8 Pile with eccentric loading

$$c = c' \cdot \frac{0.003}{0.003 - \Delta\varepsilon'} \tag{12.4}$$

and

$$a = \beta_1 \cdot c \leq h \tag{12.5}$$

Finally, $C = 0.85f'_c ab$, and the resulting axial load and moment is:

$$P_n = C - \sum T_i$$
$$M_n = C\left(\frac{h}{2} - \frac{c}{2}\right) - \sum T_i\left(\frac{h}{2} - d_i\right) \tag{12.6}$$

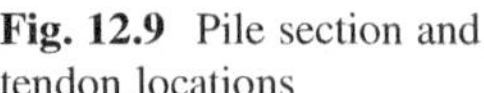

Fig. 12.9 Pile section and tendon locations

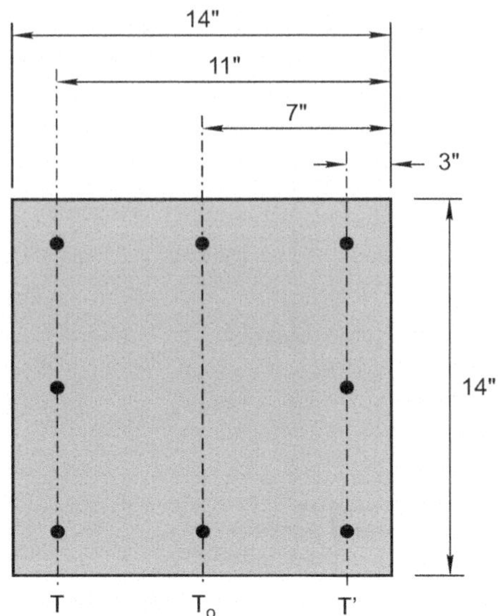

Example 12.2: Calculate the Moment and Axial Load Strength of a Pile

Calculate the moment and axial load strength of a 14-in. square pile with 8-½ in. 270 ksi low relaxation strands having an effective prestress of 175 ksi, Fig. 12.9. The concrete strength is 6000 psi.

Solution: The effective prestress is 175 ksi, giving an effective stress in the concrete of $f_{pc} = 8 \cdot 0.153 \cdot 175/(14 \cdot 14) = 1.09$ ksi and a corresponding concrete strain of 0.000248.

A location is selected for the apparent neutral axis c'. In this example, the first point is at $c' = 3$ in. A total compression strain of 0.003 is applied to the compression face and the net compression strain calculated by deleting the effective concrete compression strain due to prestress. The strain at the location of the prestressing strands is calculated. The stress in each strand is calculated using Eq. 12.3 and the true neutral axis calculated using Eq. 12.4. The nominal axial load and moment are calculated using Eq. 12.6. Finally, the net tensile strain is compared to the 0.005 limit for prestressing strand to establish a value for the strength reduction factor, and the design axial load and moment are calculated. Data for several representative points are given in Table 12.1. The ACI Building Code limits the total axial load to 0.80 times the nominal axial strength of a concentrically loaded element. This limitation is included in the calculation of design strength ϕP_n.

The data is plotted to provide an interaction diagram, Fig. 12.10. For this example, over 60 points are used and the final interaction curves for both the nominal and design conditions are given in the figure below.

The service level load on the pile given by Eq. 12.1 is:

Table 12.1 Representative interaction curve data

c'	c	a (in.)	C (kip)	T' (kip)	T_0 (kip)	T (kip)	P_n (kip)	M_n (in.-kip)	ϕ	ϕP_n (kip)	ϕM_n (in.-kip)
5.00	5.45	4.09	292	66	95	116	15	1648	0.76	11	1250
6.00	6.54	4.90	350	62	86	110	91	1785	0.67	62	1204
8.00	8.72	6.54	467	58	76	94	239	1886	0.65	200	1217
10.00	10.90	8.17	584	55	70	84	375	1815	0.65	328	1016
14.00	15.26	11.44	817	52	62	73	630	1126	0.65	381	326
16.00	17.44	13.08	934	51	60	69	754	1368	0.65	381	42
24.00	26.16	14.00	1000	49	55	61	835	48	0.65	381	29
36.00	39.24	14.00	1000	47	51	55	846	32	0.65	381	20

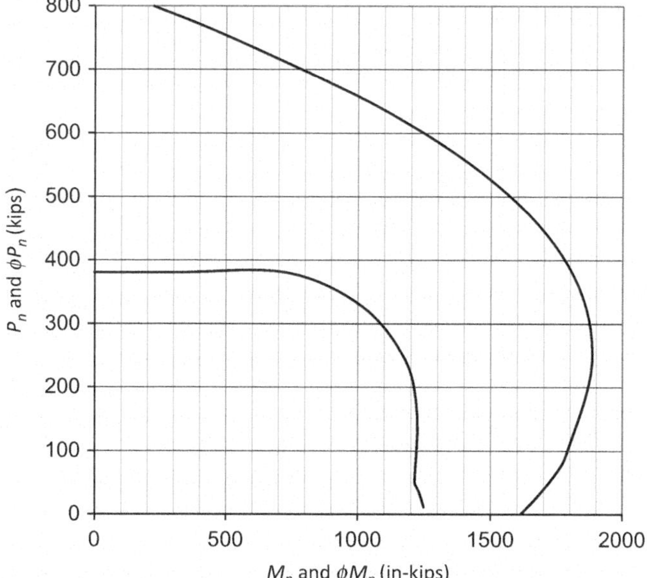

Fig. 12.10 Pile interaction diagram

$$f_{pc} = \frac{A_{ps} f_{ps}}{A_g} = \frac{8 \cdot 0.153 \cdot 175,000}{14 \cdot 14} = 1092\,\text{psi}$$

$$N = A_g\left(0.33 f'_c - 0.27 f_{pc}\right) = 14 \cdot 14 \cdot \left(0.33 \cdot 6000 - 0.27 \cdot 1092\right) = 330\,\text{kips}$$

which is 87% of the design strength.

Comment: The interaction plot is developed using strain compatibility. The presence of the pretensioning in the pile only slightly complicates the calculations. In many cases, piles are designed for vertical loads only in which case the nominal pile strength can be taken simply as $0.80\phi\left(0.85f'_c A_g - A_p f_{pe}\right)$, where the 0.80 is the ACI Building Code limit for axial load on a column to account for accidental eccentricity. In a case where only axial load is required, lateral stability of the structure is developed in alternative load paths. The above example does not address slender piles. The moment magnifier for a slender pile would be derived from the laterally unsupported length of the pile as is done in reinforced column design.

References

ACI 350-06. (2006). *Code requirements for environmental engineering concrete structures ACI Committee 350* (p. 488). Farmington Hills, MI: American Concrete Institute.

ACI 372R-13. (2013). *Guide to design and construction of circular wire- and strand-wrapped prestressed ACI Committee 372* (p. 31). Farmington Hills, MI: American Concrete Institute.

Anderson, A. R., & Moustafa, S. E. (1970, August). Ultimate strength of prestressed concrete piles and columns. *ACI Journal & Proceedings, 67*(8).

Anderson, A. R., & Moustafa, S. E. (1971). *Dynamic driving stresses in prestressed concrete piles*. New York: Civil Engineering ASCE.

Bruce Jr., R. N., & Hebert, D. C. (1974). Splicing of precast prestressed concrete piles: Part 1— Review and performance of splices. *PCI Journal, 19*(5), 70–97.

Gerwick, B. C. (1968, October). Prestressed concrete piles. *PCI Journal, 13*(5), 66–93.

Itaya, R. (1965). Design and uses of prestressed concrete columns. *PCI Journal, 10*(3), 69–76.

Lin, T. Y., & Itaya, R. (1957). A prestressed concrete column under eccentric loading. *PCI Journal, 2*(3), 5–17.

PCI. (2017). *PCI design handbook* (8th ed.). Chicago, IL: Precast and Prestressed Concrete Institute.

Zia, P., & Guillermo, E. C. (1967, June). Combined bending and axial load in prestressed concrete column. *PCI Journal, 12*(3), 52–59.

Zia, P., & Moreadith, F. L. (1966, July). Ultimate load capacity of prestressed concrete columns. *Journal of the American Concrete Institute, 63*(7), 767–788.

Chapter 13
Spliced Girders

13.1 Introduction

Precast, prestressed girders are widely used in buildings and bridges; however, one of their main limitations is the length of the span that can be shipped or the capacity of a simple span beam. AASHTO girder standards were established a half century ago. Safety requirements and higher traffic volume have caused increased bridge span length requirements to accommodate additional lanes or move abutments away from the road edge. Replacement bridges often keep the same structural profile to maintain traffic clearance and approach elevations. To increase span capacity, three strategies are employed: higher strength concrete, continuity, and girder splicing.

Increasing concrete strength is a logical first option. Increased 28-day strength allows for greater prestressing force; however, the greater force requires a higher strength at transfer. This higher transfer strength is achieved by more expensive concrete mixes or longer cure times between casting and transfer. Thus, increasing concrete strength by itself is less effective than using higher concrete strength in conjunction with other options.

The second approach to gaining structural efficiency is to use continuity. Two options for achieving continuity are illustrated in Fig. 13.1. Continuity obtained by having the cast-in-place deck reinforcement carry the negative moment is one method used in bridge design. The concept allows the superimposed dead load and live load to be carried on the continuous structure while the girders carry the self-weight and bridge deck dead load. Continuity using post-tensioning also increases efficiency (Harvey 1986). Some post-tensioning options for continuity are given in Fig. 13.2. These options include placing a tendon axially along the center of the structure to create P/A axial stresses, draping the tendon to balance the loads, using crossed tendons to double the post-tensioning over the support, and using haunched beams with either straight or draped tendons. Costs associated with the post-tensioning system and anchorage requirements are considered in the selection process. Continuity gains additional span capacity for the same beam section

© Springer Nature Switzerland AG 2019
C. W. Dolan, H. R. Hamilton, *Prestressed Concrete*,
https://doi.org/10.1007/978-3-319-97882-6_13

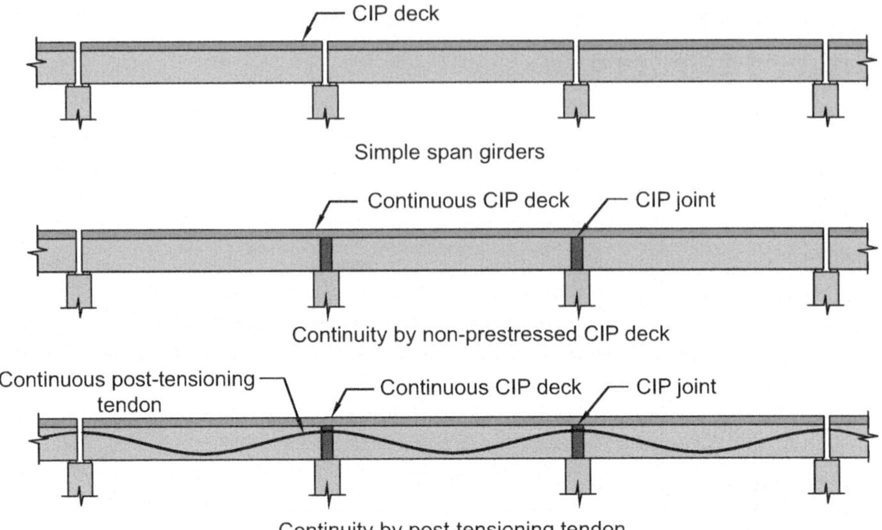

Fig. 13.1 Continuity options for prestressed girders

that would otherwise be used in a simple span configuration. Continuity does not address issues such as shipping weight and length, both potential limitations to design implementation.

To expand the reach of these girders, engineers began to **splice-**join girders at locations other than the support-girders to extend their span or load capacity. In combination with higher strength concrete and staged post-tensioning, greater efficiencies are available. To date, dozens of projects have used spliced girder systems, some having spans up to 350 ft (Castrodale and White 2004; Tadros et al. 1993; Abdel-Karim and Tadros 1992).

13.2 Concepts

Spliced girders accomplish more than making spans continuous at the supports (Geren and Tadros 1994). Splicing allows longer members to be fabricated and more complex shapes to be constructed. These benefits come at a cost of increased construction complexity. Three splicing concepts are illustrated in Fig. 13.3. The first concept uses a cap beam that splices to an end span or an extended end span with a drop-in center span. The second approach splices individual girder segments to make a longer span beam. In both cases, the splice is away from the primary structural support. The Humboldt River Bridge in Northern California uses the center span drop-in approach, Fig. 13.4. In this application, the end span is haunched to create a greater negative moment strength at the column.

The Anacis Channel East Bridge in Vancouver, British Columbia uses splice girders for a sharply curved highway bridge (Marshall and Pelkey 1986). The cost of

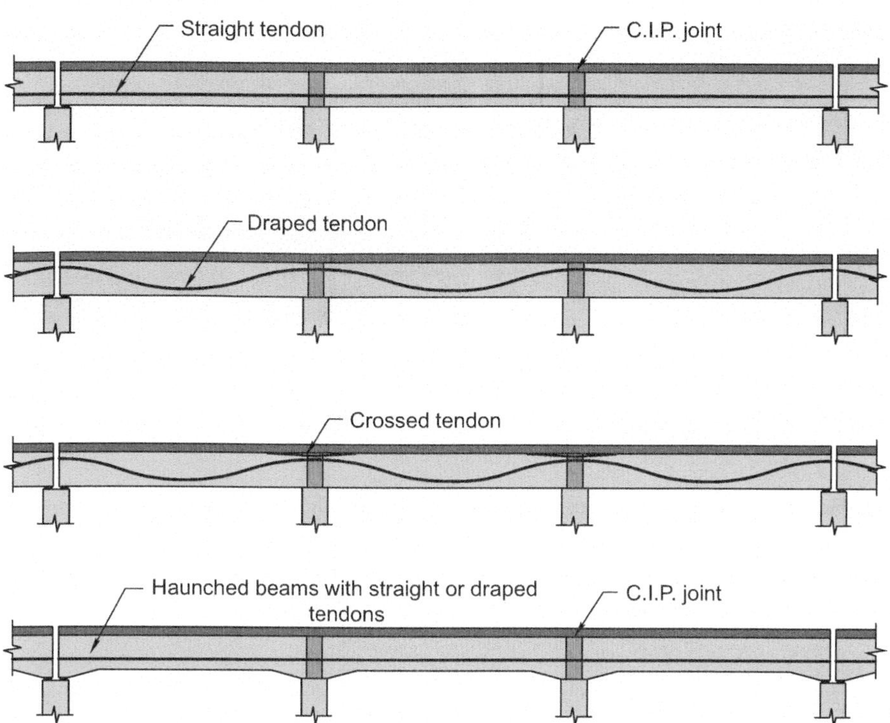

Fig. 13.2 Post-tensioning alternatives

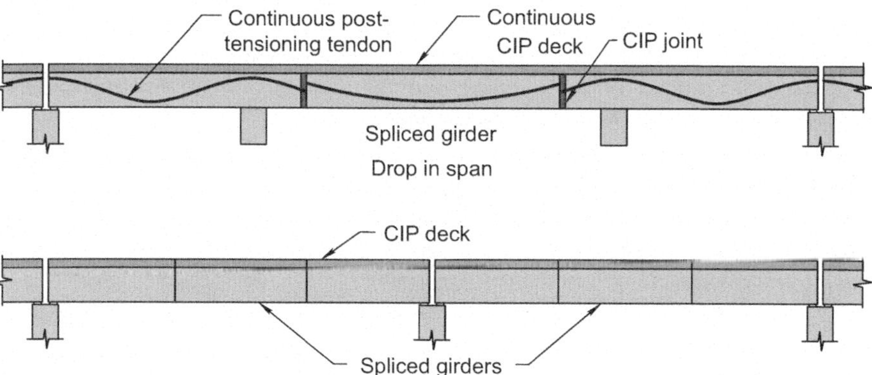

Fig. 13.3 Spliced girder concepts

the spliced girder design was approximately 20% less than the steel bridge alternative. The design used a cap beam over the column and drop-in spans on either side to create a 233 ft main span. The splices allowed the girders to be adjusted laterally to meet the 820 ft radius of curvature for the bridge.

Fig. 13.4 Humboldt River Bridge girders

13.3 Construction

Splicing requires making connections to transfer the shear and moment in the structure. Splicing girders also requires coordination with the anticipated construction methods. These construction methods address the method of splicing, the construction sequence, and post-tensioning methods and sequencing.

13.3.1 Construction Sequence

The construction sequence and loading sequence for spliced girders is similar to traditional bridges with important exceptions. Table 13.1 summarizes the loading stages and highlights the advantages and complications of spliced girder design. The time needed for field post-tensioning and for construction of any intermediate supports vary from simple span construction.

Examining the elapsed time during construction allows consideration of the time effects in the calculation of prestress losses. Creep and shrinkage losses resulting from the pretensioning are recoverable during the later splicing or system post-tensioning. Stages 4 and 5 can be reversed if the girders are designed to carry the cast-in-place deck with the field post-tensioning.

13.3.2 Splicing Options

Splicing options include dapped ends, shores, or temporary connections, Fig. 13.5. The selection of a splicing method is integral to the erection of the girders, temporary erection construction, and to the calculation of stresses. The options indicated in Fig. 13.5 present drop-in connections that minimize the use of temporary erection hardware. The same concepts are used if the girders have spliced components and

Table 13.1 Spliced girder construction sequence

Stage	Construction step	Days since precasting	Comment
1	Cast pretensioned girders	0	
2	Transfer prestress force	1	
3	Beam erection	28	28 days is somewhat arbitrary; however, there is time for sufficient creep and shrinkage to have occurred
4	Cast deck and cast-in-place joints	35	Includes time to erect several girders and coordinate multiple spans
5	Perform post-tensioning	42	See secondary moment considerations in Sect. 13.4
6	Remove temporary supports.	50	See construction notes in Sect. 13.3.3
7	Add superimposed dead load	60	Surface toppings, guard rails, mechanical equipment
8	Open bridge to traffic	70	Live load analysis considers and includes lane load distribution to girders

Fig. 13.5 Splicing options

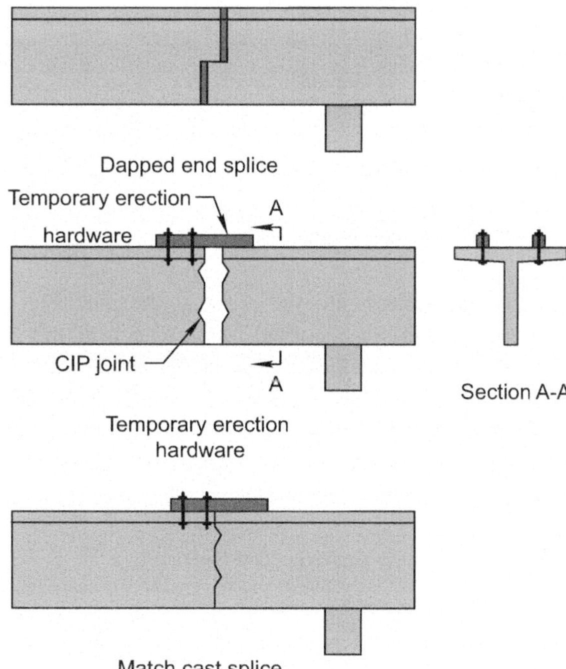

Fig. 13.6 Large spliced girder being launched. Photo courtesy of Richard J. Schmidt

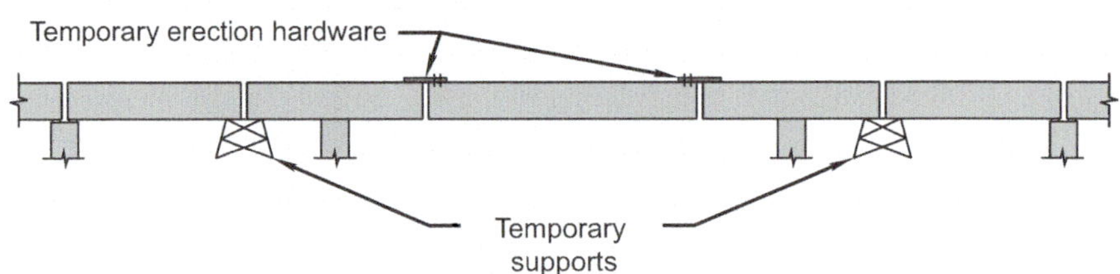

Fig. 13.7 Temporary construction support

are then launched onto the piers at the site, Fig. 13.6. The temporary erection hardware is useful on curved bridges such as the Anacis Channel because lateral diaphragms can be incorporated at the same time the joint is cast. Diaphragms cast concurrently with the cast-in-place joint provide stability for curved structural members.

13.3.3   Construction Sequence

Drop-in designs with dapped ends or temporary hardware are a method of splicing girders. Spliced girder concepts with cap beams or multiple splices still require temporary supports to balance the cap beam prior to post-tensioning, Fig. 13.7. Temporary erection hardware obviates the need for additional temporary supports to

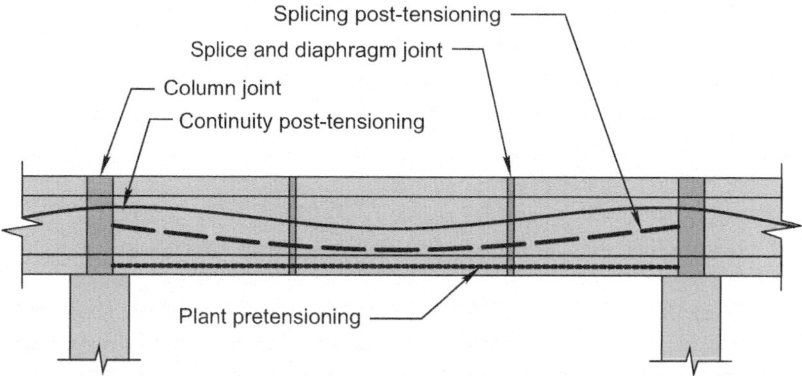

Fig. 13.8 Multiple levels of prestress in spliced girder systems

the ground. Engineers consider these temporary construction loads during design to ensure the structure behaves as intended.

Multiple levels of post-tensioning may be required. A spliced girder bridge proposed by Berger/ABAM Engineers for Snoqualmie Pass in Washington State had three levels of prestress. The project was second low bid and hence not constructed but the concepts illustrate the benefits of splicing. The individual girder segments were pretensioned. The segments would then be post-tensioned at the site along with the intermediate diaphragms to create the basic bridge unit. The bridge units would be made continuous with in-situ post-tensioning, Fig. 13.8. The basic beams had spans up to 180 ft and the end of the bridge rested on a long horizontal curve. The site environmental constraints did not allow for temporary supports. Therefore, design called for assembling the spliced girders at the construction site then use a launching truss to position the bridge units on each column.

13.4 Secondary Moments

The continuity post-tensioning generates secondary moments in the structure. Load balancing techniques are an efficient method of addressing the secondary moments. The load balanced by the post-tensioning acts as a continuous uniform load on the structure. Thus, reactions generated by the post-tensioning are accounted for in the continuity analysis.

In addition to the secondary moments generated by the post-tensioning, moments may result from differential shrinkage of the cast-in-place deck and from the temporary shoring. In shored construction, an equal and opposite force must be applied to the structure when the shore is removed. The timing of the shoring removal impacts the structure, especially in relation to the placement of the cast-in-place deck. If shoring is removed before the deck is cast, the deck is carried on the bare beam. If shoring is removed after the deck has hardened, stresses resulting from the shore removal are carried on the composite structure.

13.5 Critical Sections

Critical sections occur where loads and stresses are at maximum (or minimum) values. Initially, critical sections occur at midspan of the girder, at negative moment locations over the supports, and at the splice. The design example illustrates how these critical sections evolve as the post-tensioning changes the structural system. The splicing operations compound the number of critical sections for two reasons. First, some critical sections may occur at intermediate construction times. For example, the maximum tensile stress in a beam may be at the time of transfer of post-tensioning force or shore removal not at a later live load application. The sequential post-tensioning changes the structure response to live loads, thereby generating additional possible critical sections. The engineer establishes critical sections created due to construction, post-tensioning, and splicing details.

13.6 Design Example

This example presents the preliminary design of a four-span continuous bridge and the steps required to verify service level stresses and strength. The concrete strength and allowable stresses are given in Table 13.2. The example uses a simplified spliced girder bridge as a case study; however, the study does not examine all the bridge design parameters. It focuses on the prestressing and post-tensioning operations.

Table 13.2 Allowable stresses and steel stress

Stage	Days since precasting	Event	Precast concrete			CIP concrete			Steel stress
			f'_c (psi)	f_t (psi)	f_c (psi)	f'_c (psi)	f_t (psi)	f_c (psi)	f_{ps} (ksi)
1	0	Cast	0						216
2	1	Transfer	4000	380	2400				189
3	28	Erection	6000	465	3600				189
4	35	CIP deck	6000	465	3600				189
5	60	PT and support removal	6000	465	3600	3000	329	1800	189
6	70	Apply SDL	6000	465	3600	4000	380	2400	189
7	50 years	Open to traffic	6000	465	3600	4000	380	1600	194

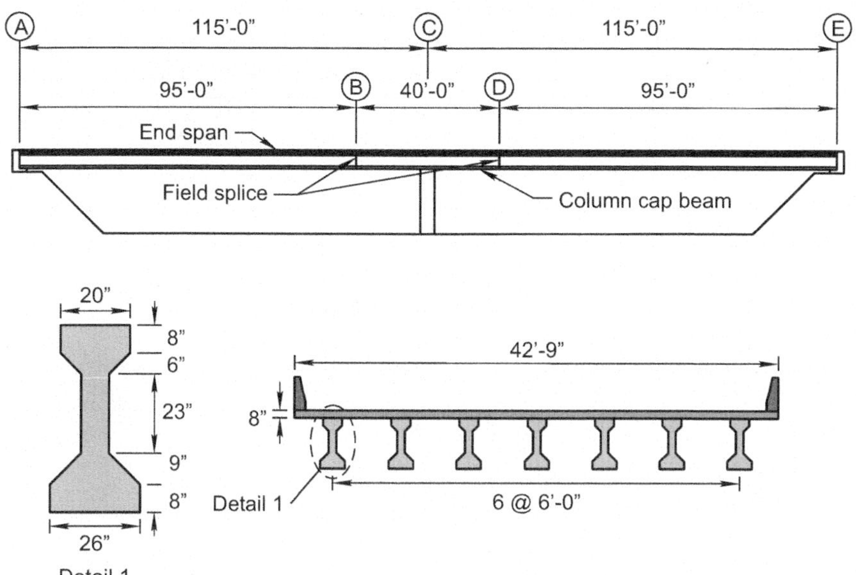

Fig. 13.9 Girder and bridge dimensions

Table 13.3 Beam section properties

Property	Non-composite	Composite stage 5	Composite stage 6 and 7
A (in.^{2})	789	1230	1298
I (in.^{4})	260,740	602,210	636,022
y_b (in.)	24.73	36.67	37.78
y_t (in.)	29.67	25.34	24.22
S_b (in.^{3})	10,543	14,564	14,393
S_b (in.^{3})	8788	21,074	22,458
S_i (in.^{3})		30,798	33,537

Allowable stresses for stages 1 through 6 are based on temporary conditions and stage 7 on final stress limits. For temporary conditions, the allowable tension is $6\sqrt{f'_c}$ and $0.6f'_{ci}$ for compression. Final stresses are $6\sqrt{f'_c}$ for tension and $0.4f'_c$ for compression. The prestressing steel stresses are $0.8 f_{pu}$ jacking, $0.7 f_{pu}$ immediately after anchor set and $0.72 f_{pu}$ final stress. The bridge is fabricated from AASHTO Type IV girders and uses a dapped end splice connection, Fig. 13.9.

The bridge section properties are given in Table 13.3, and the final beam section is given in Fig. 13.10.

At stage 5, f'_c for the deck concrete strength is 3000 psi and the modular ratio is $n = \sqrt{3/6} = 0.707$ resulting in an effective flange width of 55.15 in. For stages 6 and 7 the deck concrete strength is $f'_c = 4000\,\text{psi}$ and the modular ratio is $n = \sqrt{4/6} = 0.816$ with a corresponding effective flange width of 63.69 in. For simplicity, the properties for f'_c of 3000 psi are used for stages 4 through 7. This

Fig. 13.10 Final
composite beam

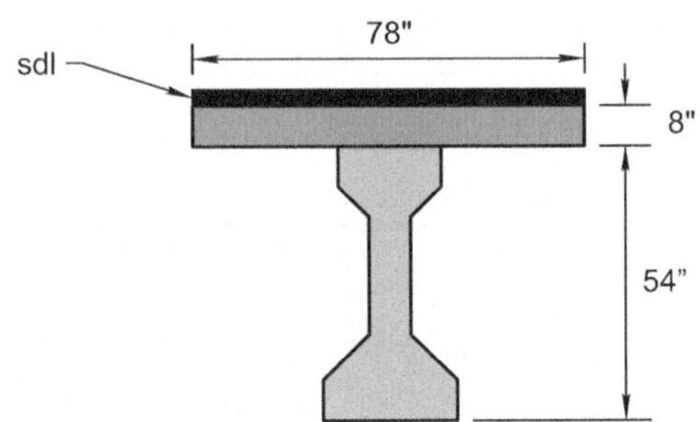

simplification of the effective deck width would be revisited if the allowable stresses
are exceeded in Stage 7. A superimposed dead load (sdl) of 130 plf is included for a
2-in. thick asphalt overlay. The live load is an AASHTO HS20-44 lane load
simplified so that each girder carries 0.66 times the lane load, resulting in a uniform
live load of 422 plf per girder but ignores the AASHTO *Design Specification*
movable concentrated load. The uniform load includes consideration for girder
spacing and impact effects. Pretensioning losses and post-tensioning losses are
taken as 25 ksi. Pretensioning considers a two-point harping arrangement. This
harping creates a potential critical section at the harp point, which is not checked
in this example.

Service level stresses are calculated at the precast beam midspan and at the face of
the center column. The continuous structure has a maximum positive moment at 3/8
the total span length. This is sufficiently close to the midspan of the end beam that
the stresses are simply summed, which is conservative. A final design analysis would
examine if a nearby point, for example the harp point, has higher stresses. The beam
nominal bending strength is checked at these same two locations. The magnitude of
the shear at the column is checked to verify that it falls within allowable limits and
the horizontal shear transfer reinforcement is calculated.

13.6.1 Stage 1 and 2

In stage 1, the pretensioning for the end and cap beam is designed. The girder weight
is 849 plf based on a concrete unit weight of 150 pcf. The precast end span is 95 ft.
Looking ahead to the addition of the deck, the beam carries its simple span self-
weight plus the dead weight of the cast-in-place slab, which is 650 plf resulting in
$M_g = 11,126$ in-kips and $M_{slab} = 8799$ in-kips. Using load balancing with third point
harping, the trial value for the pretensioning is based on the moment due to the full
girder weight plus the moment due to one quarter the slab weight. The one quarter of
the slab load came by trial. Balancing half the deck load resulted in excess prestress.
An eccentricity of 19.23 in. is selected, which is equal to the distance from the beam
centroid to the bottom of the beam less 1.5 in. cover, 0.5 in. stirrup diameter, and
3.5 in. to the center of the tendon. The strands are stressed to an initial stress of

Table 13.4 Summary of stresses in end span for stages 1–3

Properties		End beam Stresses					
		Initial stresses			Final Stresses		
789	in.^{2}		Top	Bottom		Top	Bottom
260,740	in.^{4}		(psi)	(psi)		(psi)	(psi)
29.67	in.	P_i/A	776	776	P_e/A	679	679
24.78	in.	P_iey/I	−1478	1235	P_eey/I	−1294	1080
21.23	in.	M_gy/I	1266	−1057	M_gy/I	1266	−1057
		Sub			Sub		
95	ft	total	563	953	total	651	702
6	in.				M_{slab}/I	1001	−836
					Subtotal	1652	−135
					$M_l y/I$	0	0
					Total	1652	−135

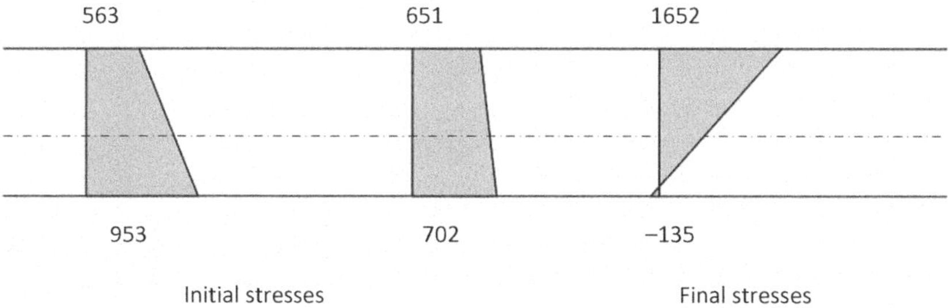

563	651	1652
953	702	−135

Initial stresses   Final stresses

$f_{ci} = 0.6 f'_{ci} =$ 2400 psi $f_c = 0.40 f'_c =$ 3600 psi

$f_{ti} = 6\sqrt{f'_{ci}} =$ −379 psi $ft = 6\sqrt{f'_c} =$ −465 psi

200 ksi, giving a pretensioning force of 30.6 kips per ½ in. diameter low relaxation strand. The trial number of strands is:

$$n = \frac{M_g + M_s/4}{P_i \cdot e} = \frac{11{,}497 + 1100}{30.6 \cdot 21.35} = 21.6$$

The trial suggests 22 strands. A final selection of 20 strands is eventually chosen to minimize the prestress in the beam. The 22 strands result in an initial prestressing force of 612 kips and a final prestressing force of 536 kips based on a final stress of 175 ksi after losses. The stresses in the simply supported beam are summarized in Table 13.4.

The column dimension is 4 ft in the direction of the bridge. The cap beam cantilevers 18 ft from the face of the column in each direction. The tendon eccentricity over the column is the distance from the neutral axis to the top fiber less 4 in. cover and clearance or 25.67 in. The cap beam carries its self-weight, the reaction from the end span, and the weight of the cast-in-place deck. The resulting girder

Table 13.5 Summary of stresses at column face for stages 1–3

Prestress

			Properties			**Cap beam stresses**					
						Initial stresses			**Final Stresses**		
$n =$	12	Strands	$A=$	789	in.^{2}		Top	Bottom		Top	Bottom
$P_i =$	367	kip	$I =$	260740	in.^{4}		(psi)	(psi)		(psi)	(psi)
$P_e =$	321	kip	$y_t =$	29.67	in.	P_i/A	465	465	P_e/A	407	407
			$y_b =$	24.78	in.	$P_i ey/I$	804	−671	$P_e ey/I$	703	−587
Loads			$e =$	−19.23	in.	$M_g y/I$	−182	152	$M_g y/I$	−182	152
						Sub total	1087	−54	Sub total	928	−28
			$L =$	18	ft				$M_{end\ girder}/I$	−662	553
$M_g =$	−1598	in-kip	$e_{end} =$	6	in.						
$M_{end\ girder}$ =	−5814	in-kip							Subtotal	267	524
$M_{slab} =$	−3966	in-kip							$M_{slab} y/I$	−451	377
$w_g =$	822	plf							Total	−184	901

Concrete

$f'_{ci} =$	4000	psi
$f'_c =$	6000	psi
$\gamma_{girder} =$	150	pcf
$\gamma_{deck} =$	150	pcf
$w_{deck} =$	650	plf

1087	928	267	−184
−54	−28	524	901

Initial stresses   Final stresses

Allowable stresses

$f_{ci} = 0.6\,f'_{ci} =$	2400	psi	$f_c = 0.40\,f'_c =$	3600	psi
$f_{ti} = 6\sqrt{f'_{ci}} =$	−379	psi	$f_t = 6\sqrt{f'_c} =$	−465	psi

moment is $wl^2/2 = 0.821 \text{ klf} \cdot 18^2 \cdot 12 \text{ in./ft}/2 = 1598$ kip in plus the reaction of the end span times the cantilever length $= 0.821 \cdot 95/2 \cdot 18 \cdot 12 \text{ in./ft} = 8432$ in-kip. The trial number of strands is

$$n = \frac{M_g + V_{end} \cdot l/2}{P_i \cdot e} = \frac{1598 + 8432}{30.6 \cdot 25.67} = 11.6$$

Select 12 strands for an initial trial giving an initial prestressing force of 367 kips and a final prestress force of 321 kips. The stresses in the cap beam are summarized in Table 13.5.

13.6.2 Stage 3 Erect Drop-in Precast Beams

Temporary shoring is installed and the cap beam is erected. The precast end span is set onto the cap beam and the end abutment. A dapped end connection is used so the temporary shoring only needs to balance the cap beam until both beams are placed. This support can be shoring under the end of the first beam to be placed, a tie down on the end opposite the first beam to be placed, or both. The shoring primarily addresses the unbalanced load until the final beam is set. The placement of the end beam results in a moment of 5813 in-kip at the center of the beam. The stresses due to beam placement are summarized in Table 13.5. There is no change in the moments in the end beam resulting from this stage of the construction.

13.6.3 Stage 4 Cast Deck

Two options exist for casting the deck. First, the deck can be cast on the bare beams followed by post-tensioning. Alternatively, the structure may be post-tensioned before the deck is cast. In this example, it is assumed that the deck is cast prior to post-tensioning. This option allows the deck and the closure joint to be post-tensioned. The post-tensioning provides precompression on the entire structure and potentially improves service life by reducing cracking. Because the deck is post-tensioned, removal of the deck for future repairs could be more complicated. A membrane sealer between the asphalt overlay and the deck would be prudent practice. The unit weight of the deck concrete is 150 pcf.

The end beam carries the deck weight as a simple span beam because the joint that provides continuity is not yet complete. The cap beam carries the added reaction of the end beam and deck plus the self-weight of the deck on the cap span. Casting the deck generates a midspan moment of 8800 in-kip on the end span and 3966 in-kip at the column face. The resulting stresses are summarized in Table 13.4 for the end beam and Table 13.5 for the cap beam.

13.6.4 Stage 5 Post-tension the Structure

A continuous tendon runs the length of the structure. The distance from the composite neutral axis to the tension face is 36.66 in. The pretensioning tendons are located 5.5 in. above the bottom soffit. Allowing 6 in. for the duct and clearance to the pretensioning tendon allows a maximum eccentricity for the drape at midspan of the simple end beam of 36.67 in. $-$ 5.5 in. $-$ 6.0 in. $=$ 25.17 in. A final eccentricity of 21.23 in. is selected to reduce the total design moment strength and facilitate lower horizontal shear requirements. The duct is placed entirely within the precast beam. The eccentricity over the support is the distance to the top of the composite

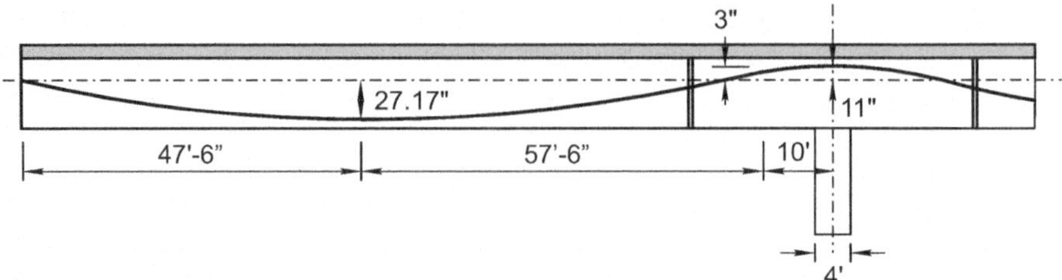

Fig. 13.11 Post-tension tendon profile

beam, less the thickness of the deck, less 6 in. to the duct, giving an eccentricity of 25.34 in. $-$ 8 in. $-$ 6 in. $=$ 11.33 in. and clears the pretensioned strand. For load balancing, the total drape is taken as the sum of the two eccentricities, Fig. 13.11. This ignores the reverse curvature and provides a total drape of 27.17 + 11.33 = 38.5 in. The tendon is located at the composite neutral axis at the end of the beam.

A trial prestress for the beam uses load balancing to compensate for the deck slab weight of 650 plf. The trial prestress force is

$$P_{\text{trial}} = \frac{w_{\text{slab}}l^2}{8e_{\text{drape}}} = \frac{0.650 \cdot 115^2}{8 \cdot 35.8/12} = 360\,\text{kip}$$

The post-tensioned tendon is initially stressed to 216 ksi and anchored at 200 ksi. Elastic shortening losses reduce the stress at the anchors to the $0.70\,f_{\text{pu}}$ limit. The final stress after losses is 175 ksi. Frictional losses are 3.7 ksi at midspan and 7.2 ksi over the support. Based on the final stresses of approximately 175 ksi per ½ in. diameter strand, this requires 13.5 tendons. Select two 6 strand tendons, which provide an initial prestress of force of 12·(200–3.7)·0.153 = 360 kips at midspan of the end beam and an effective prestress force of 314 kips. The prestress forces at the column are initially 353 kips and after losses 307 kips. Twelve strands are selected because the critical section is negative moment at the column. The maximum moment at the face of the column is used for design, and this moment is less than that at the center of the column.

The balance load on the beam due to post-tensioning is calculated using the final prestress of 360 kips giving an initial equivalent load of 663 plf and a final equivalent load of 579 plf. The positive moment is 9/128 wl^2 based on the two-span continuous structure and the negative moment is $wl^2/8$. This results in a final positive moment of -6455 in-kip and a negative moment of 11,476 in-kip. The negative sign on the positive moment indicates that the balanced load is upward. The corresponding stresses in the end span are summarized in Table 13.6 and stresses at the column face in Table 13.7. The tables summarizing the loads include the stresses locked in due to girder self-weight and deck weight. These stresses are on the beam at the composite interface. The benefit of post-tensioning is seen at the column face where the deck remains in compression once the superimposed dead load is applied.

13.6.5   *Stage 6 Superimposed Dead Load*

The superimposed dead load of 130 plf is applied to the continuous structure. The positive moment is based on 9/128 wl^2 and the negative moment is $wl^2/8$ resulting in a final positive moment of 1450 in-kip and a negative moment of 2579 in-kip. The resulting beam stresses are given in Tables 13.6 and 13.7.

13.6.6   *Stage 7 Live Load*

The superimposed live load is based on an AASHTO lane load adjusted for girder spacing and including impact factor. This produces an equivalent uniform load of 422 plf per girder and is applied to the continuous structure. The positive moment is based on 9/128 wl^2 and the negative moment is $wl^2/8$ resulting in a final positive moment of 2619 in-kip and a negative moment of 8379 in-kip at the center of the column. The final beam stresses are given in Table 13.7 and the negative moment could be reduced to the face of the column.

Table 13.6 End span summary of stresses for stages 4–6

Prestress			Properties				Initial stresses	Top	Mid	Bottom	Final Stresses	Top	Mid	Bottom
$P_i =$	360	kip	$A =$	1230	in.^{2}			(psi)	(psi)	(psi)		(psi)	(psi)	(psi)
$P_e =$	314	kip	$I =$	602210	in.^{4}									
			$S_c =$	34735	in.^{3}		Carry over stresses		1652	−135			1652	−135
			$y_t =$	25.34	in.		P_i/A	293	293	293	P_e/A	255	255	255
			$y_b =$	36.66	in.									
Loads			$e =$	21.23	in.									
			$L =$	115	ft		$M_{bal}y/I$	−275	−188	398	$M_{bal}y/I$	−275	−188	398
$M_{bal} =$	−6544	in-kip	$e_{end} =$	0	in.		Sub total	17	1757	557	Sub total	−20	1719	519
$M_{sdl} =$	1451	in-kip									M_{sdl}/I	61	42	−88
$M_l =$	2578	in-kip									Subtotal	41	1761	431
											$M_l y/I$	108	74	−157
$f'_{ci} =$	4000	psi									Total	149	1835	274
$f'_c =$	6000	psi												

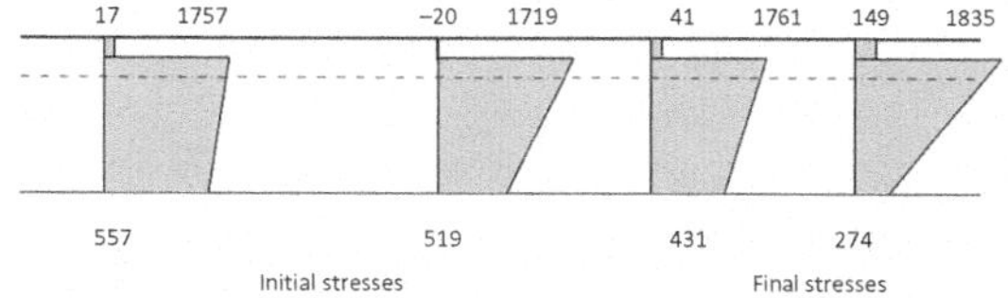

Allowable stresses

$f_{ci} = 0.6 f'_{ci} =$	2400	psi		$f_c = 0.40 f'_c =$	3600	Psi
$f_{ti} = 6\sqrt{f'_{ci}} =$	−379	psi		$f_t = 6\sqrt{f'_c} =$	−464	Psi

Table 13.6 (continued)

Prestress				Loads				Concrete		
$P_i =$		360	kip	$M_{bal} =$	-6544		in-kip	$f'_{ci} =$	4000	psi
$P_e =$		314	kip	$M_{sdl} =$	1451		in-kip	$f'_c =$	6000	psi
				$M_l =$	2578		in-kip			

Properties			End span stresses				Final Stresses			
$A=$	1230	in.^{2}		Top	Mid	Bottom		Top	Mid	Bottom
$I =$	602210	in.^{4}		(psi)	(psi)	(psi)		(psi)	(psi)	(psi)
$S_c =$	34735	in.^{3}								
$y_t =$	25.34	in.	Carry over stresses		1652	-135			1652	-135
$y_b =$	36.66	in.	P_i/A	293	293	293	P_e/A	255	255	255
$e =$	21.23	in.								
$L =$	115	ft	$M_{bal}y/I$	-275	-188	398	$M_{bal}y/I$	-275	-188	398
$e_{end} =$	0	in.	Sub total	17	1757	557	Sub total	-20	1719	519
							M_{sdl}/I	61	42	-88
							Subtotal	41	1761	431
							$M_l y/I$	108	74	-157
							Total	149	1835	274

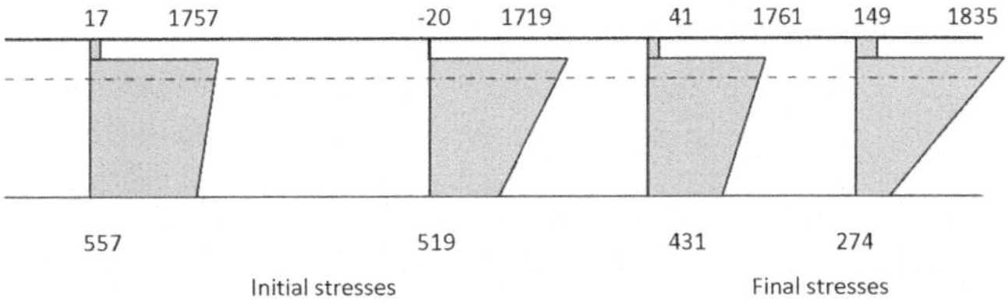

<table>
<tr><td>17</td><td>1757</td><td>-20</td><td>1719</td><td>41</td><td>1761</td><td>149</td><td>1835</td></tr>
<tr><td>557</td><td></td><td>519</td><td></td><td>431</td><td></td><td>274</td><td></td></tr>
<tr><td colspan="4" align="center">Initial stresses</td><td colspan="4" align="center">Final stresses</td></tr>
</table>

Allowable stresses

$f_{ci} = 0.6 f'_{ci} =$	2400	psi		$f_c = 0.40 f'_c =$	3600	Psi
$f_{ti} = 6\sqrt{f'_{ci}} =$	-379	psi		$f_t = 6\sqrt{f'_c} =$	-464	Psi

13.6.7 Flexural Strength

The flexural strength is based on the factored live loads acting on the continuous structure. The factored load is $w_u = 1.2(w_g + w_{slab} + w_{sdl}) + 1.7\, w_{live} = 2.72$ kips/ft. The positive moment is based on $9/128\, w_u l^2$ and the negative moment is $w_u l^2/8$ resulting in a positive moment of 30,356 in-kip and a negative moment of 53,966 in-kip.

The ACI formula for tendon strength (Eq. 5.7) is used. The positive moment calculation assumes the compressive stress block is in the deck so β_1 is 0.85 for $f'_c = 4000$ psi concrete. Low-relaxation strand has a γ_p of 0.28 and a nominal strength of 270 ksi. The positive moment section has a total of 32-½ in. diameter strands at an average depth of 53.4 in. With the compression block in the top flange, the prestressing reinforcement ratio is 32 strands divided by the flange width times the depth or 0.0014. The resulting nominal strand stress is

Table 13.7 Summary of stresses at column face for stages 4–6

Prestress			Properties				Initial stresses			Final Stresses			
n =	12	strands	A =	1230	in.^{2}		Top		Bottom		Top		Bottom
P_i =	353	kip	I =	602210	in.^{4}		(psi)		(psi)		(psi)		(psi)
P_e =	307	kip											
						Carry over stresses							
Concrete			y_t =	25.34	in.			−184	901			−184	901
f'_{ci} =	4000	psi	y_b =	36.66	in.	P_i/A	287	287	287	P_e/A	250	250	250
f'_c =	6000	psi	e =	−9.33	in.	$P_i ey/I$				$P_e ey/I$			
Loads			L =	115	ft	$M_{bal}y/I$	483	330	−698	$M_{eq}y/I$	483	330	−698
M_{bal} =	11467	in-kip	S_c =	34735	in.^{3}	Sub total	770	433	490	Sub total	732	395	453
M_{sdl} =	−2579	in-kip								M_{sdl}/I	−109	−74	157
M_l =	−8379	in-kip								Subtotal	624	321	610
										M_l/I	−353	−241	510
										Total	271	80	1120

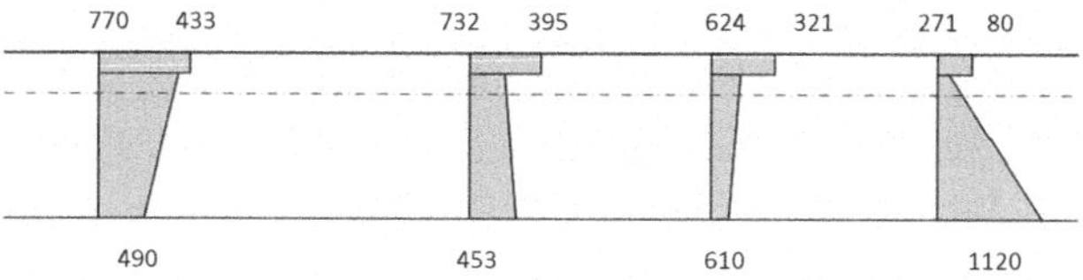

Allowable stresses

Initial stresses			Final stresses		
$f_{ci} = 0.6 f'_{ci}$ =	2400	psi	$f_c = 0.40 f'_c$ =	3600	psi
$f_{ti} = 6\sqrt{f'_{ci}}$ =	−379	psi	$f_t = 6\sqrt{f'_c}$ =	−464.8	psi

$$f_{ps} = f_{pu}\left(1 - \frac{\gamma_p}{\beta_1}\frac{\rho_p f_{pu}}{f'_c}\right) = 270\left(1 - \frac{0.28}{0.85}\frac{0.0014 \cdot 270}{4}\right) = 261.4\,\text{ksi}$$

A check indicates that the compression stress block has a depth of 4.1 in. placing it within the deck thickness. The nominal moment strength is then

$$M_n = nA_{ps} f_{ps}(d - a/2) = 32 \cdot 0.153 \cdot 261.4\left(1 - \frac{4.1}{2}\right) = 65,778\,\text{in-kip}$$

The design moment is the nominal moment times the strength reduction factor of 0.9 giving 59,200 in-kip, which is greater than the factored moment of 30,356 in-kip, indicating the design is satisfactory.

Calculation of the negative moment follows the same format with the following revisions: β_1 is 0.80 because the compression block is at the base of the precast beam, f'_c is 6000 psi, the average depth of the tendon is 54.2 in., and the prestress reinforcement ratio is 0.0035 based on the 26-in. wide beam stem. These revisions result in a nominal strand stress of 255.2 ksi, a compression stress block 8.01 in. deep, and a nominal moment of 66,311 in-kip. The stress block is equal to the 8-in. depth of the bottom of the beam, thereby verifying the initial assumption. The 0.01 in. additional depth of the equivalent stress block is well within the margin of acceptance. The design moment is 0.9 M_n = 59,680 in-kip which is larger than the factored moment of 53,966 in-kip, concluding that the negative moment strength is adequate.

13.6.8   Check Transverse Shear Strength

The check of the shear strength examines whether the factored shear stress is within allowable limits. In this case, the shear is checked against the provisions of the ACI Building Code requirements. The ACI Building Code does not require shear reinforcement if the shear stress is less than ½ v_c where v_c is $2\sqrt{f_c'}$. Further, the ACI Building Code does not allow the factored shear stress to exceed $10\sqrt{f_c'}$. The shear is carried in the precast beam with concrete compressive strength of 6000 psi. Using this information, factored shear stresses less than 77.5 psi do not require shear reinforcement and factored shear stresses greater than 774 psi require redesign of the beam.

The maximum shear in the structure occurs the column centerline and is 5/8 $w_u l$ or for this case 195.5 kips. The final factored shear is taken at a distance d from the column face and is less than the above value. The beam web width is 8 in. Using d from the negative moment section, the factored stress in the beam is

$$v_c = \frac{V_u}{b_w \cdot d} = \frac{195.5}{8 \cdot 54.2} = 451\,\text{psi}$$

The factored shear stress of 451 psi lies within the Code limits. Therefore, the beam requires shear reinforcement. For this example, the critical shear condition is web cracking V_{cw} and No. 4 grade 60 stirrups at 11 in. on center at h from the column face is adequate. The stirrup spacing is additionally dependent on the horizontal shear transfer.

13.6.9   Horizontal Shear Transfer

The depth of the equivalent rectangular stress block is 4.1 in. The factored horizontal shear on the pinned end of the span for the 4000 psi concrete is then

$$V_{uh} = 0.85 f_c' b_f a = 0.85 \cdot 4.0 \cdot 78 \cdot 4.1 = 1088\,\text{kips}$$

Using shear friction and a coefficient of friction equal to 1.0 for concrete cast onto a roughened surface, the total area of shear reinforcement is

$$A_{vf} = \frac{V_{uh}}{\phi f_{yt} \mu} = \frac{1088}{0.75 \cdot 60 \cdot 1.0} = 24.2\,\text{in.}^2$$

The maximum positive moment occurs at approximately 3/8 of the span length. The reinforcement for horizontal shear is then distributed along 3/8 the length of the beam or 43.1 ft. The resulting shear reinforcement is 0.56 in.^{2}/ft or No. 4 Grade

60 stirrups at 8.5 in. Therefore, No. 4 stirrups at 8.5 in. are used for the first 3/8 of the beam in lieu of the vertical shear requirement. No. 4 stirrups at 14 in. are required for the 5/8 portion of the beam for horizontal shear strength and would reduce to 11 in. near the column to meet the demand for vertical shear reinforcement.

13.7   Comments on Example

The example presented in Sect. 13.6 illustrates the complexity of spliced girder calculations and the steps for designing structures with multiple stages of prestressing. The negative moment strength check indicates that the total number of strands is adequate. The end span has ample positive moment strength, Table 13.6. The concentrated load required by the AASHTO code is not included, but the excess strength and the ability to lower the post-tensioning tendon provide reserve for this loading. Examining the initial stresses, there is considerable compression in the end span. This resulted in the number of pretensioning strands in the end span being reduced with no reduction in serviceability. The "over-design" in the end span results from balancing the girder and carrying the slab load on a simple span structure. A follow-up exercise would be to calculate the stresses to better optimize the total prestress or to evaluate the effects of shoring and carrying the deck on the composite structure.

13.8   Crossed Tendon Post-tensioning

Crossed tendons as seen in Fig. 13.2 increase the negative moment strength and have less total frictional loss than multi-span tendons. The crossed tendons provide a further advantage over using a short "cap" tendon due to the longer length and lower effects of anchorage losses.

The anchorage locations are needed to calculate the prestressing effects. Examining the moment envelope for a beam in a series of continuous beams suggests that a minimum moment location occurs between 0.1 and 0.2 L, Fig. 13.12. An anchor location of 0.15 L is selected for development of the prestress forces. This location is adequate for development of prestressing effects even though it is not an exact minimum moment and it is away from maximum shear locations.

As seen in Chap. 9, continuity alters the effective eccentricity of a post-tensioning tendon due to the reactions generated by the post-tensioning. Therefore, an effective eccentricity is established to calculate the stresses due to prestress.

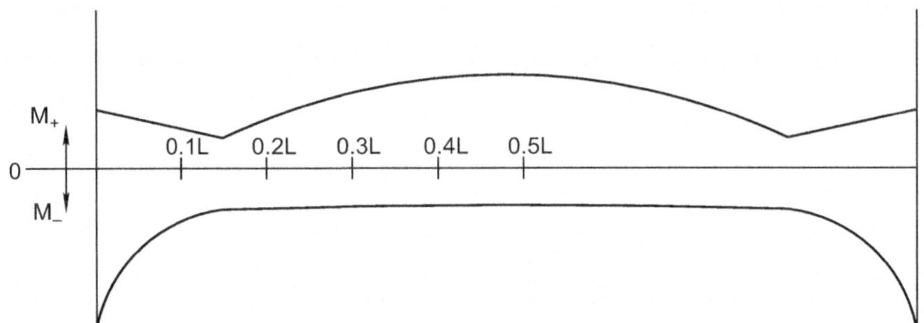

Fig. 13.12 Representative moment envelope for a beam in a continuous structure

13.8.1   Determination of Effective Eccentricity for Interior Beams

In a structure with multiple equal spans, the internal stresses redistribute to create a condition such that no end rotation can occur. That is, each beam is considered to have a fixed end condition. Within each span there are two components of the crossed tendon, the draped segment and the cap segment, Fig. 13.13. Each component is considered separately and the results combined.

The basic theory for the behavior of an interior beam is that the end rotation is zero. Assuming that the beam is symmetric, this condition is satisfied when

$$\int_0^{l/2} \frac{M}{EI} dx = 0 \tag{13.1}$$

where E and I are variables. Assume a draped tendon with a parabolic sag s and a beam of a constant cross section so that E and I are constant, Fig. 13.14. where

e = effective eccentricity
s = total sag
y = eccentricity of the tendon at any point x
M = internal moment due to prestress = P_y, and
P = post-tensioning force

If the tendon is defined by a parabolic shape then

$$y = ax^2 + bx + c \tag{13.2}$$

where the constants a, b, and c are to be determined. It is convenient to define the curve at the centerline of the beam. The resulting boundary conditions are then,

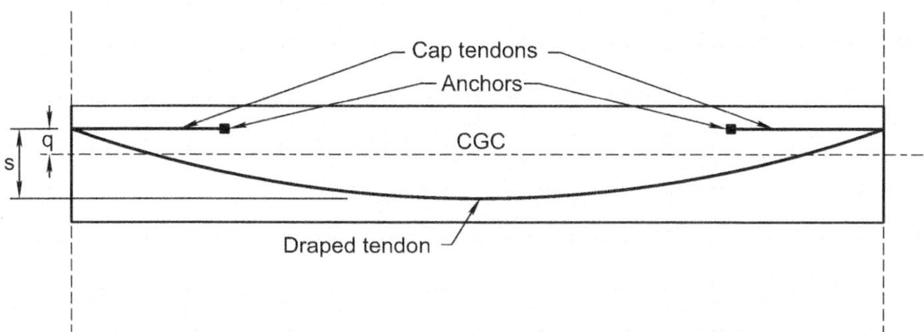

Fig. 13.13 Crossed tendon configuration

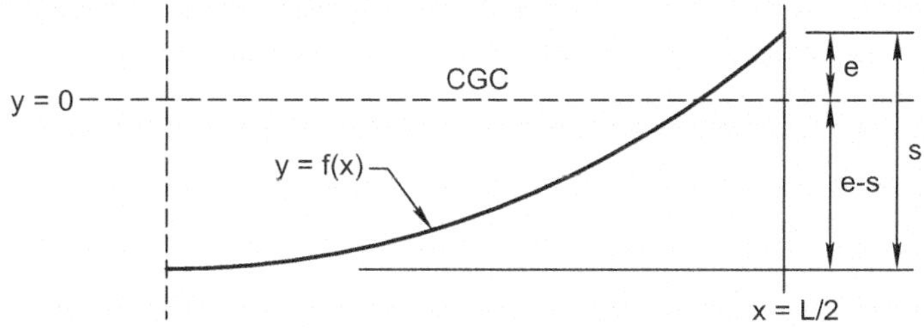

Fig. 13.14 Draped tendon geometry

$$x = 0, y = e - s \tag{13.3}$$

$$x = 0, y' = 0 \tag{13.4}$$

$$x = L/2, y = e \tag{13.5}$$

Solving using Eq. 13.2 for the conditions in Eq. 13.3 gives $c = e - s$. Similarly, solving using Eq. 13.2 for conditions in Eq. 13.4 gives $b = 0$ and finally solving Eq. 13.2 for conditions in Eq. 13.5 gives $a = 4\ s/L^2$. For simplicity, assume that the post-tensioning force is constant along the beam and substituting into Eq. 13.1 gives

$$\int_0^{l/2} \frac{M}{EI} dx = \frac{1}{EI} \int_0^{L/2} Py\, dx = \frac{P}{EI} \int_0^{L/2} \left[\frac{4s}{L^2} x^2 + (e - s) \right] dx = 0 \tag{13.6}$$

Integrating results in

$$\left[\frac{P}{EI}\left[\frac{4sx^3}{3L^2} + ex - sx\right]\right] = 0 \text{ evaluated from } 0 \text{ to } L/2 \qquad (13.7)$$

which, after applying the integration limits gives

$$\left[\frac{4sL^3}{24L^2} + eL/2 - sL/2\right] = 0 \qquad (13.8)$$

Solving for e gives $e = 2/3\ s$. Thus, regardless of the location of the anchor, the effective eccentricity at the end of the draped tendon is 2/3 of the total drape.

The cap tendon is resolved in a similar fashion or may be solved using the principles of area-moment. The real tendon end eccentricity is q and the corresponding end moment is Pq. Using the assumed anchor location of 0.15 L, the area-moment diagram is given in Fig. 13.15 with no end restraints as is consistent with zero deflection.

Algebraically balancing the M/EI diagram gives

$$\frac{Pe}{EI}0.15L - \frac{P(q-e)}{EI}0.35L = 0 \qquad (13.9)$$

Solving for e gives $e = 0.70q$.

Combining the two conditions gives the effective eccentricity in Eq. 13.10 at the beam end and an effective eccentricity in Eq. 13.11 at midspan for a beam of constant cross section.

$$e_{\text{end}} = \frac{2}{3s} + 0.70q \qquad (13.10)$$

$$e_{\text{mid}} = s/3 + 0.30q \qquad (13.11)$$

These effective eccentricities are used to calculate the stresses due to post-tensioning in the cap beam.

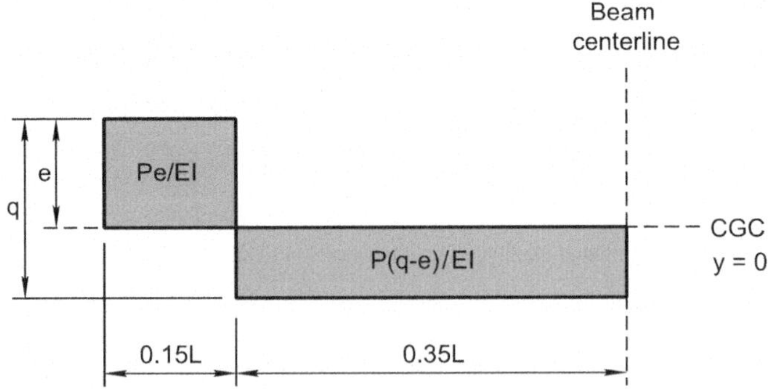

Fig. 13.15 Area-moment diagram for cap tendon

13.8.2 *Determine Effective Eccentricities for the End Beam*

The end span differs from the interior span because the end rotation at the simple support is not zero. Furthermore, the sag in the tendon s' may vary, as may the end eccentricity q'. The total drape in the end span s' is set to maintain equilibrium at the first interior support B, Fig. 13.16.

The parabolic draped tendon produces a uniform loading on the beam of

$$w = 8P's'/L^2 \tag{13.12}$$

The notation P' is used to anticipate that the post-tensioning force in the end span may differ from the interior spans. And the resulting M/EI diagram for the end span is shown in Fig. 13.17.

The ratio of the end moment to the total moment is

$$\frac{wL^2/8}{\dfrac{wL^2}{8}+\dfrac{9wL^2}{128}} = \frac{\dfrac{1}{8}}{\dfrac{1}{8}+\dfrac{9}{128}} = \frac{16}{25}$$

By analogy to the interior beam, the effective eccentricity at the interior support is $e = 16/25\ s'$.

The cap tendon effective eccentricity is obtained using a conjugate beam analysis. In a conjugate beam, there is no support at the interior end since the M/EI must be zero and the pinned end is replaced with a pin, Fig. 13.18. M_f is the fixed end moment resulting from the conjugate reaction R_A at A.

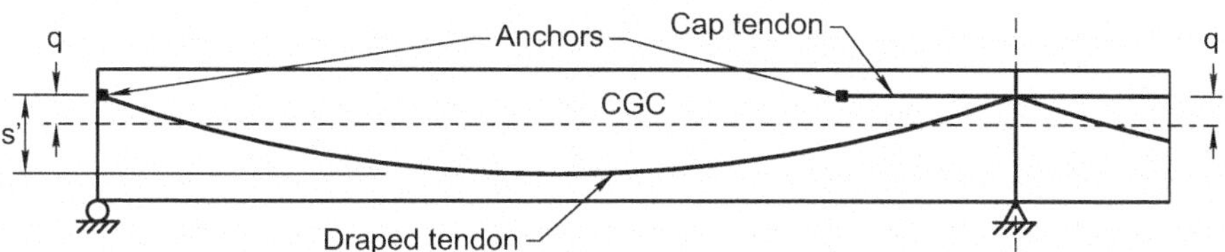

Fig. 13.16 Tendon geometry of end span

Fig. 13.17 *M/EI* diagram for end span

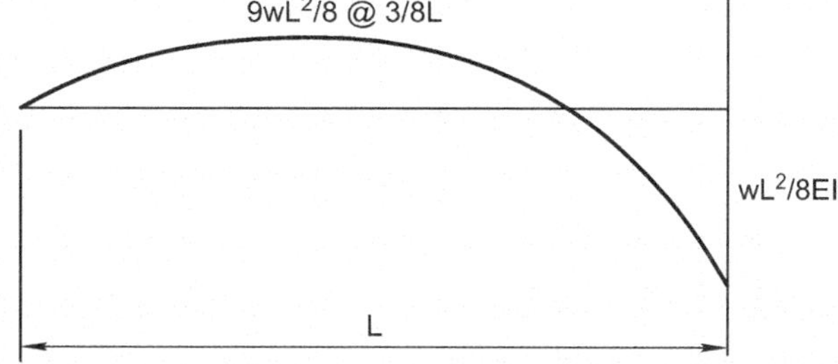

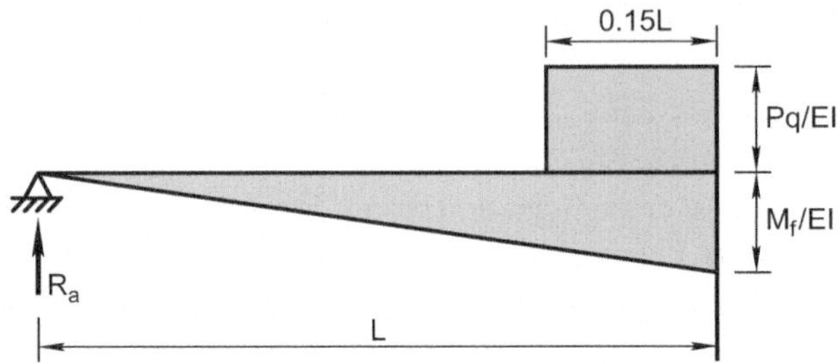

Fig. 13.18 Conjugate beam for end span cap tendon

For equilibrium, the sum of the moment on the conjugate beam about A is zero, thus

$$\sum \frac{M_A}{EI} = 0 = \frac{Pq \cdot 0.15L \cdot 0.925L}{EI} - \frac{M_f \cdot L/2 \cdot 2L/3}{EI}$$

resulting in $M_f = 0.416\,Pq$. The effective eccentricity is then obtained by noting that $Pe = Pq - M_f$ resulting in an effective eccentricity of $e = 0.584q$.

The above derivation assumes a zero eccentricity at the pinned end, that is, $q' = 0$, or the post-tensioning anchor is at the center of gravity of the concrete section. If the end eccentricity is not zero, the moment at the pinned end is $M = P'q'$. Again, assuming a constant section beam the effective eccentricity can be found. This time, the most common tool for finding the relation between an applied concentrated end moment and the fixed end moment on a propped cantilever is used. Look it up in a book! The result is the fixed end moment for an applied moment M_f is equal to $-M/2$. Thus.

$$P'e = -\frac{1}{2}P'q'$$

resulting in $e = -q'/2$. Finally, the sag of the end span is adjusted to that the total moment at the end of the propped cantilever is equal to the fixed end moment of the interior beam. For this solution, the end span post-tension force is carried as a variable giving

$$P'\left(\frac{16s'}{25} + 0.548q - \frac{q'}{2}\right) = P\left(\frac{2s}{3} + 0.70q\right)$$

Solving for s' gives the required sag for a given eccentricity at the pinned end

$$s' = 1.04\frac{Ps}{P'} + q\left(1.09\frac{P}{P'} - 0.9125\right) + 0.781q' \qquad (13.13)$$

For the case where the pinned end eccentricity is zero and the same tendon and post-tensioning force is used in the end span as in the interior spans, the required drape of the tendon in the end span is

$$s' = 1.04s + 0.178q' \qquad (13.14)$$

13.8.3 Discussion and Detailing Considerations

The determination of effective eccentricities illustrates linear transformation of post-tension forces in an indeterminate structure. This is a result of the so called "secondary moments" that generate changes in the beam reactions. Use of the effective eccentricities eliminates the need to calculate the secondary moments.

The crossed tendon solution requires close attention to the details at the interior anchors. A block-out for jacking is required if the tendon runout is horizontal, as indicated in the above figures. A detailed examination of the stresses around the block-out is needed to control localized stresses due to the anchor force transfer into the concrete. A localized thickening of the flange at the anchor can produce eccentric loads in front of the thickened section resulting in localized bending and cracking. The area behind the anchor is reinforced to transfer forces due to shear lag back into the beam.

To facilitate jacking the end of an interior tendon, the tendon is turned upward so that jacking can be completed above the flange at an angle to the top flange. Leonhardt's advice from Chap. 1 becomes relevant as reinforcement is required to restrain the anchor and the duct from unstitching during jacking. Finally, some box beams move the anchors toward the center of the beam. This requires a horizontal displacement of the anchors, which, in turn, create in-plane forces in the flange. Reinforcement for these in-plane forces is part of the final detailing.

References

Abdel-Karim, A. M., & Tadros, M. K. (1992). Design and construction of spliced I-girder bridges. *PCI Journal, 37*(4), 114–122.

Castrodale, R. W., & White, C. D. (2004). *Extending span ranges of precast prestressed concrete girders* (No. 517). Transportation Research Board.

Geren, K. L., & Tadros, M. K. (1994). The NU precast/prestressed concrete bridge I-girder series. *PCI Journal, 39*(3), 26–39.

Girgis, A. M., Hennessey, S. A., Drews, D., & Tadros, M. K. (2004). Value engineering of a spliced girder bridge produces pretensioned concrete records. In *The 2004 Concrete Bridge Conference.*

Harvey, D. I. (1986). Spliced segmental precast concrete bridges using staged post-tensioning. In *ACI Special Publication 93 Structures in Transportation* (pp. 721–736).

Marshall, S. L., & Pelkey, R. E. (1986). Production, transportation, and installation of spliced prestressed concrete 'I'Girders for the Annacis Channel East Bridge. In *ACI Special Publication 93 Structures in Transportation* (pp. 737–768).

Tadros, M. K., Ficenec, J. A., Einea, A., & Holdsworth, S. (1993). A new technique to create continuity in prestressed concrete members. *PCI Journal, 38*(5), 30–37.

Chapter 14
Strut-and-Tie Method

14.1 Introduction

Beam theory, including the assumption that plane sections remain plane, does not apply at points closer than the distance h to a discontinuity in applied load or geometry. This leads to the identification of **discontinuity regions** within reinforced concrete members near concentrated loads, openings, or changes in cross section. Because of their geometry, the full volume of deep beams, dapped ends, and column brackets qualify as discontinuity regions. Thus, structural concrete members are divided into regions where beam theory is valid, often referred to as ***B-regions***, and regions where discontinuities affect member behavior, known as ***D-regions***. Several representative D-regions are illustrated in Fig. 14.1.

At low stresses, when the concrete is elastic and uncracked, the stresses within D-regions may be calculated using finite element analysis or elasticity theory. When concrete cracks, the strain field is disrupted, causing a redistribution of the internal forces. Once cracked, the internal forces within discontinuity regions are represented using a statically determinate truss, referred to as a **strut-and-tie method**. This allows a complex design problem to be simplified, producing a solution that satisfies statics. Strut-and-tie models consist of concrete compression **struts**, steel tension **ties**, and joints that are referred to as **nodal zones** or **nodes**, Fig. 14.2. Struts are represented by dashed lines, ties are represented by solid lines, and nodes are shaded triangles.

The strut-and-tie method is used in several ways during the design process. At the conceptual design level, sketching a strut-and-tie model provides insight into structural behavior and detailing requirements. Conceptual design assists in the development of connection details. Strut-and-tie models may be used to validate design details, such as special reinforcement configurations. Finally, strut-and-tie models may form the basis for detailed design of a member.

Strut-and-tie models evolved in the early 1980s in Europe (Schlaich et al. 1987; Schlaich and Schäfer 1991; Marti 1985a, b). Their use is permitted in Section 6.2.4

© Springer Nature Switzerland AG 2019
C. W. Dolan, H. R. Hamilton, *Prestressed Concrete*,
https://doi.org/10.1007/978-3-319-97882-6_14

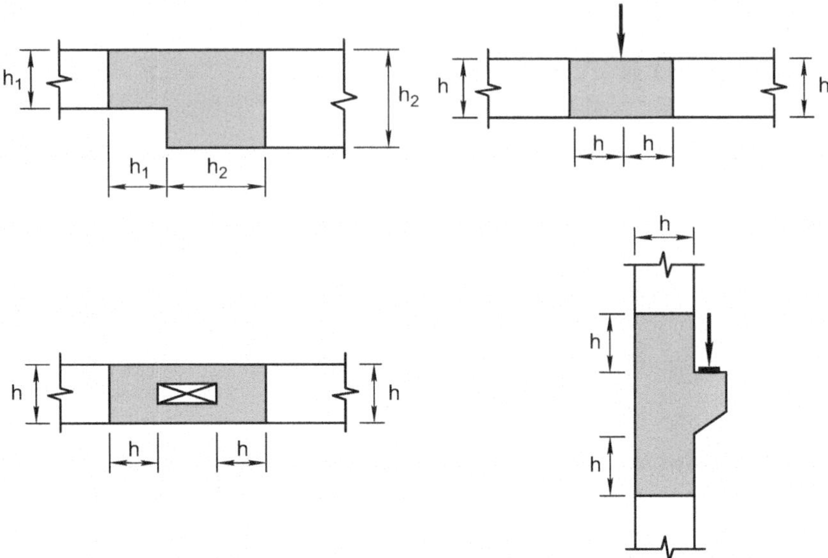

Fig. 14.1 Representative discontinuity areas

Fig. 14.2 Schematic strut-and-tie model

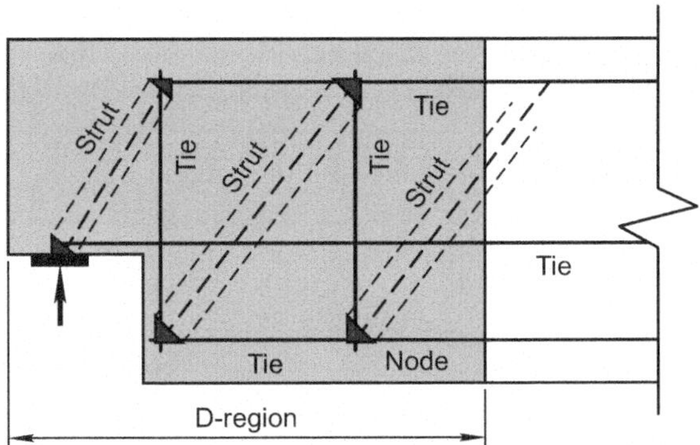

of ACI 318-14 (2014) and defined in Chapter 23 of the ACI Building Code. The portion to be designed by strut-and-tie method lies within a distance equal to the member height h from a force or geometric discontinuity. Models consist of struts and ties connected at nodal zones that are capable of transferring loads to the supports or adjacent B-regions. The cross-sectional dimensions of the struts and ties are designated as thickness and width. Thickness b is perpendicular to the plane of the truss model, and width w is measured in the plane of the model.

Design and detailing connections and attachments requires attention to the flow of forces in the structure. Strut-and-tie models assist in visualizing the behavior of the joint and assist to assure that proper reinforcement is detailed. In addition to identifying load paths, the strut-and-tie method identifies the forces to be transferred and where reinforcement needs to be developed. Anchorage to concrete provides insight to the behavior of embedded steel elements. Inserts may pullout at loads lower than the tensile strength of the insert. The anchorage methodology allows visualization of the failure planes and correct insert design (Reineck 2002).

14.2   Struts

A strut is an internal compression member. It may consist of a single element, parallel elements, or a fan-shaped compression field. An **edge** strut occurs along a free edge of a member and is considered rectangular (Nowak and Sprenger 2002). **Interior** or ***bottle-shaped*** struts allow the compression field to spread laterally between nodal zones (Uribe and Alcocer 2002). For design purposes, a strut is typically idealized as a prismatic member between two nodes but may also be idealized as a uniformly tapered compression member if the geometry requires different widths at the strut ends. The dimensions of the cross section of the strut are established by the contact area between the strut and the nodal zone. Interior struts are wider at the center than at the ends and form when the surrounding concrete permits the compression field to spread laterally. As the compression zone spreads along the length of interior struts, tensile stresses perpendicular to the axis of the strut may result in longitudinal cracking, Fig. 14.3. For simplicity, the strength of a strut is a function of the effective concrete compressive strength, which is affected by transverse stresses within the struts. Because of longitudinal splitting, interior struts are weaker than rectangular struts, even though they possess a larger cross section at mid-length. Transverse reinforcement is needed to control longitudinal splitting.

14.3   Ties

A tie is a tension member within a strut-and-tie model. Ties consist of reinforcement, prestressed or nonprestressed, plus a portion of the concrete that is concentric with and surrounds the axis of the tie. The surrounding concrete defines the tie area and

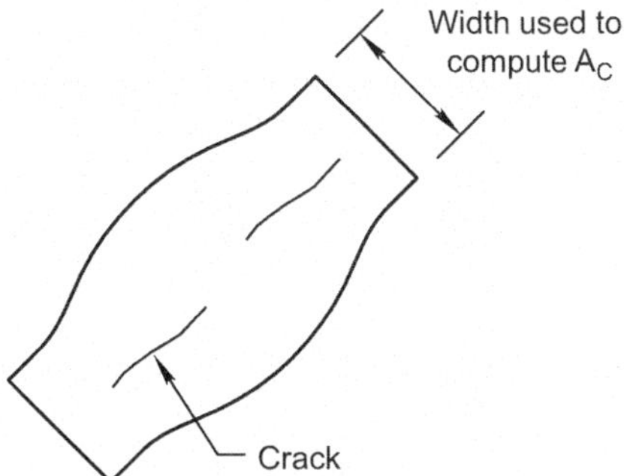

Fig. 14.3 Interior or bottle-shaped strut

the region available to anchor the tie. For design purposes, it is assumed that the concrete within the tie does not carry any tensile force. Even though the tensile strength of the concrete is not used in design, it assists in reducing tie deformation at service load.

14.4   Nodal Zones

Nodes are points within strut-and-tie models where the axes of struts, ties, and concentrated loads intersect. A nodal zone is the volume of concrete around a node where force transfer occurs. A nodal zone may be treated as a single region or may be subdivided into smaller zones to equilibrate forces. For equilibrium, at least three forces must act on a node and nodes are classified by the sign of these forces, Fig. 14.4. Thus, a *C-C-C* node resists three compressive forces, and a *C-C-T* node resists two compressive forces and one tensile force. Both tensile and compressive forces place nodes in compression because tensile forces are treated as if they pass through the node and apply a compressive force on the far side, or anchorage face. Within the plane of a strut-and-tie model truss, nodal zones are considered in compression. If the nodal zone dimensions w_{n1}, w_{n2}, and w_{n3} in Fig. 14.4 are proportional to the applied compressive forces and a state of *hydrostatic* compression exists. The dimension of one side of a nodal zone is often determined based on the contact area of the load, such as a bearing.

The length of a hydrostatic zone is often not adequate to allow for anchorage of tie reinforcement. For this reason, an ***extended nodal zone***, defined by the intersection of the nodal zone and the associated strut is used, Fig. 14.5. An extended nodal zone

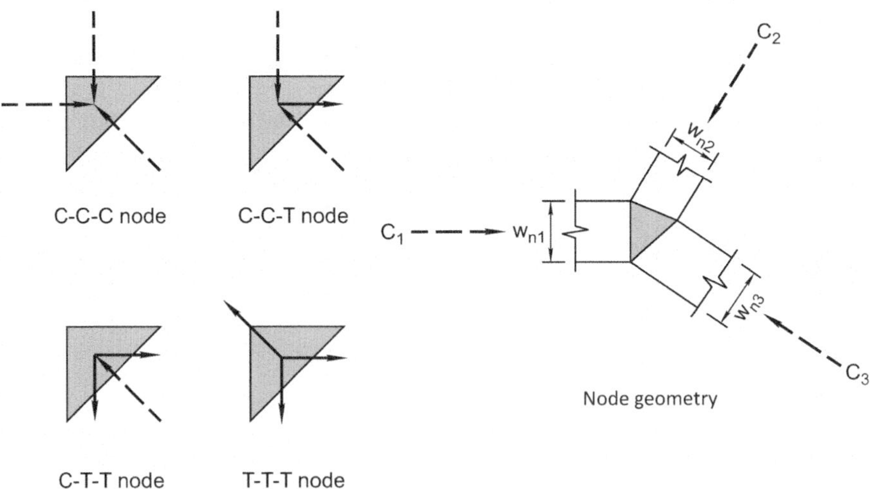

Fig. 14.4 Node nomenclature and geometry

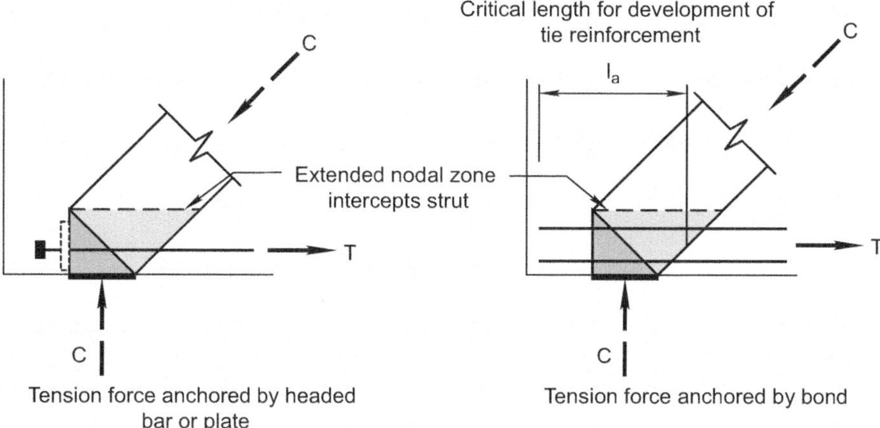

Fig. 14.5 Extended nodal zones

is regarded as the portion of the overlap region between struts and ties that is not already counted as part of a primary node. It increases the length within which the tie tensile force from the tie is transferred to the concrete and, thus, defines the available anchorage length for ties. Ties may be developed outside of the nodal and extended nodal zones if space is available, as shown to the left of the node in Fig. 14.5.

14.5   ACI Provisions for Strut-and-Tie Method

Chapter 23 of ACI 318-14 provides guidance for sizing struts, nodes, and ties. The strength of struts and nodes consists of a basic concrete compressive strength equal to $0.85f'_c$ modified by a factor β. The factor β accounts for the effects of cracks caused by spreading compressive resultants, confining reinforcement in struts, and anchorage of ties in nodal zones. The factor β is specified in the ACI Building Code as $\boldsymbol{\beta_s}$, $\boldsymbol{\beta_t}$, and $\boldsymbol{\beta_n}$ for struts, ties, and nodes respectively. A strength reduction factor $\phi = 0.75$ is used for all struts, ties, nodal zones, and bearing stresses within the strut-and-tie model.

14.5.1   Strength of Struts

The strength of a strut is based on the strength of the concrete in the strut and the strength of the nodal zones at the ends of the strut. The nominal compressive strength of a strut F_{ns} is given as

Table 14.1 β_s values for struts

Condition	β_s
Edge strut	1.0
Interior strut with the width at midsection larger than the width at the nodes	0.75
Struts in tension members or in tension zone of members	0.40
All other cases	0.60

$$F_{ns} = f_{ce}A_{cs} \tag{14.1}$$

where f_{ce} is the effective compressive strength of the concrete in a strut or nodal zone and A_{cs} is the cross-sectional area at one end of the strut, which is equal to the product of the strut thickness and the strut width. The effective strength of concrete in a strut is

$$f_{ce} = 0.85\beta_s f'_c \tag{14.2}$$

Values of β_s range from 1.0 for a strut with a uniform cross-sectional area over its length to 0.4 for struts in tension members or the tension flanges of members. Table 14.1 summarizes β_s values. Intermediate values include 0.75 for struts with a width at midsection that is larger than the width at the nodes (bottle-shaped struts) and crossed by transverse reinforcement to resist the transverse tensile force resulting from the compressive force spreading in the strut and 0.60λ for bottle-shaped struts without the required transverse reinforcement, where λ is the correction factor related to the unit weight of concrete, described in Chap. 3. The value for β_s is 0.60λ for all other cases, as when parallel diagonal cracks due to flexure or shear divide the web struts or when the diagonal cracks are likely to turn and cross a strut.

Compression reinforcement may be added to increase the strength of a strut, so that

$$F_{ns} = f_{ce}A_{cs} + A'_s f'_s \tag{14.3}$$

where f'_s is based on the strain in the concrete at peak stress. For Grades 40 and 60 reinforcement, $f'_s = f_y$. For compression reinforcement to be effective Section 23.6 of ACI 318-14 requires compression reinforcement be properly anchored, oriented parallel to the axis of the strut, located within the strut, and enclosed by ties or spirals; just as required for columns.

14.5.2 Minimum Transverse Reinforcement

Transverse reinforcement is required to prevent splitting and permits the assumption that the compressive force in the struts spreads. For $f'_c \leq 6000\,\mathrm{psi}$, the transverse

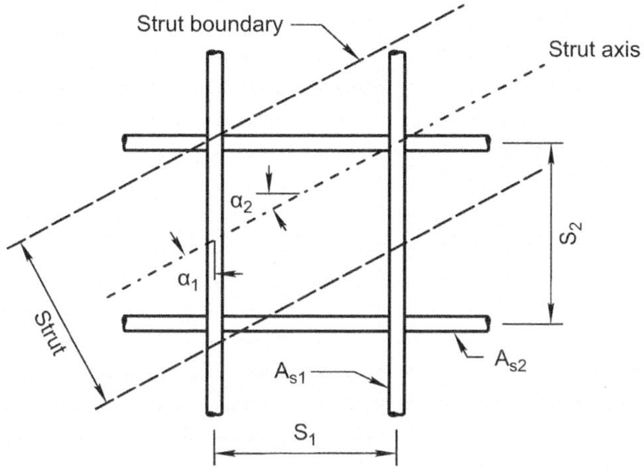

Fig. 14.6 Definition of reinforcement crossing a strut

reinforcement requirement is satisfied if the strut is crossed by a layer of orthogonal reinforcement that satisfy Eq. 14.4.

$$\frac{A_{si}}{b_s s_i} \geq 0.0025 \tag{14.4}$$

where A_{si} is the total area of reinforcement at spacing s_i in a layer of reinforcement with bars at an angle α_1 to the axis of the strut, and b_s is the thickness of the strut. The spacing of the reinforcement is limited to 12 in. The reinforcement may be perpendicular to the strut axis or may be placed in an orthogonal grid pattern. The subscript i denotes the layer of reinforcement. The values s_i and α_1 are shown in Fig. 14.6. Equation 14.4 assumes that the transverse reinforcement is placed in each face of the member. The minimum transverse reinforcement placed in a single face is $A_{si} = \sin^2\alpha_1$.

14.5.3   Strength of Nodal Zones

The nominal compressive strength of a nodal zone is

$$F_{nn} = f_{ce}A_{nz} \tag{14.5}$$

where f_{ce} is the effective strength of the concrete in the nodal zone and A_{nz} is (1) the area of the face of the nodal zone taken perpendicular to the line of action of the force from the strut or tie or (2) the area of a section through the nodal zone taken perpendicular to the line of action of the resultant force on the section. The latter

Table 14.2 β_n values for struts

Nodal zone condition	Node classification	β_n
Bounded by struts or bearing area	C-C-C	1.00
Anchoring one tie	C-C-T	0.80
Anchoring two or more ties	C-T-T or T-T-T	0.60

condition occurs when multiple struts intersect a node. The effective concrete strength in a nodal zone is

$$f_{ce} = 0.85\beta_n f'_c \tag{14.6}$$

where f'_c is the compressive strength of the concrete in the nodal zone and β_n is a factor that reflects the degree of disruption in nodal zones due to the incompatibility of tensile strains in ties with compressive strains in struts and are summarized in Table 14.2. Section 23.9.3 of ACI 318-14 permits the strength of a node to be increased above the value given in Eq. 14.6 if the node contains confining reinforcement and the effect of that reinforcement is demonstrated by tests and analysis. Unless compression reinforcement is used in the struts, the lower value of f_{ce} from Eqs. 14.2 and 14.6 govern and should be used to design both the node and the adjoining struts.

14.5.4   Strength of Ties

The nominal strength of ties F_{nt} is the sum of the strength of the reinforcing steel and prestressing steel within the tie.

$$F_{nt} = A_{ts}f_y + A_{tp}\left(f_{pe} + \Delta f_p\right) \tag{14.7}$$

where A_{ts} is the area of reinforcing steel, f_y is the specified yield strength of reinforcing steel, A_{tp} is the area of the tie prestressing steel, if any, f_{pe} is the effective stress in prestressing steel, and Δf_p is the increase in prestressing steel stress due to factored load. The sum $f_{pe} + \Delta f_p$ must be less than or equal to the yield stress of the prestressing reinforcement f_{py}, and A_{tp} is zero for nonprestressed members. The value of Δf_p may be found by analysis; or, in lieu of formal analysis, Section 23.7.3 of ACI 318-14 allows a value 60,000 psi to be used for bonded tendons and 10,000 psi to be used for unbonded tendons.

The effective width of a tie w_t depends on the distribution of the tie reinforcement. If the reinforcement in a tie is placed in a single layer, the effective width of a tie may be taken as the diameter of the largest bars in the tie plus twice the cover to the surface of the bars. Alternatively, the width of a tie may be taken as the width of the tie anchor plates. The practical upper limit for tie width $w_{t,max}$ is equal to the width corresponding to the width of a hydrostatic nodal zone, given as

$$w_{t,\max} = \frac{F_{nt}}{b_s f_{ce}} \tag{14.8}$$

where f_{ce} is the effective nodal zone compressive stress given in Eq. 14.6 and b_s is the thickness of the strut, Fig. 14.5.

Ties must be anchored before they leave the extended nodal zone as shown in Fig. 14.5. If the combined lengths of the nodal zone and extended nodal zone are inadequate to provide for development of the reinforcement, additional anchorage is obtained by extending the reinforcement beyond the nodal zone, using 90° hooks, headed bars, or mechanical anchors. If the tie is anchored with a 90° hook, the hooks should be confined by reinforcement extending into the beam from supporting members to avoid splitting of the concrete within the anchorage region.

14.6 Strut-and-Tie Design

Strut-and-tie models, which are based on strength, do not specifically address serviceability. To this end, the maximum spacing of the transverse reinforcement should be less than 12 in. Section 9.9.2 of ACI 318-14 limits the nominal shear strength of deep beams to $10\sqrt{f'_c}b_w d$. This shear stress limit applies to strut-and-tie models and should be checked at the beginning a detailed design. Deep beams having a shear span a_v less than the structural depth have steep struts and the $10\sqrt{f'_c}b_w d$ limit may not be sufficient. Zsutty (1971) reported that the shear span ratio a_v/d is an inverse ratio to the strength. A better predictor of the ability of a deep beam to develop its nominal capacity is given in Eq. 14.9; however, some consideration should be given to the maximum value of V_u/ϕ as a_v becomes small. Test data by Zsutty suggests that a lower bound for a_v/d of 0.5 may be reasonable. Additional adjustment for the depth of the member may be required.

$$\frac{V_u}{\phi} \leq \frac{10\lambda\sqrt{f'_c}b_w d}{a_v/d} \tag{14.9}$$

Application of a detailed strut-and-tie method involves completion of the following steps.

- Define and isolate the D-regions.
- Calculate the force resultants on each D-region boundary.
- Select a truss model to transfer the forces across a D-region.
- Select dimensions for strut-and-tie nodal zones.
- Verify the strength of the node and the strut, the latter both at mid-length and at the nodal interface.
- Design the ties and the tie anchorage.

- Prepare design details and check minimum reinforcement requirements.
- The design process requires interaction between these steps.

According to Section 23.3.1 of ACI 318-14, design using a strut-and-tie model requires that

$$
\begin{aligned}
\phi F_{ns} &\geq F_{us} && \text{for struts} \\
\phi F_{nt} &\geq F_{ut} && \text{for ties} \\
\phi F_{nn} &\geq F_{un} && \text{for nodes}
\end{aligned}
\tag{14.10}
$$

where F_{ns}, F_{nt}, and F_{nn} are the nominal strengths of strut, tie, and nodal zone, respectively F_{us}, F_{us}, and F_{us} factored forces acting in strut, tie, and nodal zone, respectively, and ϕ is the strength reduction factor.

14.6.1   The Truss Model

The truss representing the strut-and-tie model must fit within the envelope defined by the D-region. Selection of struts and ties is made at the discretion of the engineer, and, therefore, multiple solutions are possible. Truss member axes are positioned to coincide with the centroids of the tension and compression fields. The resulting geometry is used to calculate the forces in the members. Truss model layout is constrained by the geometric requirement that struts must intersect only at nodal zones. Ties may cross struts. An effective model represents a minimum energy distribution through the D-region (Schlaich et al. 1987; Marti 1985a, b); that is, within the model, forces should follow the stiffest load path. Because struts are typically much stiffer than ties, a model with a minimum number of tension ties is preferable.

14.6.2   Selecting Dimensions for Struts and Nodal Zones

The struts, ties, and nodal zones within the truss that represents a strut-and-tie model have finite widths that must be considered when selecting the dimensions of the truss. The width of each truss member depends on the magnitude of the forces and the dimensions of the adjoining elements. An external element, such as a bearing plate or column, serves to define a nodal zone. If the bearing area is too small, a high hydrostatic pressure results, and the corresponding width of the node or struts are not sufficient to carry the applied load. The solution in this case is to increase the size of the bearing surface and reduce the contact pressures. Some engineers intentionally select struts and nodes that are large enough to keep the compressive stresses low; in which case, only the tension ties require detailed design. To minimize cracking and to reduce complications that may result from incompatibility in the deformations due to struts shortening and ties elongating in nearly the same plane, the angle between struts and ties at a node should be at least 25°. Engineers, however, often prefer to use an angle of at least 40° because forces in the struts and ties are often unacceptably high at lower angles.

The design of nodal zones assumes that the principal stresses within the intersecting struts and ties are parallel to the axes of these truss members. The widths of the struts and ties are, in general, proportional to the magnitude of the force in the elements. Examination of Tables 14.1 and 14.2 indicates that β_s is less than β_n for most applications. That observation suggests selecting strut dimensions first and then checking nodes. If two or more struts converge on the same face, it is generally necessary to resolve the forces into a single force and to orient the face of the nodal zone so that it is perpendicular to the combined force. Some geometric arrangements preclude establishing a purely hydrostatic node. In these cases, the width of the strut is determined by the geometry of the bearing plate or tension tie.

The thickness of the strut, tie, and nodal zone is typically equal to the thickness of the member. If the thickness of the bearing area is less than the thickness of the member, it may be necessary to add reinforcement perpendicular to the principal plane of the member for confinement and to prevent splitting parallel to the plane of the truss. In this instance, a strut-and-tie model perpendicular to the plane of the truss may be used to determine the requirements for transverse reinforcement.

14.6.3   Strength of Struts

Strut strength is based on both the strength of the strut itself and the strength of the nodal zone. If a strut does not have sufficient strength, the design is revised by providing compression reinforcement or by increasing the size of the nodal zone. This may, in turn, affect the size of the bearing plate or column. Increasing the nodal zone is preferred providing the geometry allows the larger area.

14.6.4   Design of Ties and Anchorage

To control cracking in a D-region, ties are designed so that the stress in the reinforcement is below yield at service loads. The geometry of the tie is selected so that the reinforcement fits within the tie dimensions and is fully anchored.

Anchorage for ties is provided within the nodal and extended nodal zones plus regions on the far side of the node that may be available based on the geometry of the member. Figure 14.9a illustrates an extended nodal zone and the length available for anchorage of ties ℓ_a. In this case, the tie is extended to the left of the nodal zone to allow for full development of the reinforcement. The shape of the extended nodal zone is a function of the strut angle θ and the width of the tie w_t. Figure 14.5 illustrates the geometry and dimensions of a *C-C-T* node with a tension tie that contains multiple layers of reinforcement. Figure 14.9b shows a *C-T-T* nodal zone. If insufficient length is available to anchor the reinforcement within the nodal and extended nodal zones, the reinforcement must extend beyond the node or a hook or headed bar must be used to fully develop the reinforcement.

14.6.5 *Design Details and Minimum Reinforcement Requirements*

A complete design includes verification that (1) tie reinforcement can be placed in the section, (2) nodal zones are confined by compressive forces or tension ties, and (3) minimum reinforcement requirements are satisfied. Reinforcement within ties must meet the ACI Building Code requirements for bar spacing and fit within the overall width and thickness of the tie. Tie details should be reviewed to ensure that ties are adequately developed by tension development length, hooks, headed bars, or mechanical anchorage. Shear reinforcement requirements are satisfied by ensuring that the factored shear is less than the ACI Building Code maximum, as described in Chap. 7, longitudinal cracking of bottle-shaped struts is controlled, or the minimum reinforcement requirements described in Sect. 14.5.2 are met.

14.7 Dapped Beam Ends

Precast and prestressed concrete beams often have **dapped** or notched ends, as shown in Fig. 14.2. The notch allows structural overlap between the main beams and the floor beams and is advantageous in controlling building floor-to-floor height. Dapped ends create two structural challenges. First, the shear at the end of the beam is carried by a much smaller section. Second, the mechanism of load transfer through the notched zone is difficult to represent using conventional design techniques. As a result, dapped-end beams lend themselves to strut-and-tie design methods.

Dapped ends have been problematic since their inception and studies indicated that careful detailing is required (Mattock and Theryo 1986). A PCI research and development study examined and tested several commonly used dapped end details (Klein et al. 2015). This and other studies found that if the remaining portion of the dapped end is less than 60 percent of the overall depth, a shear failure is likely. Two of the more successful designs from the Klein report are shown in Fig. 14.7 with a corresponding strut-and-tie model. Detail *a* developed 100% of the theoretical design load and detail *b* developed 125% of the design moment. Both designs address the critical issue of transferring the inclined tension tie force to the horizontal reinforcement at the bottom of the beam. Other details in the PCI report called attention to details around **curved bar nodes**, nodes with larger curved reinforcement present as in Fig. 14.7a (Klein 2008).

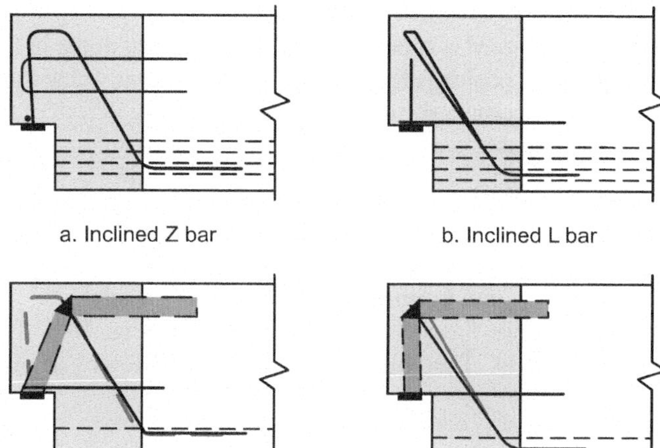

a. Inclined Z bar   b. Inclined L bar

Fig. 14.7 Strut-and-tie models for dapped end

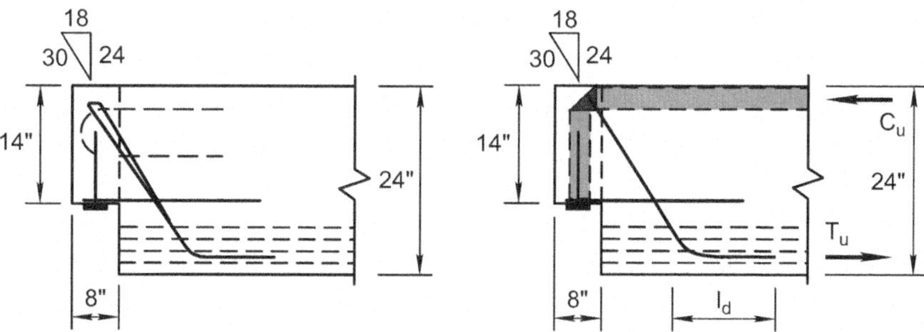

Fig. 14.8 Detail of dapped beam end

Dapped beam ends are especially troublesome when all the strands are below the dap as seen in Fig. 14.7. The strands tend to pull the lower portion of the beam inward from the nib. The compression leads to the propensity to form a diagonal crack at the intersection of the nib and the full beam depth.

Example 14.1: Design of a Dapped Beam End

A 24-in. deep precast concrete T beam has an 8-in. thick web that carries factored end reactions of 53.2 kip in the vertical direction and 11 kip in the horizontal direction, as shown in Fig. 14.8. The beam end is notched 10 in. vertically and 8 in. along the beam axis. The load is transferred to the support through a 4 × 8 in. bearing plate. Design the end reinforcement, using $f'_c = 5000\,\text{psi}$ and $f_y = 60,000\,\text{psi}$.

Solution. The combination of the concentrated load and the geometric discontinuity suggests the use of a strut-and-tie solution. The 14 in. nib depth is close to the 60% desired for successful implementation.

Definition of D-region

The D-region for this beam is approximately one structural depth in from the end of the notch. The bearing plate has longitudinal reinforcement welded to it to allow for horizontal load transfer and a vertical bar for shear-friction design of the end. The effective depth at the notch is taken as 13.0 in. The maximum allowable shear strength is $V_u \leq \phi V_n = \phi 10 \sqrt{f'_c} wd = 0.75 \cdot 10 \cdot \sqrt{5000} \cdot 8 \cdot 13 = 55.2$ kips. This exceeds the 53.2 kip applied load, so the section is adequate to proceed with the design.

Force resultants on D-region boundaries and the truss model

The truss model follows the inclined L configuration, Fig. 14.8. The assumed depth of the truss is 18 in. The tensile and compressive forces at the D-Region interface are

$$T_u = C_u = V_u \cdot 24/18 = 70.9 \text{ kip}$$

Selecting dimensions for strut and nodal zones

The nodal zone stress is established at the bearing plate. The stress at the node and the strut is $V_u/bw = 53.2/4 \times 8 = 1.66$ ksi.

Strength of struts

The strut strength for a bottle shaped strut with $\beta_s = 0.75$ is $\phi 0.85 f'_c = 2.39$ ksi. This exceeds the applied load of 1.66 ksi, so the strut is adequate. The horizontal strut matching the D-region boundary is assumed to be 4 in. deep and 8 in. wide. The stress at the boundary is $T_u/w \times b = 70.9/(4 \times 8) = 2.12$ ksi, which is less than the 2.39 ksi allowed.

Design ties and anchorage

The tension tie force is 30/24 times the vertical reaction or $T_u = 53.2 \times 30/24 = 66.5$ kip. The area of reinforcement required is

$$A_s = \frac{T_u}{\phi f_y} = \frac{66.5}{0.75 \cdot 60} = 1.48 \text{ in.}^2$$

Two No. 6 hairpin bars provide $A_s = 4 \times 0.44 = 1.76$ in.^{2}

Design details and minimum reinforcement requirements

The horizontal force of 11 kip is taken by a bar welded to the bearing plate. The area of steel is $A_s = 11/(0.75 \times 60) = 0.24$ in.^{2} indicating one No. 5 bar is sufficient. To prevent a shear friction failure at the bearing plate the area of reinforcement for a friction factor of $\mu = 1.4$ is

$$A_{sv} = \frac{V_u}{\phi f_y \mu} = \frac{53.2}{0.75 \cdot 60 \cdot 1.4} = 0.84 \text{ in.}^2$$

Use 2 – No. 6 bars or 2-¾ in. diameter headed studs.

The design is completed by detailing the inclined L bar to have a development length along the length of the strand to assure that the tensile force is transmitted into the longitudinal reinforcement.

Comment: The example in this section illustrates the methodology of strut-and-tie design and the importance of understanding the detailing requirements needed to transfer forces at nodes. Failure to appreciate the need to provide anchorage for the tie or to supply thrust resistance for the struts can lead to failure. In the examples, the contact area was used to establish the hydrostatic nodal pressure. As discussed, an equally acceptable solution would have been to select the maximum stress for one of the struts. The diagonal L bar is a loop to ensure anchorage at the top node. The remaining strut and tie widths would then be adjusted accordingly.

Example 14.2: Problem: Design a Column Bracket

Design the column bracket for a factored load of 35-kip vertical and 7-kip horizontal resting on a 3 in. by 6 in. bearing plate. Material properties are f'_c is 5000 psi and f_y is 60,000 psi. Dimensions are given in Fig. 14.9.

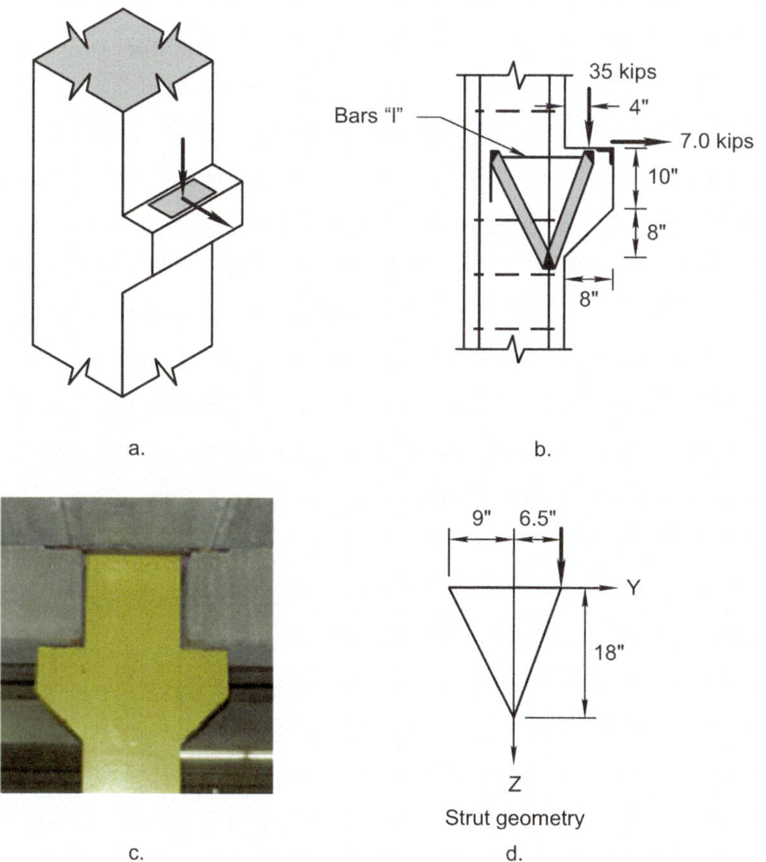

Fig. 14.9 Column bracket details

Solution: The truss geometry is given in Fig. 14.9d. Assume bottle-shaped struts and $\beta_s = 0.75$. The stress under the bearing pad is $V_u/(3 \text{ in.} \times 6 \text{ in.}) = 35/3 \times 6 = 1.94 \text{ ksi}$. By similar triangles, the force in the strut under the load is 37.2 kip and the stress in the strut is 2.07 ksi. The allowable stress in the strut is $0.85 f'_c \beta_s = 0.85 \times 5 \times 0.75 = 3.61 \text{ ksi}$, which exceed the strut stress concluding that the strut is OK.

The force in the tie $x - y$ is 19.6 kip and the required area of reinforcement is $A_s = 19.6/(0.75 \times 60 \text{ ksi}) = 0.44 \text{ in.}^2$ This is provided by 2 – No. 5 bars with $A_s = 0.62 \text{ in.}^2$

Comment: In this example, the calculation is more complex than the shear-friction solution in Example 15.2, but the total amount of reinforcement is less. Assuming an edge strut means that additional reinforcement in the bracket is required to control cracking. Adding 2 No. 4 hoops brings the total amount of reinforcement about equal. The solution for an elastomeric bearing pad would be identical and the strut area would increase to the full width of the bracket. Therefore, the stress in the strut would reduce, keeping the design within Code limits.

Example 14.3: Design General Zone Reinforcement for Post-tensioning Anchorage

A 12-in. wide by 30 in. deep rectangular beam is prestressed with 8-1/2 in. 270 ksi strand tendons centered on the beam centroid 12 in apart, Fig. 14.10, such that the anchor end eccentricity $e = 6$ in. The concrete strength at transfer is 5000 psi.

Solution: Two checks must be made: the tie requirements when both anchors are stressed, and the tie requirements when one anchor is being jacked. The cross-sectional area of the anchor is taken as an 8 in. by 8 in. based on an anchor of approximately 8 in. diameter and adjusting for the local zone. The length of the general zone is taken as h for the disturbed region.

For both anchors stressed, the anchor force is taken as $0.70 f_{pu} = 189$ ksi giving an initial prestressing force of 231 kip. The struts are symmetrical and extend from the anchor to the centroid of the P_i/A_g in of the beam height giving an angle of the strut of $16.7°$ from the horizontal. This results in the following: $F_{ab} = 241.5$ kip, $F_{bd} = 66.5$ kip. The stress in the strut is 214.5 kip/96 in.^{2} or 2520 psi. The allowable

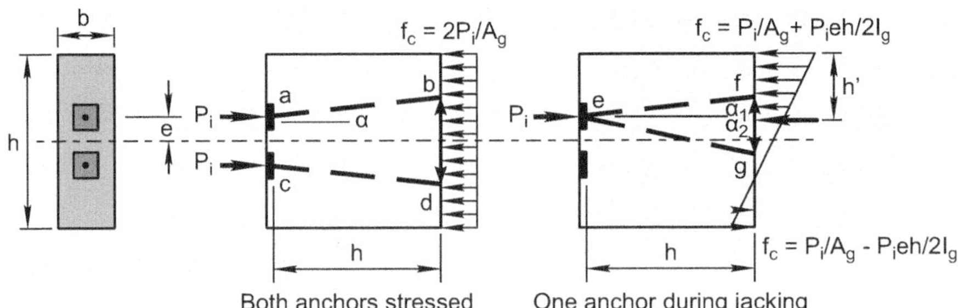

Fig. 14.10 Anchor and strut-and-tie geometry

stress in a strut with a β_s of 0.75 and 5000 psi transfer stress is 3190 psi, confirming that the strut is acceptable. The tension tie force of 66.5 kip is provided using Grade 60 reinforcement. The area of steel required for the tie is 1.48 in.^{2} or 3—No. 5 closed stirrups. The stirrups are centered at 30 in. in from the end of the beam.

The second condition occurs when only one anchor is being stressed. For this case, the full jacking force of 251 kip is used as this is the force applied to the anchor. The jacking force is divided into two struts and the struts react with the linearly varying stress at a distance h in from the end. By trial, equilibrium is obtained at 0.30 h down from the top of the beam. This results in the following geometry and forces: $\alpha_1 = 9.1°$, $\alpha_2 = 11.1°$, $F_{ef} = 140°\text{kip}$, $F_{eg} = 114.7$ kip, and $F_{fg} = 21.8$ kip. By inspection the tie force F_{fg} is less than the tie force for both anchors stressed so no revision of reinforcement is required. The stress in strut F_{ef} is 2930 psi based on half the strut area being available at the anchor. This is larger than the strut for both anchors stressed so it the critical condition but is still less than the allowable stress for the strut.

References

ACI 318-14. (2014). *Building code requirements for structural concrete and commentary*. Farmington Hills, MI: American Concrete Institute.

Klein, G. J. (2008). Curved-bar nodes. *Concrete International, 30*(9), 43–47.

Klein, G. J., Andrews, B. M., & Holloway, K. P. (2015, July 23). *Development of rational design methodologies for dapped ends of prestressed concrete thin-stemmed members, Draft final report*, PCI R&D committee, pp 218.

Marti, P. (1985a). Truss models in detailing. *Concrete International, 7*(12), 66–73.

Marti, P. (1985b). Basic tools for reinforced concrete design. *ACI Journal, 82*(1), 46–56.

Mattock, A., & Theryo, T. (1986, September–October). Strength of precast prestressed members with dapped ends. *PCI Journal 31*(5), 58–75.

Nowak, L. C., & Sprenger, H. (2002). Example 5: deep beam with opening. In K.-H. Reineck (Ed.), *Examples for the design of structural concrete with strut-and-tie models, ACI SP 208* (pp. 129–144). Farmington Hills, MI: American Concrete Institute.

Reineck, H. K. (Ed.). (2002). *Examples for the design of structural concrete with strut-and-tie models, ACI SP 208*. Farmington Hills, MI: American Concrete Institute.

Schlaich, J., & Schäfer, K. (1991, March 13). Design and detailing of structural concrete using strut-and-tie models. *Structural Engineer, 69*(6), 13 pp.

Schlaich, J., Schäfer, K., & Jennewein, M. (1987, May–June). Toward a consistent design of structural concrete. *Journal of PCI, 32*(3), 74–150.

Uribe, C. M., & Alcocer, S. (2002). Example 1a: Deep beam design in accordance with ACI 318-2002. In K.-H. Reineck (Ed.), *Examples for the design of structural concrete with strut-and-tie models, ACI SP 208* (pp. 65–80). Farmington Hills, MI: American Concrete Institute.

Zsutty, T. C. (1971, Feb 2). Shear strength prediction for separate categories of simple beam tests. *ACI Journal, Proceedings, 68*, 138–143.

Chapter 15
Connections and Anchoring to Concrete

15.1 Introduction

Structures require members to be connected and loads to be transferred. These loads can be simple bearing, cast-in-place continuous joints, or mechanical attachments. Three tools assist in connection design: **shear friction**, **anchorage to concrete**, and **strut-and-tie methods**. Each tool has its benefits. Shear friction is simple, straightforward, and beneficial for loads causing potential failure in a known plane through the concrete. Shear friction applications include column brackets, wall corbels, and horizontal shear in composite members. Anchorage to concrete addresses the resistance that may be obtained if short steel elements are embedded in the concrete. The strut-and-tie method addresses complex load paths and the response of structures where the internal stresses do not follow the assumption of plane sections remaining plane. Shear friction and anchorage to concrete are presented in Subheadings 15.2 and 15.3, and strut-and-tie methods are presented in Chap. 14.

Simple bearing connections are designed to ensure that the bottom portion of the beam does not break off using tools such as shear friction, Fig. 15.1. The dapped end of the beam in Fig. 15.1d requires establishment of an internal load path using tools such as strut-and-tie methods, described in Chap. 14, to move the vertical reaction into the main body of the beam.

Attachments consist of steel plates, steel weldments, or precast components that are fastened to the supporting concrete using **anchors**. Anchors are reinforcements either cast into the concrete, or mechanically post-installed, or adhesively bonded. Examples of simple connections and attachments include column base plates, column brackets, wall corbels, stair supports, and mechanical equipment supports. A precast column bracket with the anchor reinforcement extended the bracket cast into a column; steel cast-in brackets and post-installed and steel corbels are examples of concrete anchors, Fig. 15.1. Figure 15.2 shows mechanical equipment fastened to the soffit of a precast prestressed concrete slab.

© Springer Nature Switzerland AG 2019

C. W. Dolan, H. R. Hamilton, *Prestressed Concrete*,

https://doi.org/10.1007/978-3-319-97882-6_15

a) Precast column bracket   b) Placement in precast column   c) Steel corbel in precast element   d) Steel corbel post-installed to a cast-in-place wall

Photographs a) though c) courtesy Rocky Mountain Prestress Company   *Photograph d) courtesy Hilti Corporation*

Fig. 15.1 Anchor-based connections. (**a**) Precast column bracket. (**b**) Placement in precast column. (**c**) Steel corbel in precast element. (**d**) Steel corbel post-installed to a cast-in-place wall

Fig. 15.2 Mechanical equipment post-installed in a prestressed precast concrete slab

15.1.1   *Loads*

Returning to Leonhardt's commandments in Chap. 1, prestressed structures move over time because of temperature changes and creep and shrinkage. This movement is accommodated either by sliding relative to the bearing or by deformation of the structure itself. An examination of some unenclosed parking garages reveals horizontal cracks in the columns resulting from the expansion and contraction of the beams. These cracks are prevalent in the columns supporting the roof where it is subjected to solar input and ambient temperature. Since the roof does not have columns above, the column restraint is half of that of the floors, which aggravates the problem.

Loads on connections are derived from the structural analysis and are affected by the relative stiffness of the various components in the structure. In addition to the

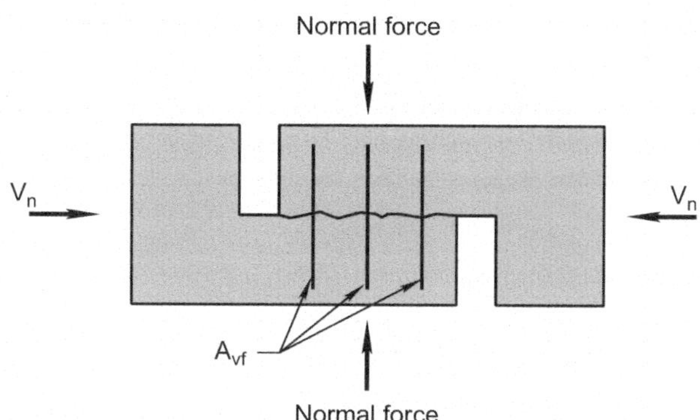

Fig. 15.3 Shear friction model

loads generated by the structural analysis, precast beams are often supported by elastomeric or Teflon coated bearing pads. In Fig. 15.1b the elastomeric pad is placed between the concrete and the steel support. These connections are intended to minimize the horizontal restraint of the supported element and are typically analyzed as rollers or pins. Shrinkage, creep, and temperature changes that result in the longitudinal movement of the member, however, still generate longitudinal forces resulting from friction between the member and the bearing pad or by deformation of the bearing pad. Industry practice is to include a horizontal shear force on the bearing at least equal to 0.2 times the factored vertical shear. This horizontal force affects the connection and the beam end design and details.

15.2 Shear Friction

Shear friction principles address the possibility of a failure plane forming through the concrete or between concrete and steel. The basic principle says that the shear strength of the connection is a function of the normal force and the coefficient of friction at the failure interface, Fig. 15.3. The failure plane is indicated with a dashed line and the normal force is provided by the reinforcement labeled A_{vf}.

Formulation of the nominal shear strength then follows directly from friction behavior:

$$V_n = A_{vf} f_y \mu \qquad (15.1)$$

where V_n is the nominal shear strength, A_{vf} is the area of anchored reinforcement crossing the shear failure plane, f_y is the specified yield strength of the reinforcement, and μ is the coefficient of friction. The coefficient of friction for various conditions is given in Table 15.1. The maximum shear stress that can be developed across an assumed shear plane of area A_c is given in Table 15.2 based on the limits established in the ACI Building Code. The yield stress of the reinforcement is limited to 60 ksi, even if a higher strength steel is used. Shear friction design often uses headed studs,

Table 15.1 Coefficients of friction

Contact surface condition	Coefficient of friction μ
Concrete placed monolithically	1.4 λ
Concrete placed against hardened concrete that is clean, free of laitance, and intentionally roughened to a full amplitude of approximately ¼ in.	1.0 λ
Concrete placed against hardened concrete that is clean, free of laitance, and not intentionally roughened.	0.6 λ
Concrete placed against as-rolled structural steel that is clean, free of paint, and with shear transferred across the contract surface by headed studs or by welded deformed bars or wires.	0.7 λ

Notes: λ is the correction for lightweight concrete given in Chap. 3. Adapted from ACI 318-14 (2014)

Table 15.2 Maximum V_n allowed across the assumed shear plane

Condition	Maximum V_n	
Normal-weight concrete placed monolithically or placed against hardened concrete intentionally roughened to a full amplitude of approximately ¼ in.	Least of:	$0.2 f'_c A_c$
		$(480 + 0.08 f'_c) A_c$
		$1600 A_c$
Other cases	Lesser of:	$0.2 f'_c A_c$
		$800 A_c$

A_c is the area of concrete subject to shear friction resistance. Adapted from ACI 318-14 (2014)

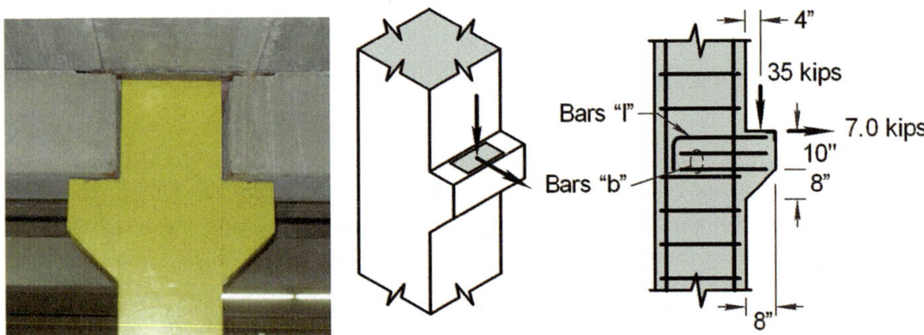

Fig. 15.4 Column bracket

especially when calculating the shear strength between a steel beam and a concrete slab. The yield strength of headed studs is 51 ksi.

Example 15.1

Problem: Design column bracket using shear friction. The column bracket shown in Fig. 15.4 carries a factored load of 35 kips and a factored horizontal load of 7.5 kips.

Solution 1: a narrow stem rests on a 3 in. by 6 in. steel bearing plate. Reinforcement material properties are f_y is 60,000 psi, ϕ is 0.75, and μ is 1.4 for a monolithic concrete connection.

The required shear friction reinforcement is

$$A_{vf} = \frac{V_u}{\phi f_y \mu} = \frac{35}{0.75 \cdot 60 \cdot 1.4} = 0.56 \text{ in}^2$$

and the required tension reinforcement is

$$A_s = \frac{T_u}{\phi f_y} = \frac{7}{0.75 \cdot 60} = 0.16 \text{ in}^2$$

Detailing the bracket includes one No. 4 bar (bar l in Fig. 15.4), $A_s = 0.20$ in.^{2}, welded to the bearing plate and 2 No. 4 closed hoops (bars b in Fig. 15.4), $A_s = 4 \times 0.20 = 0.80$ in.^{2}, below the welded bar to provide the shear friction reinforcement.

Solution 2: the beam seated on a full width elastomeric pad as shown in the photo at the left of Fig. 15.4. In this case, the reinforcement requirement would be identical but is provided by 3 – No. 4 hoops, $A_s = 6 \times 0.20 = 1.20$ in.^{2} and there is no welding or embedded plates.

Comment: The photograph shows a column with a double bracket and elastomeric pads. In this case, the reinforcement would serve both brackets simultaneously. As a practical matter in Option 1, 2 – No. 4 bars would be welded to the embedded plate for stability in handling the insert and to resist any eccentric loading.

15.3 Anchorage to Concrete

15.3.1 Behavior of Anchors

Anchors are steel elements that are cast into the concrete, adhesively bonded, or mechanically post-installed in hardened concrete. The **effective embedded depth** h_{ef} is shown in Fig. 15.5.

The variety of possible anchors and their varied failure modes require the engineer to determine the strength associated with each possible failure mode and base the design on the lowest strength to assure overall structural safety (Fuchs et al. 1995). Anchor failures are evaluated for steel strength, concrete breakout intension, concrete pullout in tension, side face blowout, pryout in shear, combined tension and shear, and bond failure of adhesive.

The failure modes result from tension, shear, or combined loading. Figure 15.6 illustrates several failure modes. Except for the adhesive anchors, headed studs are

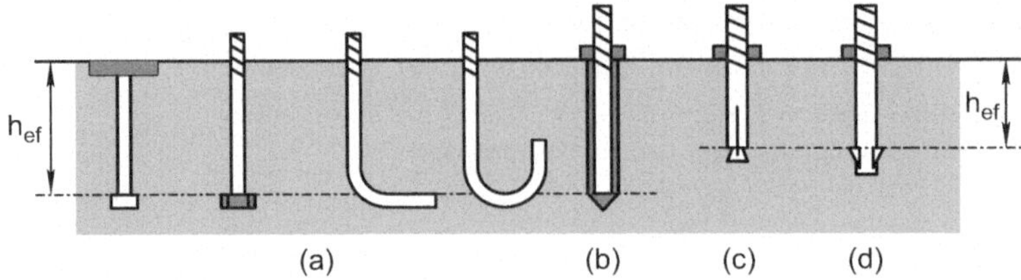

Fig. 15.5 Anchor types. Cast-in anchors: (a) headed studs and bolts, and post-installed anchors: (b) adhesive anchor, (c) drop-in type displacement-controlled expansion anchor undercut anchor, and (d) torque-controlled expansion anchors

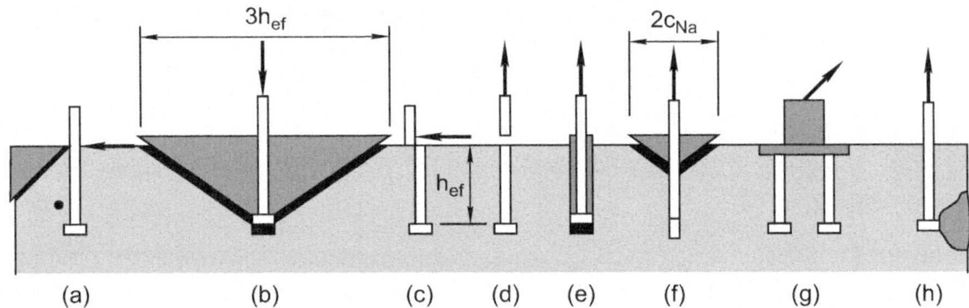

Fig. 15.6 Anchor failure modes: (a) concrete breakout in shear; (b) concrete breakout in tension; (c) steel failure in shear; (d) steel failure in tension; (e) concrete pullout in tension; (f) bond failure of adhesive; (g) combined tension and shear; and (h) side face blowout

illustrated. Any design considers eccentric or unequal loading of individual anchors. The characteristics of the failure modes are described in sections a though g below.

(a) **Steel strength in shear or tension**—The shear or tensile strength of an anchor (Fig. 15.6 c and d) is based on the specified tensile strength of the steel. Strength is based the net area of the anchor when threads are present. Tensile strength rather than yield strength is used to calculate anchor strength to allow for direct comparison with the concrete breakout strength. Individual anchors are assumed to yield and the load distributed among adjacent anchors.

(b) **Concrete breakout in shear or tension**—Concrete breakout occurs when an anchor in tension generates tensile stresses on the surface of a prism of concrete radiating out from the head of the anchor, Fig. 15.6b. Shear failure prisms form between the anchor and the edge of the concrete, Fig. 15.6a. When the stresses are high enough for the concrete to fracture, an anchor prism of concrete separates from the surrounding concrete, Fig. 15.7. Attachments with multiple anchors, such as shown in Fig. 15.6g, may generate overlapping breakout regions. Nominal concrete breakout strength is modified to account for the overlap.

(c) **Anchor pullout in tension**–As the head size on cast-in anchors diminishes, the anchor can pull out by creating a cylinder of concrete directly above the head Fig. 15.6e. The pullout strength is a function of the area of the head less the area

Fig. 15.7 Single stud breakout prism indicating a failure angle of approximately 35°

of the shaft. Normally proportioned headed studs and bolts have sufficient head area that steel rupture or concrete breakout occurs before concrete pullout can develop.

(d) **Side face blowout**–Anchors with deep embedment and thin side cover can fail by concrete spalling on the side face around the embedded head with no major breakout occurring at the top concrete surface, Fig. 15.6h.

(e) **Pryout in shear**–If an attachment, like that shown in Fig. 15.6*g*, has short anchors, is located away from an edge, and is subjected to high shear load, the plate may bend and the anchors on the back side of the attachment may rotate upward, leading to a pryout failure. Longer anchors are less prone to pryout failures.

(f) **Combined tension and shear** Anchors on some attachments, as shown in Fig. 15.6*g*, are subjected to both shear and tensile loading. Determination of the strength of these attachments requires examining the interaction between the effects of the shear and the tensile loads.

(g) **Bond failure of adhesive**—Adhesive anchors develop a bond between the anchor, the epoxy, and the concrete. A typical bond failure results in the anchor pulling out due to a concrete breakout around the portion of the anchor near the surface combined with an adhesive failure along the lower portion of the anchor, as shown in Fig. 15.6*f*.

15.3.2   Concrete Breakout Strength

The variation between historical predicted behavior and test results led to the development of the **concrete capacity design (CCD)** method for determination of concrete breakout strength (Fuchs et al. 1995; ACI 349 2001). The CCD method correlates strength with key material and geometric parameters. The research concluded that a prism with a failure angle of approximately 35° provides a statistically more reliable strength prediction than earlier models. The strength of an anchor with a 35° failure cone correlates to the embedment depth of the anchor raised to the 1.5 power $h_{ef}^{1.5}$.

The ACI Building Code assumes that the concrete cracks at some point in its service lifetime. Behavior in cracked concrete is critical to anchors and especially for anchors placed in negative moment regions and in structures subjected to earthquake loading (Eligehausen and Balogh 1995; Cook and Klingner 1992a, b). Cracked concrete specimens exhibit a loss of strength ranging up to 30% for cast-in-place bolts and headed studs and 40–60% for post-installed anchors compared to uncracked concrete specimens. The greater loss in post-installed anchors is attributed to a variation in the ability of the mechanical expansion devices on the anchor to fully engage the concrete. In evaluation tests, anchors expanded to only 50% of their specified limits confirmed this loss. These observed anchor strength reductions in cracked concrete led to the conclusion that cracked concrete should be the basis for design unless analysis indicates that the concrete will remain uncracked throughout its service life.

Precast and prestressed members are often assumed to remain uncracked during removal from the forms, lifting and handling, and throughout their design life. An industry study on headed stud anchors in uncracked concrete confirmed the applicability of the concrete capacity design approach and provided an alternative set of equations for the *PCI Design Handbook* 2017 (Anderson and Meinheit 2005). The PCI equations provide the same strength predictions for anchors in cracked concrete if adjusted to include cracked concrete factors. Research sponsored by PCI concluded that the strength prediction equations were valid for concrete strengths as low as $f'_c = 1000$ psi (Winter and Dolan 2014).

ACI 355.2 *Qualification of Post-Installed Mechanical Anchors in Concrete and Commentary* (2007) provides test acceptance criteria for the certification of post-installed anchors in cracked concrete and ACI 355.4 *Qualification of Post-Installed Adhesive Anchors in Concrete* (2010) provides acceptance criteria for adhesive anchors. Manufacturers use these criteria to establish the anchor strength and installation conditions associated with their proprietary anchors.

Analyses of tests led to the conclusion that the statistically significant variables for predicting the CCD anchor strength in tension are the tensile strength of the concrete and the anchor embedment depth. The nominal breakout strength of a single anchor is established at the **5% fractile**.[1] Using the 5% fractile, rather than the mean strength, provides a level of structural reliability consistent with other equations in the Code.

---

[1]Nominal strength for which there is 90% confidence that there is a 95% probability that it will be exceeded by the actual strength.

15.3.3   Anchor Design

Attachments may include multiple anchors. The design of attachment anchors involves the calculation the strength of individual anchors in the group based on their possible failure modes and then uses the lowest individual strength as the basis for the nominal strength of the **anchor group**, individual anchors functioning together. The strength reduction factor corresponding to the failure mode associated with the lowest strength anchor failure mode and is used to calculate the design strength for comparison with the factored loads. Tensile and shear failure modes must be checked and, if both tension and shear are present, verification of the strength due to the interaction of the two loads is required.

The overlap of the concrete breakout prisms for anchors in tension that are placed less than $3h_{ef}$ apart on an attachment or anchors with an edge distance less than $1.5h_{ef}$ results in a concrete breakout strength that is less than the sum of the individual anchor breakout capacities. The calculated strength of anchor groups placed near an edge requires further modification because concrete may split near individual anchors and the total failure breakout prism may not be mobilized.

15.4   ACI 318-14 Provisions for Concrete Breakout Strength

Chapter 17 of ACI 318-14 includes equations for possible failure modes. The ACI Building Code allows the use of any method that is in substantial agreement with test results. The equations in the ACI Building Code are deemed to meet this requirement. Temporary lifting attachments, specialty attachments, and anchors subject to high cycle fatigue or impact loading are outside the scope of the Code. The Code addresses cast-in headed studs, bolts and hooked rods, post-installed bolts, and adhesively bonded anchors. The ACI Building Code addresses the concrete breakout strength of the cast-in and adhesive anchors. The capacity of post-installed mechanical, undercut, and expansion anchors requires manufacturers' certification and is addressed in the Code only through the selection of strength reduction factors.

Reinforcement placed within the anchor breakout prism affects anchor performance. **Supplementary reinforcement** is added to reduce spalling and control crack width. Supplementary reinforcement restrains the potential concrete breakout but is not designed to transfer the full design load from the anchors into the structural member. **Anchor reinforcement** transfers the full anchor load into the structural member and is discussed in Sect. 15.4.7. When anchor reinforcement is provided, the calculation of concrete breakout strength is not required.

The design strength of anchors, the product of the nominal strength S_n and the strength reduction factor ϕ, must exceed the factored load U or $\phi S_n \geq U$. The strength reduction factor is a function of the type of anchor, type of loading, the presence of supplementary reinforcement, and the installation conditions. Strength

Table 15.3 Summary of strength reduction factors for anchors[a]

Strength condition		ϕ for tension	ϕ for shear
Ductile steel element		0.75	0.65
Brittle steel element[b]		0.65	0.60
Concrete breakout, for cast in studs, headed bolts, or hooked bolts	Condition A: Supplementary reinforcement present, except pryout and pullout strength	0.75	0.75
	Condition B: Supplementary reinforcement not present. Includes pryout and pullout strength	0.70	0.70
Post-installed anchors with supplementary reinforcement present except for pryout and pullout strength[c]	Category 1—low sensitivity to installation and high reliability	0.75	0.75
	Category 2—medium sensitivity to installation and medium reliability	0.65	0.65
	Category 3—high sensitivity to installation and low reliability	0.55	0.55
Post-installed anchors without supplementary reinforcement present and for pryout and pullout strength[b]	Category 1—low sensitivity to installation and high reliability	0.65	0.65
	Category 2—medium sensitivity to installation and medium reliability	0.55	0.55
	Category 3—high sensitivity to installation and low reliability	0.45	0.45

[a]Adapted from ACI 318-14 (2014)

[b]Brittle steel elements have an elongation less than 14% when tested by their appropriate ASTM method

[c]Strength and sensitivity conditions for post-installed anchors are established by the ACI 355.2. The effects of variability in anchor torque during installation, tolerance on drilled hole size, and energy level used in setting anchors are considered; for expansion and undercut anchors approved for use in cracked concrete, increased crack widths are considered. ACI 355.4 tests for sensitivity to installation procedures determine the category for a particular adhesive anchor system considering the influence of adhesive mixing and the influence of hole cleaning in dry, saturated and water-filled/underwater bore holes

reduction factors listed in Table 15.3 are derived from analysis of test results. Cast-in-place anchors have higher strength reduction factors than post-installed or adhesive anchors. The strength reduction factors for post-installed anchors reflect both the difficulty of installation and the ability of the post-installed anchor to perform properly. Post-installed anchor details are often proprietary and vary between anchor suppliers.

To reduce local cracking and splitting, anchors placement must satisfy the minimum and critical edge distances given in Table 15.4. These distances are based on both the embedment depth h_{ef} and the anchor diameter d_a. Use of an edge distance below that given in Table 15.4 is allowed if the lower distance is validated by tests according to ACI 355.2 or ACI 355.4 and supplementary reinforcement is present. In all cases, attachments with anchors loaded in shear require a reinforcing bar between the anchor and the edge, Fig. 15.6.

Table 15.4 Minimum and critical edge distance and anchor spacing[a]

Anchor type	Minimum edge distance[b]	Critical edge distance[c] c_{ac}	Minimum anchor spacing
Cast-in	Minimum cover[d]		$4\,d_a$
Adhesive	$6\,d_a$	$2\,h_{ef}$	$6\,d_a$
Undercut or expansion[e]	$6\,d_a$	$2.5\,h_{ef}$	$6\,d_a$
Torque controlled	$8\,d_a$	$4\,h_{ef}$	$6\,d_a$
Displacement controlled	$10\,d_a$	$4\,h_{ef}$	$6\,d_a$

[a]Minimum and critical edge distances can be reduced if validated by tests following ACI 355.2 or ACI 355.4 protocols
[b]Minimum edge distance is the greater of the required cover or the value in the table
[c]Critical edge distance, c_{ac}, is required to develop the basic strength as controlled by concrete breakout or bond of a post-installed anchor in tension in uncracked concrete without supplementary reinforcement
[d]Cast-in anchors to be torqued require $6\,d_a$ edge distance
[e]h_{ef} is the lesser of (a) the actual h_{ef}, (b) 2/3 of the slab thickness, h_a, and (c) the slab thickness less 4 in.

15.4.1   Steel Strength

Steel tensile and shear strengths assume that the anchors yield and distribute the load to all anchors in the group. The calculated nominal strength of each anchor is based on the specified tensile strength of the steel f_{uta} and the net cross-sectional area of the anchor A_{se}. A subscript extension on the area denotes whether the area is for tension $A_{se,N}$ or shear $A_{se,V}$.

The specified tensile strength is used because the steel in many anchors does not have a well-defined yield point. Tensile strength is higher than yield strength, the value most often used in reinforced concrete design. The strength reduction factors for anchors in tension given in Table 15.3 are lower than those associated with steel yielding because the nominal tensile strength is the basis. The final strength reduction factors are compatible with the tensile strength and with the load factors given in ASCE/SEI 7-16 (2016). Section 17.4.1.2 of ACI 318-14 limits the allowable tensile strength to 1.9 times the specified yield strength of an anchor but not more than 125,000 psi. This limit is to assure sufficient yield strength and prevent the use of high strength steel with very low yield strength. The nominal capacities for steel tension and shear strengths are summarized in Table 15.5 an anchor group with n anchors.

Example 15.2
Problem: Determine the nominal steel tensile strength of the anchor group for the attachment shown in Fig. 15.8. The attachment is anchored with 6 ½ in. diameter headed studs, each having a specified yield strength $f_{ya} = 51$ ksi.

Table 15.5 Nominal steel anchor strength

Anchor	Failure mode	
	Tension	Shear
Headed studs	$N_{\mathrm{sa}} = n\,A_{\mathrm{se},N}\,f_{\mathrm{uta}}$	$V_{\mathrm{sa}} = n\,A_{\mathrm{se},V}\,f_{\mathrm{uta}}$
Headed bolts[a] or hooked bolt anchors	$N_{\mathrm{sa}} = n\,A_{\mathrm{se},N}\,f_{\mathrm{uta}}$	$V_{\mathrm{sa}} = n\,0.6\,A_{\mathrm{se},V}\,f_{\mathrm{uta}}$
Post-installed	Per manufacturer's specifications	Per manufacturer's specifications

[a]ANSI/ASME B1.1 (2003) defines $A_{\mathrm{se},N}$ and $A_{\mathrm{se},V}$ for threaded bolts and hooked bolts as
$\frac{\pi}{4}\left(d_{\mathrm{a}} - \frac{0.9743}{n_{\mathrm{t}}}\right)^{2}$ where d_{a} is the diameter of the anchor and n_{t} is the number of threads per inch

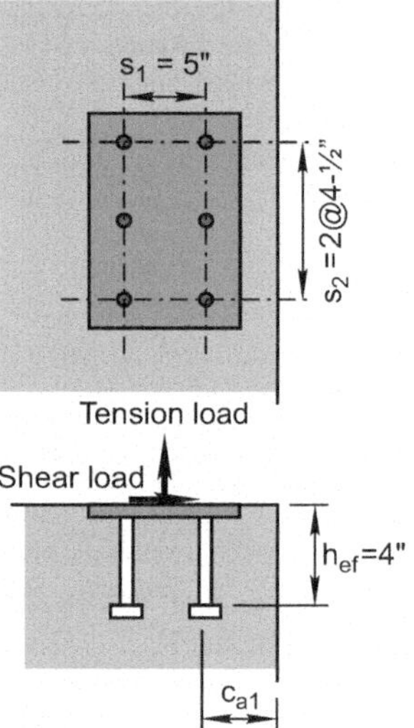

Fig. 15.8 Attachment details for Examples 15.2 through 15.9

Solution: The area of a ½ in. diameter headed stud is 0.20 in.^{2} and for a yield strength of 51 ksi, the specified tensile strength of a headed stud is 65 ksi. Based on the equations in Table 15.3, the six studs shown in Fig. 14.8 then have a nominal tensile strength of

$$N_{\mathrm{sa}} = nA_{\mathrm{se},N}f_{\mathrm{uta}} = 6 \times 0.20 \times 65 = 87.0\,\mathrm{kips}$$

Comment: This example uses the specified tensile strength rather than $1.9f_{\mathrm{ya}}$ to obtain the tensile strength of the anchor. The ASTM specification for headed studs specifies the tensile strength as 65 ksi, which is less than the value from the Code calculation. Using $f_{\mathrm{uta}} = 1.9\,f_{\mathrm{ya}} = 116.3$ ksi and results in a tensile strength of 116.3 kips, which would exceed the anchor strength.

Example 15.3
Problem: Determine the nominal steel shear strength of the anchor group for the attachment shown in Fig. 15.8 if the attachment is anchored with six ¾ in. diameter

A307 threaded bolts, each with ten threads per inch and specified yield strength of 36 ksi and a tensile strength of 65 ksi. There is no grout pad under the plate.

Solution: From Table 15.3, the effective area of a ¾ in. diameter headed bolt is

$$A_{se,V} = \frac{\pi}{4}\left(d_a - \frac{0.9743}{n_t}\right)^2 = \frac{\pi}{4}\left(3/4 - \frac{0.9743}{10}\right)^2 = 0.334 \text{ in.}^2$$

For a specified yield strength $f_{ya} = 36$ ksi, the tensile strength is 65 ksi but ranges between 58 and 80 ksi.

For the six bolts shown in Fig. 15.8 and using Table 15.3 for headed bolt strength, the nominal shear steel strength is

$$V_{sa} = n0.6A_{se,V}f_{uta} = 6 \times 0.6 \times 0.334 \times 65 = 78.1 \text{ kips}$$

Comment: While the Code allows this solution, the specified tensile strength of some grade steel bolts differ from A-36 steel. In such cases, the actual specified tensile strength is used for design, and checked against the $1.9f_{ya}$ and 125 ksi limits.

15.4.2   Concrete Breakout Strength of Single Cast-In and Post-installed Anchors

The concrete breakout strength of an anchor group is based on the breakout strength of a single anchor in cracked concrete and then adjusted for the effect of anchors in a group. The individual anchor strength is further modified to account for concrete cracking, the distance to the edge of the member from the closest anchors, eccentricity of the load on the attachment, anchor pullout, and anchor pryout effects.

Tensile Breakout Strength of a Single Anchor

The tensile breakout strength of a single anchor N_b in cracked concrete is

$$N_b = k_c \lambda_a \sqrt{f'_c} h_{ef}^{1.5} \tag{15.2}$$

where k_c is 24 for cast-in anchors, 17 for post-installed anchors, and λ_a is the modification factor for lightweight concrete in the anchor zone. The value of λ_a is 1.0λ for cast-in and undercut anchors, and 0.8λ for expansion and adhesive anchors. Equation 15.2 is limited to anchors with diameters of 4 in. or less due to the lack of test data on larger bolts. Section 17.4.2.2 of ACI 318-14 allows the value of k_c to increase above 17 if supported by tests based on ACI 355.2 evaluation protocols for

Fig. 15.9 Single anchor breakout prism projections shear and tension

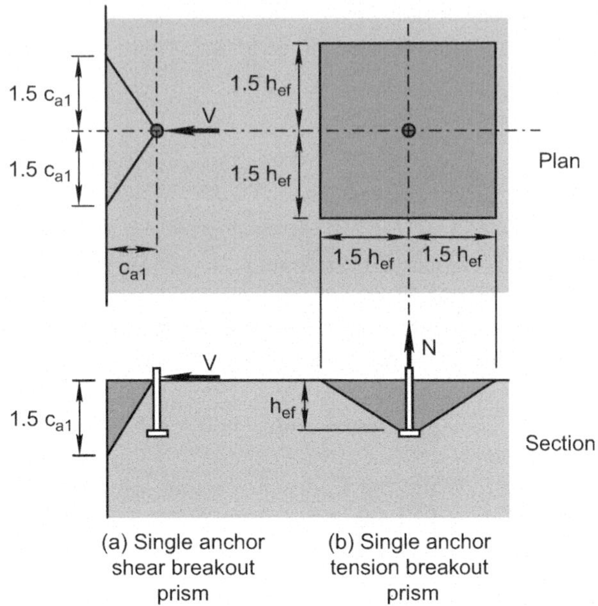

post-installed anchors that justify the higher value. In no case is k_c allowed to exceed 24.

Test data for bolts with deep effective embedment lengths indicate a greater strength than predicted by Eq. 15.2. Thus, the basic breakout strength of headed studs or threaded bolts with h_{ef} between 11 and 25 in. may be calculated as

$$N_b = k_c \lambda_a \sqrt{f'_c} h_{ef}^{5/3} \tag{15.3}$$

Shear Breakout Strength of a Single Anchor

The calculated concrete shear breakout strength of a single anchor V_b in concrete is the lesser of Eqs. 15.4 and 15.5.

$$V_b = \left[7 \left(\frac{l_e}{d_a} \right)^{0.2} \sqrt{d_a} \right] \lambda_a \sqrt{f'_c} (c_{a1})^{1.5} \tag{15.4}$$

where d_a is the diameter of the anchor and c_{a1} is the distance from the edge of the concrete to the first anchor, Fig. 15.9a. The value for l_e, the load-bearing length of the anchor for shear, is equal to h_{ef} for anchors with constant stiffness over their full length of embedment or $2d_a$ for torque-controlled expansion bolts separated from the expansion sleeve. In all cases, l_e must be $\leq 8d_a$. For headed studs, headed bolts, or hooked bolts continuously welded to steel plates with a thickness of at least 3/8 in., the value of 7 in Eq. 15.4 may be increased to 8, providing the strength does not exceed V_b calculated as

$$V_b = 9_a \sqrt{f_c'} (c_{a1})^{1.5}$$
(15.5)

Concrete Breakout Strength of Anchor Groups

The tension breakout strength of a single anchor is based on the $35°$ breakout prism. This results in a prism of concrete extending $1.5h_{ef}$ from the center of the anchor, as shown in Fig. 15.9b. The tension breakout prism for a single anchor has a single anchor projected area $A_{Nco} = (3h_{ef})^2 = 9h_{ef}^2$. The shear breakout prism extends from the anchor downward toward the free edge, giving a shear breakout area for a single anchor A_{Vco} on the front face of the concrete of $A_{Vco} = 2(1.5c_{a1}$ wide$)\times(1.5c_{a1}$ deep$) = 4.5\ c_{a1}^2$, as shown in Fig. 15.9a and Fig. 15.10e.

Anchors spaced less than $3h_{ef}$ apart have overlapping breakout prisms. Experimental evaluation of these groups indicates that the strength of groups in tension can be accounted for by multiplying the breakout strength of a single anchor N_b by the ratio of the projected breakout area of the anchor group A_{Nc} to the projected breakout area of a single anchor A_{Nco} (Cook and Klingner 1992a, b). Calculation of the anchor group breakout prism projected area is influenced by both the spacing of the anchors and the distance to the edge of the concrete. Calculation of tensile strength is a function of the anchor embedment depth h_{ef} the distances to the edges c_{a1} and c_{a2} and the anchor spacing s_1. Typical conditions for anchors in tension are illustrated in Fig. 15.10a through c.

The shear strength of individual anchors in a group is obtained by multiplying the shear strength of a single anchor by the ratio of the projected area of the shear anchor group A_{Vc} to the projected area for a single anchor A_{Vco}. The calculation of the projected shear breakout area is limited in a thin concrete section and the placement of the shear anchors. Anchors placed perpendicular to the edge of the concrete offer two possible failure modes. The first mode is based on the anchor nearest the edge carrying half the load and failing first, Fig. 15.10f. The second mode is based on the anchor closest to the edge "riding with the failure breakout" with the entire load carried by the anchor farthest from the edge, Fig. 15.10g. Where the load is placed symmetrically on two or more anchors, the projected area is a function of the slab thickness h_a, the anchor spacing s_1, and the distance from the edge c_{a1}, Fig. 15.10h. For slabs with a thickness greater than c_{a1}, h_a is replaced with c_{a1}, as shown in Fig. 15.10d.

Anchor strength is further modified for the condition of the concrete immediately around the individual anchor. Modification factors for anchor concrete breakout strength are designated as ψ, followed by subscripts defining the condition and direction of loading. The subscript c indicates modification for cracked or uncracked concrete, the subscript ed for edge distance, the subscript ec for load eccentricity, the

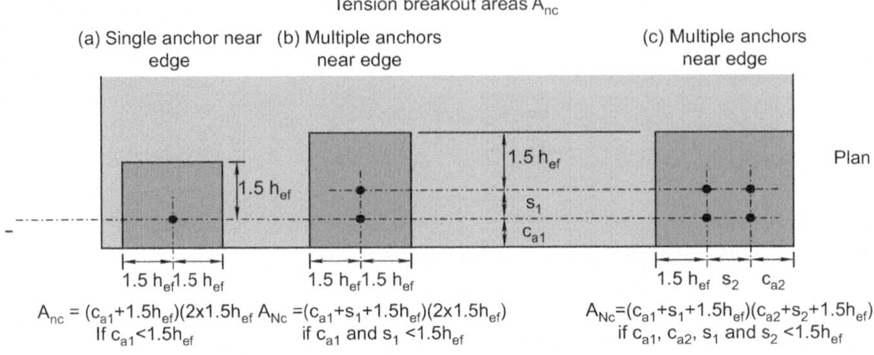

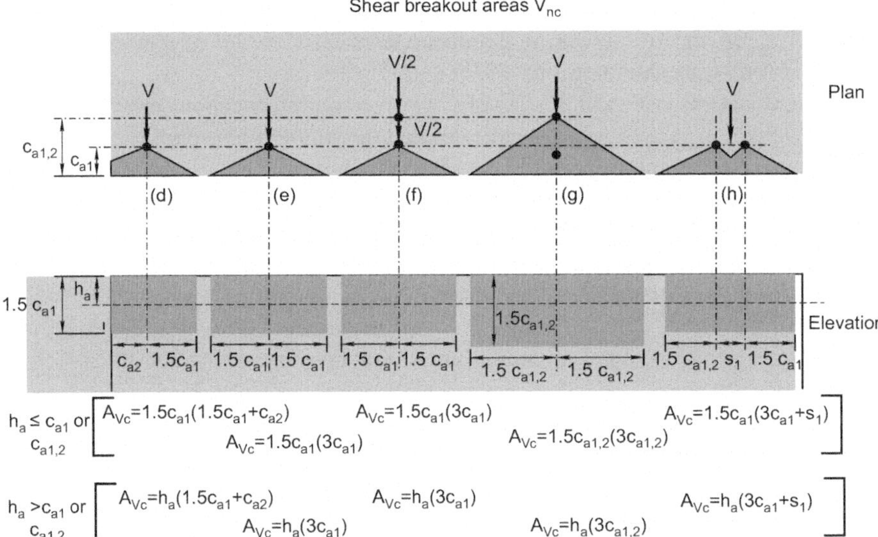

Fig. 15.10 Anchor group breakout projected areas for tension and shear near edges

subscript cp for anchor pullout, and the subscript h for the slab thickness. Combining these effects, the concrete breakout strength for a single anchor in tension is

$$N_{cb} = \frac{A_{Nc}}{A_{Nco}} \psi_{ed,N} \psi_{c,N} \psi_{cp,N} N_b \qquad (15.6)$$

and for an anchor group in tension,

$$N_{\mathrm{cbg}} = \frac{A_{Nc}}{A_{Nco}} \psi_{\mathrm{ec},N} \psi_{\mathrm{ed},N} \psi_{\mathrm{c},N} \psi_{\mathrm{cp},N} N_{\mathrm{b}} \qquad (15.7)$$

The concrete breakout strength in shear for a single anchor loaded *perpendicular* to the edge is

$$V_{\mathrm{cb}} = \frac{A_{Vc}}{A_{Vco}} \psi_{\mathrm{ed},V} \psi_{\mathrm{c},V} \psi_{h,V} V_{\mathrm{b}} \qquad (15.8)$$

and for a group of anchors loaded perpendicular to the edge,

$$V_{\mathrm{cbg}} = \frac{A_{Vc}}{A_{Vco}} \psi_{\mathrm{ec},V} \psi_{\mathrm{ed},V} \psi_{\mathrm{c},V} \psi_{h,V} V_{\mathrm{b}} \qquad (15.9)$$

Section 17.5.2.1 of ACI 318-14 permits shear forces *parallel* to an edge to be computed as twice the value determine in Eqs. 15.8 and 15.9.

For anchors near the edges of walls or other locations where there is less than 1.5 h_{ef} from three surfaces and similar instances, using the ratio of projected areas results in an overestimation of strength. In these situations, an adjusted effective embedment depth or edge distance is used in addition to the ratio of breakout areas to bring the calculated nominal strengths in line with test results. Sections 17.4.2.3 and 17.4.2.4 of ACI 318-14 provide guidance for these conditions.

Modification Factors for Concrete Cracking, Edge Distance, and Slab Thickness

Research cited in Fuchs et al. (1995) and Cook and Klingner (1992a, b) indicates that anchor strength is reduced if the concrete is cracked at service load. For this reason, Eq. 15.2 through Eq. 15.9 are based on cracked concrete. If analysis indicates that the concrete is uncracked under service load, then both the tensile and shear strength is increased. Uncracked conditions are common in precast and prestressed concrete. Modification factors for anchors placed in uncracked concrete are summarized in Table 15.6.

The strength of an individual anchor near one or more edges must be further adjusted due to localized cracking. Modification factors for anchors near free edges are summarized in Table 15.7 using the notation illustrated in Fig. 15.10. In situations where the slab thickness h_{a} is less than $1.5c_{\mathrm{a}}$, a further modification of $\psi_{h,V}$

$$= \sqrt{\frac{1.5c_{\mathrm{a1}}}{h_{\mathrm{a}}}} \leq 1.0 \text{ is required.}$$

Example 15.4
Problem: Determine the concrete tensile breakout strength of the anchor group in Fig. 15.8, given that the load is concentrically applied and the attachment is in concrete that analysis indicates is uncracked during service load. The anchors are

Table 15.6 Breakout modification factors for concrete cracking near an anchor

Cracked condition	Tension $\psi_{c,N}$ Factor	Shear[a] $\psi_{c,V}$ Factor
Cracked	1.0	1.0
Cracked		1.2[b]
Cracked		1.4[c]
Uncracked cast-in	1.25	1.4
Uncracked post-installed and k_c is 17	1.4	
Uncracked and strength determined by ACI 355.2	1.0	

[a]The ACI Building Code does not differentiate between cast-in and post-installed anchors for shear
[b]The ACI Building Code requires a No. 4 (No. 13) or larger bar between the anchors and the edge of concrete for this factor to be used
[c]The ACI Building Code requires a No. 4 (No. 13) or larger bar between the anchors and the edge and enclosed in stirrups not more than 4 in. on centers for this factor to be used

Table 15.7 Breakout modification factors for edge distance

Condition	Tension $\psi_{ed,N}$	$\psi_{ed,Na}$	Shear $\psi_{ed,V}$
$c_{a,min} \geq 1.5\, h_{ef}$ $c_{a,min} \geq 1.5\, c_{Na}$	1.0	1.0	
$c_{a,min} < 1.5\, h_{ef}$ $c_{a,min} < 1.5\, c_{Na}$	$0.7 + 0.3\frac{c_{a1}}{1.5h_{ef}}$	$0.7 + 0.3\frac{c_{a,min}}{c_{Na}}$	
$c_{a2} \geq 1.5\, c_{a1}$			1.0
$c_{a2} \leq 1.5\, c_{a1}$			$0.7 + 0.3\frac{c_{a2}}{1.5c_{a1}}$

cast in 5000 psi normalweight concrete with six - ½ in. diameter headed studs and $c_{a1} = 8$ in.

Solution. From Fig. 15.8, $s_1 = 5$ in. and $s_2 = 4.5$ in. For normalweight concrete $\lambda = \lambda_a = 1.0$. From Table 15.4 for uncracked concrete and cast in anchors, $\psi_{c,N} = 1.4$, the pullout modification factor is $\psi_{cp,N} = 1.0$, and because the value of c_{a1} is greater than $1.5\, h_{ef}$, from Table 15.4 $\psi_{ed,N} = 1.0$. The load is concentric, so $\psi_{ec,N} = 1.0$. The attachment has cast-in headed anchors resulting in a value of $k_c = 24$.

The projected area of a single anchor is

$$A_{Nco} = 9h_{ef}^2 = 9 \times 4^2 = 144 \text{ in.}^2$$

The projected area of the anchor group is

$$A_{Nc} = (3h_{ef} + s_1) \times (3h_{ef} + 2s_2) = (3 \times 4 + 5) \times (3 \times 4 + 2 \times 4.5) = 357 \text{ in.}^2$$

From Eq. 15.2, the tensile concrete breakout strength of a single anchor is

$$N_b = k_c \lambda_a \sqrt{f'_c}\, h_{ef}^{1.5} = 24 \times 1.0\sqrt{5000} \times 4^{1.5} = 13\ \text{kip}$$

Lastly, from Eq. 15.7

$$
\begin{aligned}
N_{cbg} &= \frac{A_{Nc}}{A_{Nco}} \psi_{ec,N}\psi_{ed,N}\psi_{c,N}\psi_{cp,N} N_b \\[4pt]
&= \frac{357}{144}1.0 \times 1.0 \times 1.4 \times 1.0 \times 13.58 \\[4pt]
&= 47.2\ \text{kips}
\end{aligned}
$$

Example 15.5

Problem: Determine the required length of six $\tfrac{1}{2}$ in. diameter headed studs shown in Fig. 15.8 for a factored tensile load of 35 kips and the conditions given in Example 15.3 are present.

Solution: From Table 15.3, the ϕ-factor for a cast-in headed ductile anchor is 0.75. Thus, the required nominal strength of the anchor group is $N_u/\phi = 35/0.75 = 46.7$ kips. Using the information from Example 15.3, $A_{Nco} = 9h_{ef}^2$, and after combining terms, $A_{Nc} = 9h_{ef}^2 + (6\,s_1 + 3\,s_2)h_{ef} + 2s_1s_2$. The tensile strength of a single anchor is $N_b = k_c\lambda_a\sqrt{f'_c}h_{ef}^{1.5} = 24 \times 1.0\sqrt{5000}\times h_{ef}^{1.5} = 1967\,h_{ef}^{1.5}$. Combining the terms and inserting the values for $s_1 = 5$ in. and $s_2 = 4.5$ in, gives

$$
\begin{aligned}
46,700 = \Big[\big(9h_{ef}^2 &+ (6 \times 5 + 3 \times 4.5)h_{ef} + 2 \times 5 \times 4.5\big)/9h_{ef}^2\Big] \times 1.0 \times 1.0 \\
&\times 1.0 \times 1.0 \times 1967 h_{ef}^{1.5}
\end{aligned}
$$

Using an equation solver or by trial, h_{ef} must be at least 5.85 in. Use $h_{ef} = 6$ in. because headed studs come in $\tfrac{1}{2}$ in. and 1 in. increments. An Excel or Mathcad calculation sheet with h_{ef} as a variable provides an effective trial solution tool.

Example 15.6

Problem: Determine the shear breakout strength of the anchor group shown in Fig. 15.8. The load is concentrically applied, the attachment has six $\tfrac{3}{4}$ in. diameter headed bolts cast in 5000 psi normalweight concrete that analysis indicates is uncracked during service load, and $c_{a1} = 8$ in.

Solution: From Fig. 15.8, $s_1 = 5.0$ in. and $s_2 = 4.5$ in. For normalweight concrete, λ and $\lambda_a = 1.0$. From Table 15.6 for uncracked concrete and cast-in anchors, $\psi_{c,V} = 1.2$, and because the value of c_{a2} is greater than c_{a1} by inspection, from Table 15.5 $\psi_{ed,V} = 1.0$. The load is concentric, so $\psi_{ec,V} = 1.0$, and the concrete thickness is greater than $1.5\,c_{a1}$ so $\psi_{h,V} = 1.0$.

Option 1: Assume that one-half of the load is carried by the front row of anchors. The projected area of a single anchor is

$$A_{Vco} = 4.5c_{a1}^{2} = 4.5 \times 8^{2} = 288 \quad \text{in.}^{2}$$

The projected area of the anchor group is

$$A_{Nc} = 1.5c_{a1}(3c_{a1} + 2\,s_{2}) = 1.5 \times 8 \times (3 \times 8 + 2 \times 4.5) = 408 \ \text{in.}^{2}$$

For a single anchor, $d_{a} = 0.75$ in. and $l_{e} = h_{ef}$. Then from Eq. 15.4, the shear strength of a single anchor is

$$V_{b} = \left(7\left(\frac{\ell_{e}}{d_{a}}\right)^{0.2}\sqrt{d_{a}}\right)\lambda_{a}\sqrt{f'_{c}}(c_{a1})^{1.5} = \left(7\left(\frac{4}{0.75}\right)^{0.2}\sqrt{0.75}\right)1.0\sqrt{5000}(8)^{1.5}$$

$$= 13{,}570 \ \text{lb}$$

The anchor group strength is

$$\frac{V_{cbg}}{2} = \frac{A_{Vc}}{A_{Vco}}\Psi_{ec,V}\Psi_{ed,V}\Psi_{c,V}\Psi_{h,V}V_{b}$$

$$= \frac{408}{288}1.0 \times 1.0 \times 1.2 \times 1.0 \times 13.57$$

$$= 23.1 \ \text{kips}$$

Thus, for both anchors:

$$V_{cbg} = 46.1 \ \text{kips}$$

Option 2: The strength of the anchor must also be checked for the case in which the back row of anchors carries the entire load. For this condition, the edge distance c_{a1} increases to $8 + 5$ in. $= 13$ in. With this change,

$$A_{Vco} = 4.5c_{a1}^{2} = 9 \times 13^{2} = 760.5 \ \text{in.}^{2}$$

$$A_{Nc} = 1.5c_{a1}(3c_{a1} + 2\,s_{2}) = 1.5 \times 13 \times (3 \times 13 + 2 \times 4.5) = 955.5 \ \text{in.}^{2}$$

$$V_{b} = \left(7\left(\frac{\ell_{e}}{d_{a}}\right)^{0.2}\sqrt{d_{a}}\right)\lambda_{a}\sqrt{f'_{c}}(c_{a1})^{1.5} = \left(7\left(\frac{4}{0.75}\right)^{0.2}\sqrt{0.75}\right)1.0\sqrt{5000}(13)^{1.5}$$

$$= 28 \ \text{kip}$$

$$V_{cbg} = \frac{A_{Vc}}{A_{Vco}}\Psi_{ec,V}\Psi_{ed,V}\Psi_{c,V}\Psi_{h,V}V_{b}$$

$$= 760.51.0 \times 1.0 \times 1.2 \times 1.0 \times 28.08$$

$$= 42.3 \ \text{kips}$$

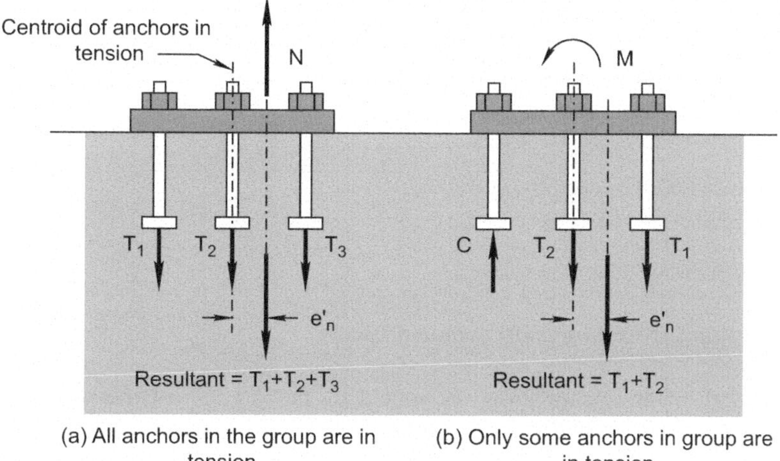

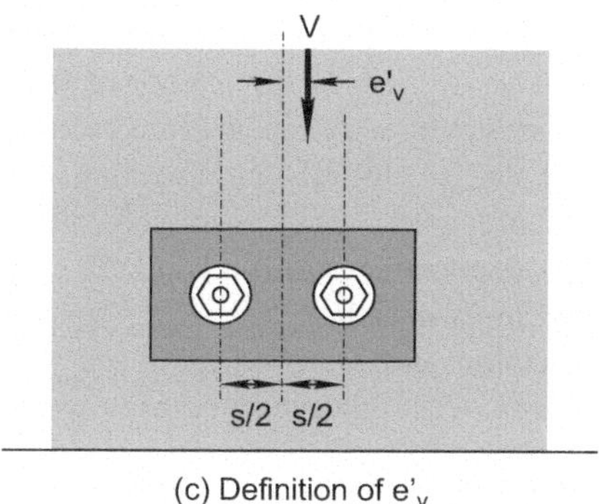

(c) Definition of e'_v

Fig. 15.11 Description of loading eccentricity (a) e'_N when all anchors are in tension, (b) e'_V for when one or more anchors are in compression, and (c) e'_V for shear loadings

In this example, the second condition limits the nominal strength of the anchor group to $V_{obg} = 42.3$ kips. The anchor group must be detailed with a No. 4 (No 13) bar or larger between the edge and the anchor to be consistent with the assumed value of $\psi_{c,v}$.

Comment: The solution to Example 15.6 satisfies the ACI Building Code requirements. At the same time, holes in the plate are designed to fit over the cast-in bolts. Tolerances are specified on the hole size and, consequently, some designers recommend only using the bolts closest to the edge to establish the anchor group strength under the assumption that the hole tolerance may not allow the back holes to completely engage the bolts. Under this restriction, the strength of the attachment in Example 15.5 would be 23.1 kips or about half of the strength calculated in the example.

Table 15.8 Breakout modification factors for eccentricity

Condition	Tension		Shear
	$\psi_{ec,N}$	$\psi_{ec,Na}$[a]	$\psi_{ec,V}$
Anchor group loaded eccentrically in tension	$\dfrac{1}{\left(1+\frac{2e'_N}{3h_{ef}}\right)} \leq 1.0$	$\dfrac{1}{\left(1+\frac{e'_N}{e_{Na}}\right)} \leq 1.0$	
Anchor group loaded eccentrically in shear			$\dfrac{1}{\left(1+\frac{2e'_V}{3h_{ef}}\right)} \leq 1.0$

[a]For adhesive anchors

Modification for Eccentrically Applied Loads

The strength of an anchor group is limited by the strength of the most severely loaded anchor. Eccentrically loaded attachments result in loads being redistributed such that some anchors are more severely loaded, Fig. 15.11. Anchors may remain in tension or be placed in compression depending on the eccentricity of the load or the magnitude of the moment.

The anchor to the right in Fig. 15.11 carries a higher load than the remaining anchors and fails first. The strength of the anchor or the anchor group in concrete breakout in tension or in shear is modified to correct for this load redistribution. Table 15.8 summarizes the eccentricity modification factors. When adjusting for eccentricity, the values of e'_N and e'_V, respectively, are calculated only for those anchors in tension or those anchors loaded toward the edge in shear. In cases where the eccentricity occurs in two orthogonal directions, the modification factor is calculated for each direction and the product of the factors from Table 15.8 are used in Eqs. 15.7 and 15.9.

15.4.3   *Pullout Strength of Anchors*

The pullout strength of an anchor in tension occurs when the head of the anchor pulls or slips through the concrete creating a cylindrical failure, Fig. 15.6e. For cast-in, post-installed, and post-installed undercut anchors, the nominal pullout strength in tension N_{pn} is

$$N_{pn} = \psi_{c,p} N_p \tag{15.10}$$

where N_p is the strength of an individual anchor given in Eq. 15.11 or Eq. 15.12 (Kuhn and Shaikh 1996; Primavera et al. 1997) and $\psi_{c,p} = 1.0$. The individual anchor strength is based on the net area of the anchor head directly bearing in the concrete A_{brg} or on the geometry of the anchor. A_{brg} is typically calculated as the area of the head of the bolt or stud less the area of the shaft. The anchor strength for a single headed stud or bolt adjusted to the 5% fractile is

$$N_p = 8A_{brg}f'_c \tag{15.11}$$

and for a hooked bolt is

$$N_p = 0.9f'_c e_h d_a \tag{15.12}$$

where e_h is the distance from the inner surface of the shaft of a J- or L-bolt to the outer tip of the J- or L-bolt and d_a is the diameter of the hooked bolt. The ACI Building Code requires that e_h, illustrated in Fig. 15.5a, meet the requirement $3d_a \le e_h < 4.5d_a$. The pullout strength of a post-installed anchor is supplied by the manufacturer based on an evaluation performed in accordance with ACI 355.2 and reduced to the 5% fractile.

The pullout strength is further modified depending whether the anchor is in cracked or uncracked concrete. For anchors located in areas where the analysis indicates the concrete is cracked, $\psi_{c,p} = 1.0$, and where the analysis indicates the concrete is uncracked, such as prestressed and precast concrete, $\psi_{c,p} = 1.4$.

Example 15.7
Problem: Calculate the pull-out strength of the anchor in Example 15.2

Solution: The diameter of the head on the headed stud is 1 in. giving a bearing area of

$$A_{brg} = \pi\left(1.0^2 - 0.5^2\right)/4 = 0.589 \text{ in.}^2$$

$\psi_{cp} = 1.0$ for cracked concrete, so the strength is

$$N_{pn} = \psi_{c,p}n\, N_p = \psi_{c,p}n8A_{brg}f'_c = 1 \times 6 \times 8 \times 0.589 \times 5000 = 141.0 \text{ kips}$$

Summary: the tensile strength of the anchor is the least of the steel strength, breakout strength, and the pullout strength from Examples 15.2, 15.4, and 15.7. The nominal breakout strength of 47.2 kips from Example 15.4 is the limiting condition for this attachment.

15.4.4   Side-Face Blowout

The anchors with deeper embedment but thinner side cover may fail by concrete spalling on the side face around the embedded head while no major breakout appearing at the top concrete surface, as illustrated in Fig. 15.6g. For a single headed anchor with deep embedment close to an edge ($h_{ef} > 2.5\, c_{a1}$), the nominal side-face blowout strength N_{sb} is given in Section 17.4.4 ACI 318-14 as

$$N_{sb} = 160\, c_{a1} \sqrt{A_{brg}} \lambda_a \sqrt{f'_c} \tag{15.13}$$

If c_{a2} for the single headed anchor is less than $3c_{a1}$, the value of N_{sb} is modified by the factor $(1 + c_{a2}/c_{a1})/4$ where the ratio c_{a2}/c_{a1} must be greater than 1 and is limited by the ACI Building Code to be less than or equal to 3. For headed anchor groups with deep embedment close to an edge ($h_{ef} > 2.5c_{a1}$) and with anchor spacing less than $6c_{a1}$, the nominal strength of those anchors susceptible to a side-face blowout failure N_{sbg} is

$$N_{sbg} = \left(1 + \frac{s}{6c_{a1}}\right) N_{sb} \tag{15.14}$$

where s is the distance between the anchors nearest the edge and N_{sb} is given in Eq. 15.13 without modification for a perpendicular edge distance.

Example 15.8
Problem: Calculate the side blowout strength of the anchor in Example 15.4.

Solution: The modification factor for the bearing group is

$$\left(1 + \frac{s_1}{6c_{a1}}\right) = \left(1 + \frac{4.5}{6 \times 8}\right) = 1.09$$

And, using the bearing area from Example 15.6, the value for N_{sb} is

$$N_{sb} = 160 c_{a1} \sqrt{A_{brg}} \lambda_a \sqrt{f'_c} = 160 \times 8 \sqrt{0.589} \times 1.0 \times \sqrt{5000} = 69.5 \;\; \text{kips}$$

Comment: The side blowout strength of a single anchor exceeds the steel tensile strength of a single anchor calculated in Example 15.2, so side blowout is not a limiting condition for this anchor group. The spacing modification factor is greater than 1.0, so no further calculations are necessary.

15.4.5   Pryout of Anchors

Pryout is a phenomenon that occurs with short anchors for attachments loaded in shear (ACI 355.2-07 2007, Anderson and Meinheit 2005). As an anchor group moves laterally, the individual anchors can fail due to a shear failure of the steel, concrete breakout, or, due to the rotation of short anchors prying out of the concrete. The nominal pryout strength for a single anchor V_{cp} is given in Section 17.5.3 of ACI 318-14 as

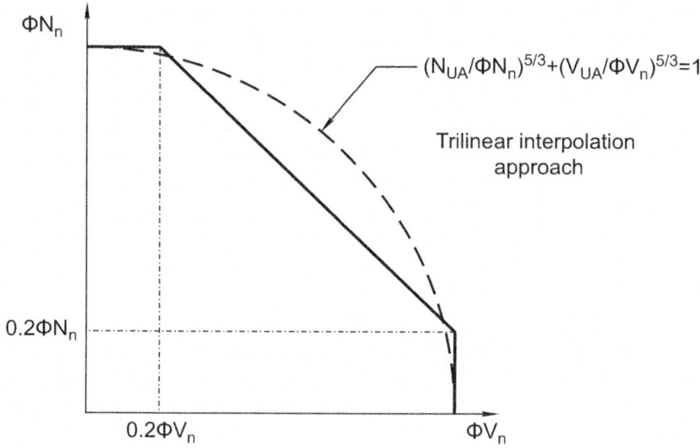

Fig. 15.12 Shear and tensile load interaction

$$V_{cp} = k_{cp}N_{cp} \tag{15.15}$$

The pryout strength for a group of anchors V_{cpg} is

$$V_{cpg} = k_{cp}N_{cpg} \tag{15.16}$$

For cast-in, expansion, and undercut anchors, N_{cp} and N_{cpg} may be taken as N_{cb} or N_{cbg} from Eqs. 15.6 and 15.7, respectively. For both single anchors and anchor groups, $k_{cp} = 1.0$ for h_{ef} less than 2.5 in. and $k_{cp} = 2.0$ for h_{ef} greater than or equal to 2.5 in.

Example 15.9
Calculate the pryout strength of the anchor group in Example 15.5.

Solution: h_{ef} is greater than 2.5 in., so $k_{cp} = 2.0$. The results from Example 15.3 provide an anchor group strength of $N_{cbg} = 33.7$ kips. The pryout strength from Eq. 15.15 is then

$$V_{cpg} = k_{cp}N_{cbg} = 2 \times 33.7 = 67.4 \, \text{kips}$$

Summary: Comparing Examples 15.3, 15.6, and 15.9, the attachment shear strength is limited by the shear breakout strength, $V_{cbg} = 42.3$ kips, from Example 15.6 or 23.1 kips if a plate with oversized holes is used.

Fig. 15.13 Steel attachment
for Example 15.10

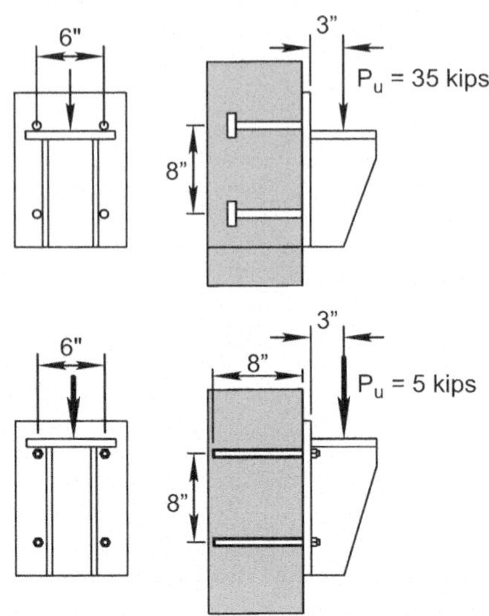

15.4.6 Combined Shear and Normal Force

Attachments such as the corbels shown in Fig. 15.1 generate both tensile and shear forces in the anchors. Experimental studies indicate that the interaction can be represented using a curvilinear relationship, such as shown in Fig. 15.12 (Lotze et al. 2001). Section 17.6 of ACI 318-14 simplifies this curvilinear relationship by using a trilinear approximation of the interaction behavior. The trilinear relationship allows the full tensile strength of the anchor to be used if $V_{ua}/\phi V_n$ is less than 0.2 and the full shear strength to be used if $N_{ua}/\phi N_n$ is less than 0.2. Between these two limits the load combination must satisfy Eq. 15.17.

$$\frac{N_{ua}}{N_u} + \frac{V_{ua}}{V_u} \le 1.2 \tag{15.17}$$

Example 15.10
Problem: Design the anchor group for the attachment shown in Fig. 15.13 using 5/8 in. diameter headed studs. The normalweight concrete has a compressive strength of 3500 psi and analysis indicates concrete will remain uncracked during the service live. The corbel carries a factored load of 35 kips. The attachment is fabricated from 3/8 in. thick plate, is located at least $10h_{ef}$ from any edge, and no supplementary reinforcement is present.

Solution: The load is resisted by four studs in shear and the tension generated by the load eccentricity is resisted by the top two studs. With no supplementary reinforcement, $\phi = 0.70$ from Table 15.3. With no edge distance issues, eccentricity, or pullout restrictions, $\psi_{ed,N}$ and $\psi_{cp,N}$ equal 1.0. The concrete is uncracked, so $\psi_{c,N}$ is 1.25 from Table 15.6.

The shear strength is controlled by the steel strength of the anchors, so the design shear strength for four anchors is

$$\phi V_n = \phi A_{se,V} \, \phi_{uta} = 0.70 \times 4 \times 0.31 \times 65 = 56.4 \text{ kips}$$

which is greater than the 35-kip applied load. Since c_{a1} is at least 10 in., pryout need not be checked.

The tensile load on the top two anchors is assumed to be resolved from a couple between the studs, which is conservative in this instance. For a stud spacing of $s = 8$ in.,

$$N_u = P_u \times \text{eccentricity}/s = 35 \times 3/8 = 13.1 \text{ kips}$$

for which, $N_N = N_u/\phi = 13.1/0.70 = 18.75$ kips.

Check if the tensile strength is controlled by tensile concrete breakout of the two top anchors. For a spacing between the studs of $s = 6$ in.,

$$A_{Nco} = 9h_{ef}^2$$

and

$$A_{Nc} = (3h_{ef} + s)\, 3h_{ef} = 9h_{ef}^2 + 18h_{ef}$$

Combining Eq. 15.2 and Eq. 15.6 gives

$$N_{cbg} = \frac{A_{Nc}}{A_{Nco}} \psi_{ed,N} \psi_{c,N} \psi_{cp,N} k_{ca} \sqrt{f'_c} h_{ef}^{1.5}$$

$$18.75 = \frac{9h_{ef}^2 + 18h_{ef}^2}{9h_{ef}^2} \, 1.0 \times 1.25 \times 1.0 \times 24 \times 1.0\sqrt{3500} \times h_{ef}^{1.5}$$

Solving by trial for the required embedment depth gives $h_{ef} = 4$ in. and $N_{cbg} = N_N = 21.3$ kips. Because both shear and tensile forces are present, the anchors must be checked for combined effects. From Eq. 15.17,

$$\frac{N_u}{N_N} + \frac{V_u}{V_N} = \frac{13.1}{0.70 \times 21.3} + \frac{35}{0.7 \times 56.4} = 1.77$$

This exceeds the code requirement of 1.2, so the embedment length must be increased. Using a 9.2 in. embedment length and the plate thickness of 5/8 in. gives $N_N = 60.3$ kips and a combined ratio from Eq. 15.17 of 1.197, less than the maximum value of 1.2. A 5/8 in. diameter headed stud has a head diameter of

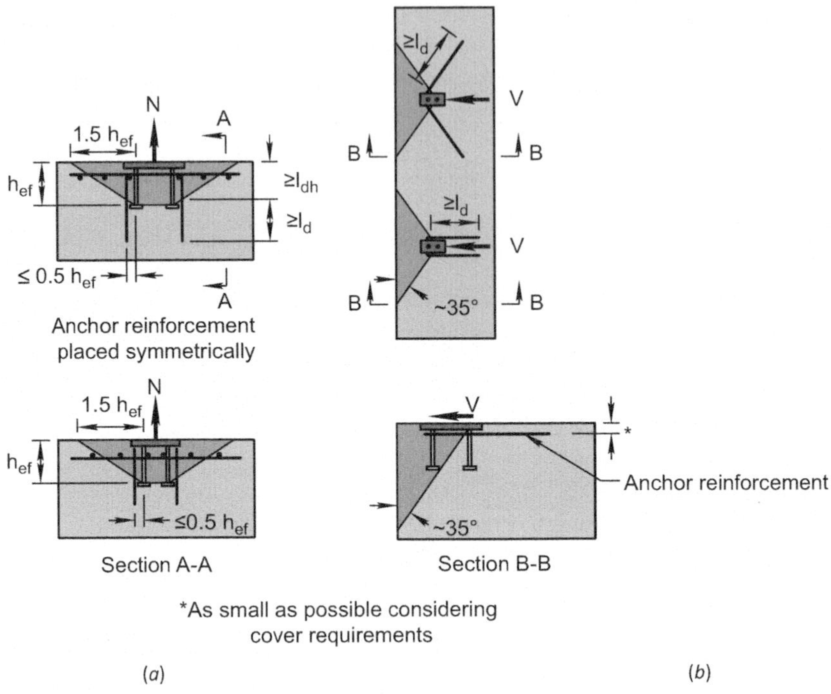

Fig. 15.14 Anchor reinforcement for (**a**) tension and (**b**) shear

1–1/4 in. A check of pullout strength can be made using Eqs. 15.10 and 15.11. $\psi_{c,p} = 1.4$ for uncracked concrete.

$$A_{brg} = \tfrac{\pi}{4}\left(1.25^2 - 0.5.25^2\right) = 0.92 \;\; \text{in.}^2 \text{ and for } n = 2 \text{ studs using Eq. 15.10,}$$

$$N_p = n\, 8A_{brg} = 2 \times 8 \times 0.92 \times 3500 = 51.5 \text{ kips}$$

Using the result in Eq. 15.10,

$$N_{pn} = \psi_{c,p}N_p = 1.4 \times 51.5 = 72.2 \text{ kips}$$

The check of pryout indicates it is not a controlling condition. A 5/8 in. diameter stud 9-½ in. long is selected. Increasing the embedment length beyond 9-½ in. could increase the anchor group strength up to the limitation of the pullout strength but the 9-½ in. depth is sufficient.

Comment: For this solution, the tensile force on the top anchors was calculated based on the distance between the top and bottom anchors. An equally valid approach would be to assume a compression centroid below the lower anchors. In either case, the anchor group strength would be established by the most highly loaded anchor.

15.4.7   Anchor Reinforcement

The ACI Building Code identifies two types of reinforcement for use with anchors. *Supplementary reinforcement* assists in controlling crack width and preventing spalling. *Anchor reinforcement* transfers the total factored load to the supporting structure. To be effective, the anchor reinforcement must be aligned with the direction of the applied load and be developed in both the concrete breakout zone associated with the anchor and in the underlying concrete, Fig. 15.14. The detailing shown in Fig. 15.14 requires that the anchor reinforcement be placed close to the surface for shear loads, commensurate with the cover requirements. Hairpins bars are often used for this purpose. Attachments in plastic hinge regions, or areas where analysis indicates substantial cracking may be present, must be detailed to include anchor reinforcement.

If anchor reinforcement is used, Section 17.2.3.4 of ACI 318-14 does not require calculation of the concrete breakout strength. In many instances, however, the addition of anchorage reinforcement is not practical. If in Fig. 15.14 for example, the anchor group is in a thin slab loaded normal to the surface, sufficient depth below the breakout prism would not be available for the development of the anchor reinforcement. A strength reduction factor of $\phi = 0.75$ is used when determining the area of the anchor reinforcement.

15.4.8   Adhesive Anchors

Following the collapse of the ceiling panels in the "Big Dig" in Boston, the National Transportation Safety Board requested ACI develop criteria for adhesive anchors (Hansen 2009). Adhesive anchors are sensitive to several factors, including installation temperature, moisture, and sustained loading. To provide uniformity of installation and use, anchor systems must be qualified in accordance with procedures described in ACI 355.4 *Acceptance Criteria for Qualification of Post-Installed Adhesive Anchors in Concrete* (ACI 355.4 2010) and must be installed by qualified technicians. The ACI Building Code requires anchors installed horizontally or in an upward sloping orientation to be subject to continuous inspection during construction.

Adhesive anchors have failure modes similar to cast-in or post-installed anchors in addition to the possibility of an adhesive bond failure (Cook et al. 1998; Eligehausen et al. 2006). Bond failures arise when the adhesive undergoes a shear failure between the cured adhesive and the concrete. This results in a bond pullout failure accompanied by a concrete breakout closer to the surface, Fig. 15.6*f*. For the performance of adhesive anchors to correlate with the qualification tests of ACI 355.4, Section 17.4.5 of ACI 318-14 places restrictions on the installation of adhesive anchors. These limitations include:

Table 15.9 Minimum characteristic bond stress

Installation and service environment [a, b]	Moisture content of concrete at time of installation	Peak in-service temperature of concrete, °F	τ_{cr}, psi	τ_{uncr}, psi
Outdoor	Dry to fully saturated	175	200	650
Indoor	Dry	110	300	1000

[a]Where anchor design includes sustained loads, the values of τ_{cr} and τ_{uncr} should be multiplied by 0.4

[b]Where the anchor design includes earthquake loads for structures assigned to Seismic Design Categories C, D, E, or F, the value of τ_{cr} should be multiplied by 0.8 and the value of τ_{uncr} should be multiplied by 0.4

The minimum concrete age is 21 days. This provision allows moisture in the concrete to be used in the hydraulic cement reaction and not be available to disrupt the adhesive cure.

A concrete strength that is equal or greater than 2500 psi. Qualification tests are conducted using concrete with compressive strengths of at least 2500 psi and data on lower strength is very limited.

Rotary impact or rock drills are used to drill the holes for adhesive anchors. These tools create a rough irregular surface to improve mechanical interlock between the adhesive and the concrete. Holes made with coring bits are smoother and have less interlock.

Installation temperature is at least 50°F. This minimum temperature is for the adhesive to cure properly.

Adhesive anchors require bond to prevent pullout and to mobilize a concrete breakout failure mode. The *characteristic bond stresses* in uncracked concrete τ_{uncr} and in cracked concrete τ_{cr} are provided by the manufacturer based on the 5% fractile results derived from the tests specified in ACI 355.4. Table 15.9 provides conservative values for the characteristic bond stresses given in the ACI Building Code. The characteristic bond stress is multiplied by 0.40 if the anchor is subjected to sustained load. Use of higher characteristic bond values in design requires that appropriate installation instructions be included in the construction documents and may require special inspection to ensure proper behavior. The characteristic bond values in Table 15.9 should be selected to be compatible with both the installation and service environments. For example, if adhesive anchors are installed before a building is enclosed, as shown in Fig. 1.15, the environment would be "outdoor."

Basic Bond Strength

The bond stress is not uniform over the embedded length and consequently, the projected area of concrete breakout strength is limited to a width of $2c_{Na}$, where c_{Na} is defined by Eq. 15.22. The formulation of the bond strength of adhesive anchors is a function of both c_{Na} and h_{ef}. The basic bond strength of an adhesive anchor is

$$N_{ba} = \lambda_a \tau_{cr} \pi d_a h_{ef} \tag{15.18}$$

where τ_{cr} is the characteristic bond stress for cracked concrete, d_a is the anchor diameter, and h_{ef} is the effective embedment depth. For the calculation of adhesive anchor bond strength, the value of λ_a is 0.6λ. The tensile breakout strength of a single adhesive anchor is

$$N_a = \frac{A_{Na}}{A_{Nao}} \Psi_{ed,Na} \Psi_{cp,Na} N_{ba} \tag{15.19}$$

and for an anchor group in tension

$$N_{ag} = \frac{A_{Na}}{A_{Nao}} \Psi_{ec,Na} \Psi_{ed,Na} \Psi_{cp,Na} N_{ba} \tag{15.20}$$

Modification factors $\Psi_{ec,Na}$ and $\Psi_{ed,Na}$, and are given in Table 15.6 and Table 15.8. The modification factor $\Psi_{cp,Na}$ equals 1.0 for $c_{a,min} \geq c_{ac}$ and equals $\frac{c_{a,min}}{c_{ac}}$ for $c_{a,min} \leq c_{ac}$. Critical edge distances are given in Table 15.7. The projected area of a single adhesive anchor A_{Nao} is

$$A_{Nao} = (2c_{na})^2 \tag{15.21}$$

where

$$c_{Na} = 10d_a \sqrt{\frac{\tau_{uncr}}{1100}} \tag{15.22}$$

and the constant 1100 carries the units of lb/in.^{2}

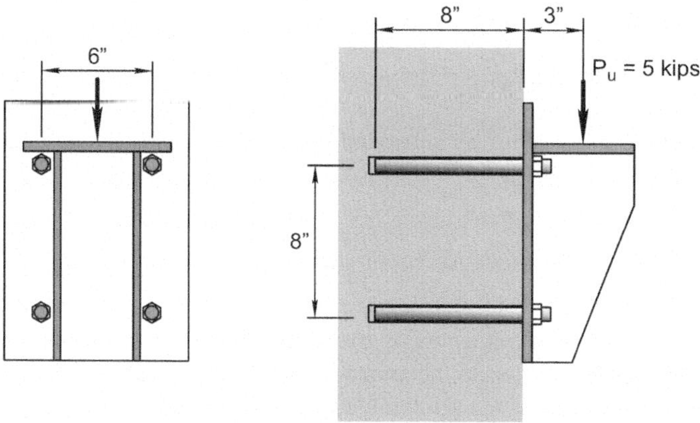

Fig. 15.15   Adhesive anchor attachment for Example 15.11

Example 15.11

Problem: Determine if the anchor group for the attachment shown in Fig. 15.15 is adequate to carry a 5-kip factored sustained load using the characteristic bond stresses from Table 15.9. The A-36 steel anchors have ten threads per inch, are ¾ in. diameter, and embedded 8 in. into an exterior concrete wall with $f'_c = 4000$ psi. The anchor group is well away from any edges, supplementary reinforcement is present, and the anchors are considered Category 2—medium sensitivity and reliability.

Solution: From Table 15.3, the strength reduction factor for a Category 2 anchor with supplementary reinforcement is 0.65. The ψ values are all 1.0, as there is no eccentricity or edge distance constraints. The shear on the anchor group is equal to the applied load and is 5 kips. To calculate the design shear strength,

$$f_{uta} = 65 \text{ ksi}$$

The net area, $A_{se,V}$,

$$A_{se,V} = \frac{\pi}{4}\left(d_a - \frac{0.9743}{10\dfrac{\text{threads}}{\text{in.}}} \right) = 0.334 \text{ in.}^2, \text{ then}$$

$$\phi V_{sa} = \phi \times n \times A_{se,V} \times \phi_{uta} = 0.60 \times 4 \times 0.334 \times 65 = 52.2 \text{ kips}$$

The tensile breakout strength of the anchor is as follows, noting that the ACI Building Code requires that τ_{uncr} to be multiplied by 0.4 for the sustained loads:

$$c_{Na} = 10 d_a \sqrt{\frac{0.4\tau_{uncr}}{1100}} = 10 \times 0.75 \sqrt{\frac{0.4 \times 650}{1100}} = 3.65 \text{ in.}$$

The basic projected area is $A_{Nao} = (2\, c_{Na})^2 = (2 \times 3.65)^2 = 53.2$ in.^{2}, and the projected area of the two top anchors is $A_{Na} = 2\, c_{Na}(2\, c_{Na} + s) = 2 \times 3.65 \times (2 \times 3.65 + 6) = 96.9$ in.^{2}

The basic bond strength of one adhesive anchor with a factor of 0.4 applied to τ_{cr} for sustained loads is

$$N_{ba} = 0.4\tau_{uncr}\pi d_a h_{ef} = 0.4 \times 650 \times \pi \times 0.75 \times 8 = 1.51 \text{ kips}$$

Again, using the 0.4 factor for sustained loads. The two top anchors then provide a strength of

$$N_a = \frac{A_{Na}}{A_{Nao}}\psi_{ec,Na}\psi_{ed,Na}\psi_{pc,Na}N_{ab} = \frac{96.9}{53.2} \times 1 \times 1 \times 1 \times 1.51 = 2.75 \text{ kips}$$

The design load on the anchors is the applied load P_u times the eccentricity divided by the distance between the anchors, and thus, $N_u = 5 \times 3/8 = 1.875$ kips, which is more than 20% of the nominal strength, so the combined loading must be checked, giving

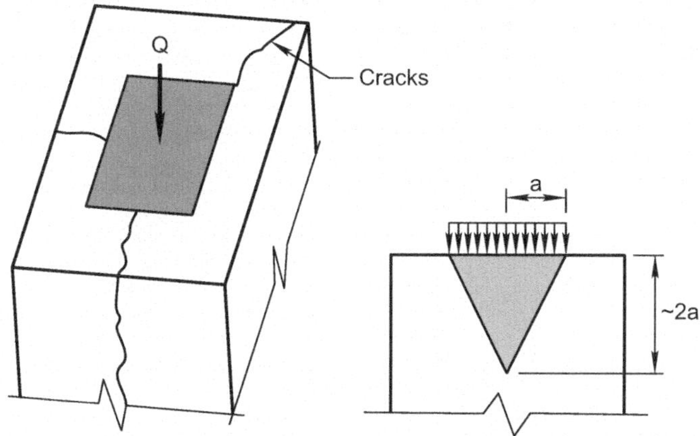

Fig. 15.16 Experimental cube and internal wedge

$$\frac{N_u}{N_a} + \frac{V_u}{V_{sa}} = \frac{1.875}{0.65 \times 2.75} + \frac{5}{0.65 \times 52.2} = 1.30 < 1.2$$

The interaction does not meet the code requirements and thus the attachment cannot carry the applied load. Increasing the anchor embedment to 9 in. results in an interaction value of 1.17, making the insert acceptable.

Comment: Comparing the anchor group in Example 15.9, the strength of the attachment with adhesive anchors is approximately 1/7th the strength of the attachment with cast-in anchors. Two conclusions can be drawn from this comparison. First, cast-in anchors are structurally more efficient than adhesive anchors when using the minimum characteristic bond stress. Second, the characteristic bond stress values in the ACI Building Code are conservative. Characteristic bond stresses for commercial adhesive anchors derived through the ACI 355.4 qualification process may provide bond stresses up to ten times the Code values.

15.5 Small Concentrated Bearing Loads

Stresses under post-tensioning anchors and bearing plates supporting concentrated loads can exceed the compression strength of the concrete. For example, an 8-strand anchor that is 5-1/4 in. in diameter and having a 2-1/2 in. diameter duct has a net area of 16.74 sq. in. The tendon force at transfer is 230 kips resulting in a compressive stress of 13.7 ksi under the anchor. Confinement of the concrete is required to support such local stresses. The following is an approximate method for checking confinement in local zones of anchor or under concentrated loads.

Hawkins (1968) published a series of reports on bearing stress tests on concrete cubes. These tests accurately predicted the failure modes and loads in unreinforced

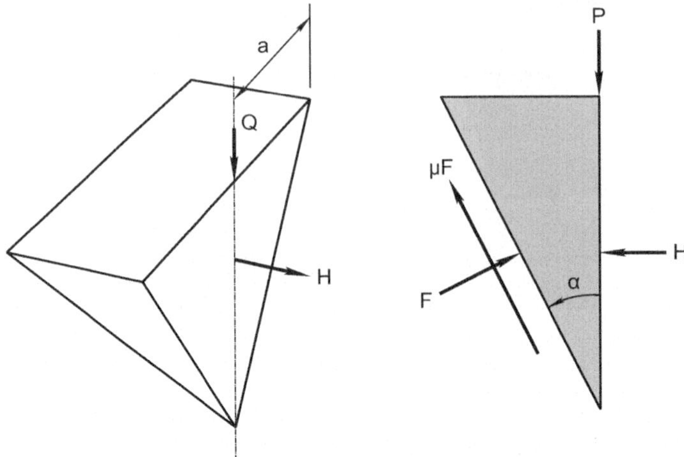

Fig. 15.17 Free body diagram of wedge

concrete cubes Fig. 15.16. Based on experimental observation, the depth of the wedge is approximately twice the half side of a square plate or the radius of a circular plate. Hawkins used a "wedge" theory that indicated that the wedge formed under the bearing plate split the concrete. Using the tensile strength of the concrete, Hawkins correlated the splitting force generated by the wedge to the failure of the cube. In his development, Hawkins assumed the load would equal the compression strength of the concrete plus the resistance of the wedge. The following solution assumes that only the splitting action of the internal wedge is active, thus resulting in a conservative solution.

Assume the wedge moves as a rigid body and that lateral confinement prevents the concrete within the wedge from crushing. That is, the following approach is independent of the strength of the concrete. The load that can be supported is a function of the horizontal force that is generated by splitting the concrete and the friction resisting the wedge from moving downward, Fig. 15.17. The frictional force is equal to the normal force restraining the wedge and the coefficient of friction along the interface. Both the horizontal force and the frictional force are a function of the normal force on the wedge surface.

Equilibrium on the wedge provides the following two relationships

$$H = F \cos \alpha - \mu F \sin \alpha$$

and

$$P = 4(F \sin \alpha + \mu F \cos \alpha)$$

where the 4 results because there is resistance on all sides of the wedge in Fig. 15.17 in addition to the face shown. Recasting the two equations as a function of the surface force F and solving for the required horizontal restraining force gives

$$H = \frac{P}{4} \frac{\cos\alpha - \mu \sin\alpha}{\sin\alpha + \mu \cos\alpha}$$

Using the results of Hawkins' experiments that the depth of the wedge is approximately twice the half width of the prism, $\cos(\alpha)$ is approximately 0.90 and $\sin(\alpha)$ is approximately 0.45. If the frictional resistance is reduced to zero, the horizontal force becomes half the vertical load.

From consideration of shear friction, the coefficient of friction would vary between 1.0 for a cold joint and 1.4 for monolithic concrete. The wedge is only active if the concrete cracks; therefore, a coefficient of friction of 1.0 is selected for comparison. This assumption results in

$$H = \frac{P}{4}\frac{\cos\alpha - \mu \sin\alpha}{\left(\cos\alpha + \mu \sin\alpha\right)} = \frac{P}{4}\frac{0.90 - 1.0 \cdot 0.45}{0.90 + 1.0 \cdot 0.45} = 0.083P$$

Thus, a restraining force between ½ and 1/8th of the applied load is required to resist the concentrated load. Hawkins original work used the tensile strength of the concrete as a resistance. Instead of relying on the concrete tensile strength to restrain the wedge, reinforcement stressed at approximately 2/3 f_y provides the horizontal force H. The area of reinforcement needed to confine the concentrated load is then

$$A_s = \frac{H}{2/3 f_y}$$

While this approach is approximate, it provides insight to the behavior of concentrated loads on a structure and behavior in the local zone of prestressing anchors.

Example 15.12

Consider a VSL monostrand anchor 4.13 in. by 2.95 in. in area with a 2.56 in. diameter trumpet beneath the anchor. The anchor is applied to the edge of a 6-in. thick slab and carries 31 kips at the time of stressing. The concrete strength is 3000 psi at the time of transfer.

Solution: The bearing area is the plate less the trumpet or $4.13 \cdot 2.95 - 2.56^2 \pi / 4 = 7.05$ in.^{2} The stress under the anchor is $31,000/(4.13 \cdot 2.95 - \pi 2.56^2 / 4) = 4400$ psi, which is greater than the strength of the concrete.

The splitting force within the slab is

$$H = 0.083P = 0.083 \cdot 31 = 2.58 \text{ kips}$$

and the required area of reinforcement using grade 60 reinforcement is

$$A_s = \frac{H}{2/3 f_y} = \frac{2.58}{2/3 \, 60} = 0.06 \ \text{in.}^2$$

If friction is ignored, the horizontal force is 15.5 kips and the required area of steel is 0.38 sq. in.

Comment: This finding supports the ACI Building Code requirement to detail 2 No. 4 bars ($A_s = 0.40$ in.^{2}) behind each anchor to control local cracking and suggests why cracking is relatively rare.

Example 15.13
The 8-strand anchor mentioned above with 230 kips at transfer has a diameter of the anchor of 5.25 in.

Solution: The form of the solution for a cone is slightly more complex than a rectangular prism. For simplicity, the rectangular prism is used and the horizontal force is

$$H = 0.083P = 0.083 \cdot 210 = 17.5 \ \text{kips}$$

The required area of reinforcement using Grade 60 reinforcement is

$$A_s = \frac{H}{2/3 f_y} = \frac{17.5}{2/3 \, 60} = 0.44 \ \text{in.}^2$$

This particular anchor uses a 3/8 in. diameter wire spiral with a pitch of 1 in. on center. The number of wraps of the spiral is

$$n = \frac{A_s}{A_{\text{wire}}} = \frac{0.44}{2 \cdot 0.11} = 1.99$$

Two wraps of the spiral are required. The factor of 2 in the denominator represents that the wire surrounds the prism twice per wrap.

Comment: The depth of the wedge is twice the diameter or about 10.5 in. Therefore ten wraps with a 1 in. pitch would be used to contain the full wedge. The required two wraps would occur in the 3 in. immediately below the anchor plate and would be roughly on the centroid of the splitting force.

References

ACI 318-14. (2014). *Building code for structural concrete and commentary*. Farmington Hills, MI: American Concrete Institute.
ACI Committee 349. (2001). *Code Requirements for Nuclear Safety Related Concrete Structures*. Farmington Hills, MI: American Concrete Institute.

ACI Committee 355.2-07. (2007). *Qualification of post-installed mechanical anchors in concrete and commentary*. Farmington Hills, MI: American Concrete Institute.

ACI Committee 355.4-10. (2010). *Acceptance criteria for qualification of post-installed adhesive anchors in concrete*. Farmington Hills, MI: American Concrete Institute.

Anderson, N. S., & Meinheit, D. F. (2005). Pryout capacity of cast-in headed stud anchors. *PCI Journal, 50*(2), 90–112.

Anderson, N. S., & Meinheit, D. F. (2007). A review of headed stud design criteria. *PCI Journal, 52*(1), 82–100.

ANSI/ASME B1.1. (2003). *Unified inch screw threads (UN and UNR Thread Form)*. New York: American Society of Mechanical Engineers.

ASCE/SEI 7-16 2016, *Minimum design loads for buildings and other structures*, ASCE/SEI 7–16. American Society of Civil Engineers, Reston, VA.

Cook R. A., & Klingner, R. E. (1992a). Behavior of ductile multiple-anchor steel-to-concrete connections with surface-mounted baseplates. *Anchors in concrete: Design and behavior*, ACI SP-130. American Concrete Institute, Farmington Hills, MI, pp. 61–122.

Cook, R. A., & Klingner, R. E. (1992b). Ductile multiple-anchor steel-to-concrete connections. *Journal of Structural Engineering, 118*(6), 1645–1665.

Cook, R. A., Kunz, J., Fuchs, W., & Konz, R. C. (1998). Behavior and design of single adhesive anchors under tensile load in uncracked concrete. *ACI Structural Journal, 95*(1), 9–26.

Eligehausen, R., & Balogh, T. (1995). Behavior of fasteners loaded in tension in cracked reinforced concrete. *ACI Structural Journal, 92*(3), 365–379.

Eligehausen, R., Cook, R. A., & Appl, J. (2006). Behavior and design of adhesive bonded anchors. *ACI Structural Journal, 103*(6), 822–831.

Fuchs, W., Eligehausen, R., & Breen, J. (1995). Concrete capacity design (CCD) approach for fastening to concrete. *ACI Structural Journal, 92*(1), 73–93. Discussion: *ACI Structural Journal, 92*(6), 1995, pp. 787–802.

Hansen, B. (2009). Investigators fault epoxy 'Creep' in big dig collapse. *Civil Engineering, 77*(9), 20–21.

Hawkins, N. M. (1968). The bearing strength of concrete loaded through rigid plates. *Magazine of Concrete Research, 20*(62), 31–40.

Kuhn, D., & Shaikh, F. (1996) *Slip-pullout strength of hooked anchors. Research Report*. University of Wisconsin-Milwaukee, submitted to the National Codes and Standards Council.

Lotze, D., Klingner, R. E., & Graves III, H. L. (2001). Static behavior of anchors under combinations of tension and shear loading. *ACI Structural Journal, 98*(4), 525–536.

PCI design handbook: Precast and prestressed concrete. (2017). MNL-120 (8th ed.). Chicago, IL: PCI.

Primavera, E. J., Pinelli, J. P., & Kalajian, E. H. (1997). Tensile behavior of cast-in-place and undercut anchors in high-strength concrete. *ACI Structural Journal, 94*(5), 583–594.

Winter J. B., and Dolan, C. W., 2014, Concrete breakout capacity of cast-in-place concrete anchors in early age concrete, PCI Journal, 59(2), pg. 114–131.

Chapter 16
Comprehensive Problems

16.1 Concept

Design of prestressed concrete is more intense than design of reinforced concrete and requires more checks and adjustment of reinforcement. The following comprehensive problems are developed to allow mastery of the topic. The problems may be assigned in parts, as a project, or as a take-home exam. They may form the basis for development of comprehensive calculation sheets using Excel or Mathcad. Development of comprehensive design aids focuses attention on the organization of the calculations and clarity of presentation.

The problems are narrowly defined to reduce the range of potential design solutions. Options exist for the student to provide supplemental information, as would occur in practice. Specified summary tables are suggested to allow efficient evaluation of submittals. As a take-home exam, these problems provide a platform for validating student calculation sheets. The problems may be extended by changing loads, spans, material properties, or adding further constraints.

16.2 Floor Beam

The section shown in Fig. 16.1 is a portion of a simple span prestressed concrete floor system. You are to provide complete design calculations including confirmation or revision of the tendon and shear reinforcement. Assume that the 120-in. flange width is the effective width. All dimensions are in inches. Report the class of your final design, U, T, or C. Stirrups are #3 bars.

$L = 30'$ Strands = ½ in. diameter, 270 ksi, low relaxation

$w_d = 20$ psf

$w_l = 60$ psf, no live load reduction factor allowed

© Springer Nature Switzerland AG 2019
C. W. Dolan, H. R. Hamilton, *Prestressed Concrete*,
https://doi.org/10.1007/978-3-319-97882-6_16

Fig. 16.1 Beam section

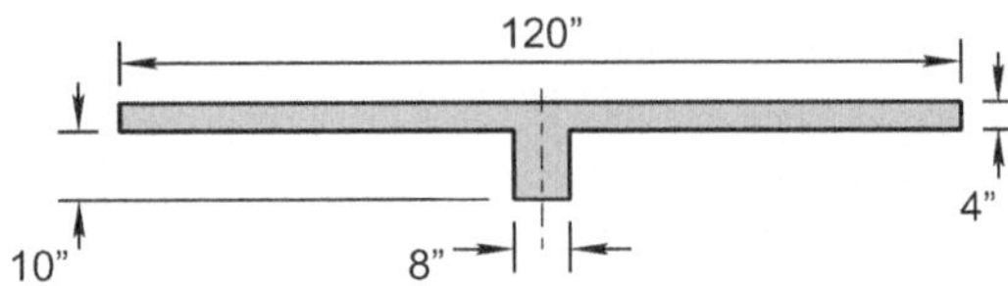

$f'_c = 6000$ psi

Number of strands = 6
End eccentricity = 1 in.
Midspan eccentricity = 8.5 in. You may use either a parabolic drape or a harped
  strand

$A = 460$ in.^{2}
$y_b = 11$ in.
$I_c = 4667$ in.^{4}

Compute or assume any additional information you need to complete the design
and report any changes and your beam classification on your cover sheet.

16.3 Pedestrian Bridge

The section shown below is suggested for a simple span pretensioned pedestrian
bridge. The superimposed dead load consists of non-structural handrails of 120 plf,
half placed on each edge of the beam. The live load is 60 psf. What is the longest
simple span that you can produce and still be within ACI Building Code limits for
flexure? (Fig. 16.2).

For your solution, provide the number of ½ in. diameter 270 ksi, low relaxation
strands, the strand profile, the stress summary, and the final strength check. Summarize your solution in a form like Table 16.1.

16.4 Post-tensioned Pedestrian Bridge

Design the pedestrian bridge girder in Sect. 16.3 using post-tensioned tendons.
Determine the number and size of the tendons assuming the bridge is cast-in-place
and spans 95 ft.

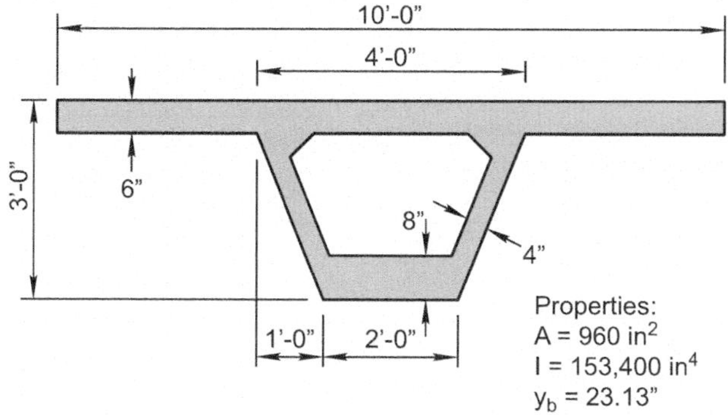

Fig. 16.2 Pedestrian beam section and properties

Table 16.1 Sample beam solution format

Item		
No. of strands		
End eccentricity	in.	Harp location:
Midspan eccentricity	in.	
Stresses	Top	Bottom
End, initial	psi	psi
Midspan, initial	psi	psi
Midspan, final	psi	psi
M_u	ft-kips	
ϕM_n	ft-kips	
Maximum span length	ft	

16.5   Torsion Design of Pedestrian Bridge

Using the information from the pedestrian bridge design in Sect. 16.3 and a span of 65 ft; double the live load on one side of the span. In addition to the prestress design, evaluate and girder for torsional effects and design the shear reinforcement accordingly.

16.6   Multistage Prestressing

A simple span beam is a decked bulb T section. The basic beam is cast with 12-straight ½ in. diameter 270 ksi low relaxation steel strands, Fig. 16.3. The strands are stressed to the ACI Building Code limits. The composite deck is then placed and a second parabolically draped post tensioning strand is stressed to ACI Building Code specification limits.

The span length is 100 ft, the superimposed dead load, exclusive of the deck weight, is 250 plf, and the live load is 550 plf. The concrete strength at transfer is

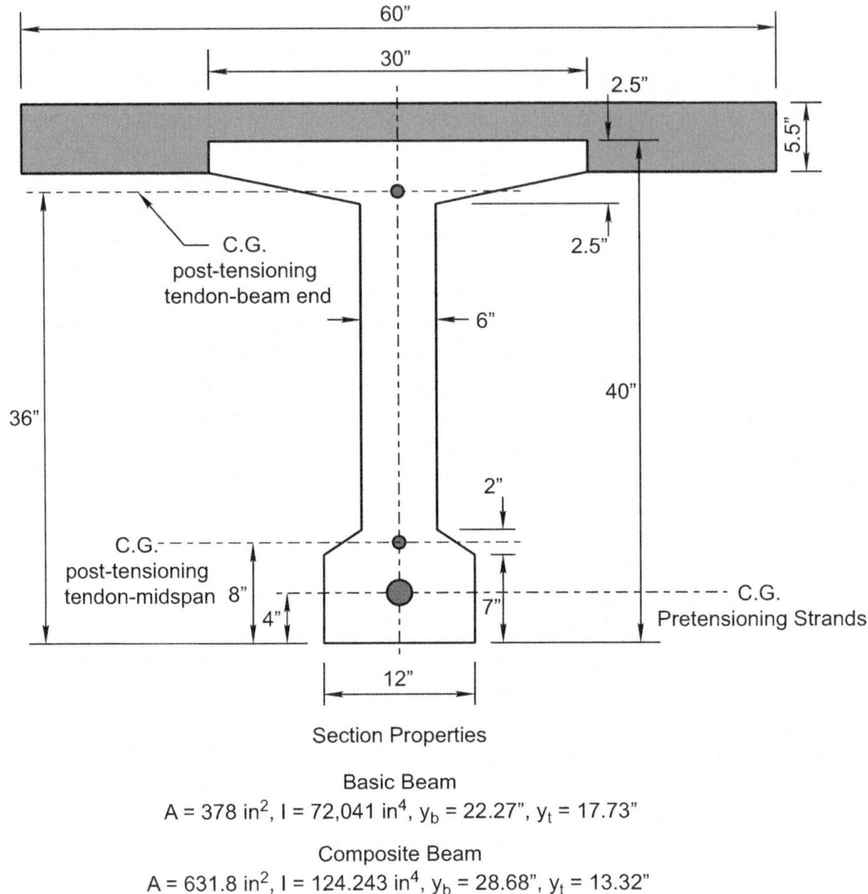

Section Properties

Basic Beam
$A = 378$ in^2, $I = 72{,}041$ in^4, $y_b = 22.27$", $y_t = 17.73$"

Composite Beam
$A = 631.8$ in^2, $I = 124.243$ in^4, $y_b = 28.68$", $y_t = 13.32$"

Fig. 16.3 Decked bulb-T beam section properties

4500 psi and the specified design strength is 6500 psi. Grade 60 reinforcement is used. Post-tensioning is Grade 270 low relaxation strand either 0.5 or 0.6 in diameter.

1. Verify that the service stresses in the basic beam meet the ACI Building Code requirements. You may add steel reinforcement if the tensile stresses are too high and you may go to $0.8 f'_{ci}$ at the ends if needed.
2. Determine the size of the post-tensioning tendon to carry the service loads. The tendon must contain a whole number of strands and may be stressed to the full ACI Building Code limit.
3. Compute the final service level stresses considering the full sequence of loads.
4. Check the nominal capacity of the beam. Adjust the post-tensioning strand or add reinforcement if needed.
5. Design shear reinforcement.

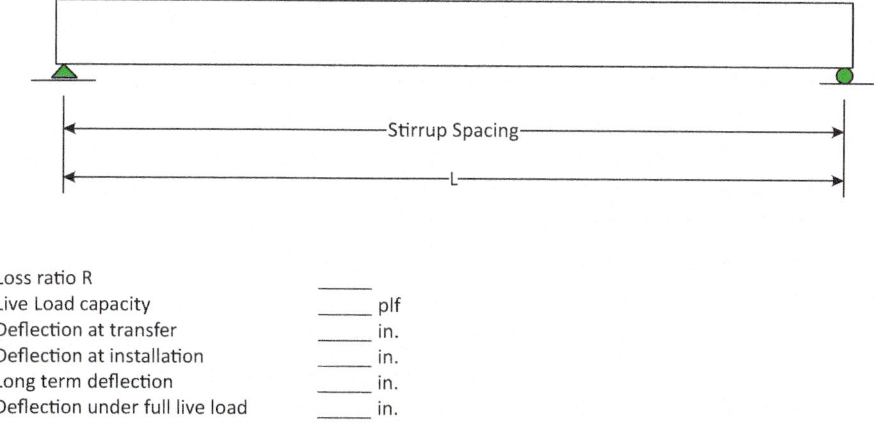

Loss ratio R   ______
Live Load capacity   ______ plf
Deflection at transfer   ______ in.
Deflection at installation ______ in.
Long term deflection   ______ in.
Deflection under full live load ______ in.

Fig. 16.4 General beam size and required information

6. Calculate initial, intermediate, and long-term deflections.

Completion of this assignment tracks of when loads are applied, which section carries each set of loads, how and when losses are applied, and what eccentricities go with each set of loadings. Submit a complete set of calculations, annotated with comments to indicate your approach and what checks have been performed.

16.7   Beam Design

Determine the stirrup requirements, service level deflections, and maximum live load that can be carried by the one of the beam sections shown below in Figs. 16.4, 16.5, 16.6, 16.7, 16.8, 16.9, and 16.10. Initial strand design is provided from the PCI Design Handbook. The maximum live load may have to be found by iteration, checking all allowable and strength limit states. Round off the live load to the nearest 5 psf or 100 plf. If needed, assume the relative humidity to be 50%. Assume the effective prestress is 85% of the initial prestress. Other loss ratios may be considered.

The beam must meet ACI Building Code service level stress and strength requirements; however, depending on your calculation of losses and live loads, you may have to adjust the tendon profile or the given concrete strength. If the section will not work, describe the Code conditions that are not met.

16.7.1   40IT32

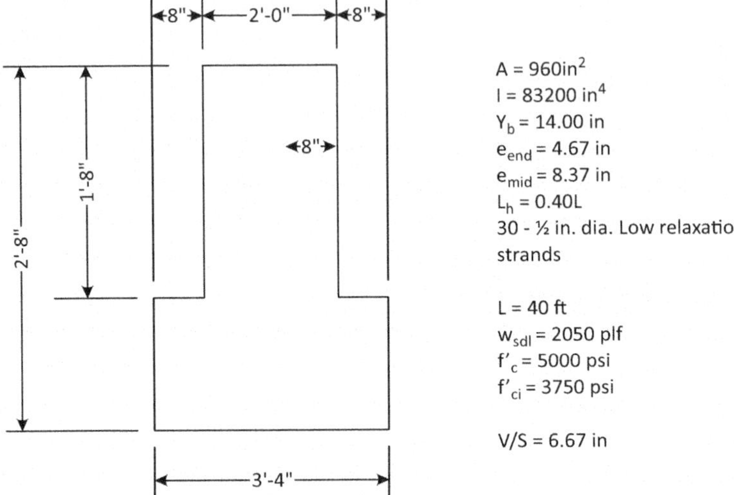

Fig. 16.5 32 in. deep inverted-T beam and properties

16.7.2   40IT48

Fig. 16.6 48 in. deep inverted-T beam and properties

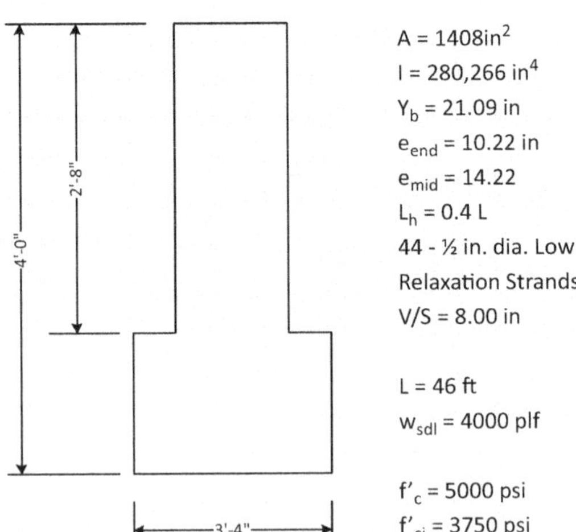

16.7.3   10DT34-68

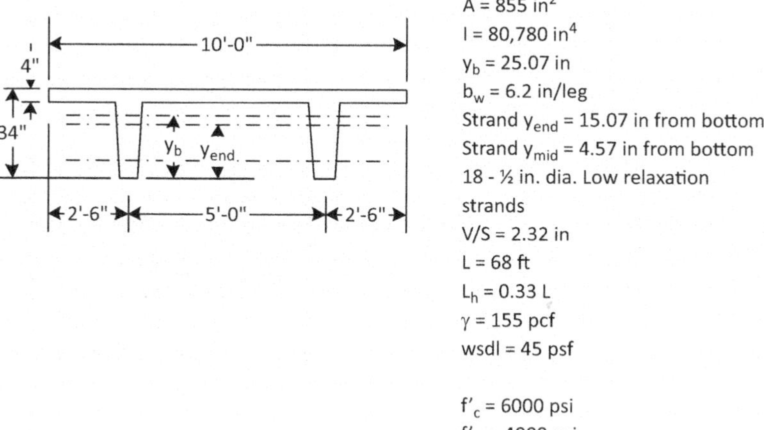

Fig. 16.7   34 in. deep double-T beam

16.7.4   10DT34LW-68

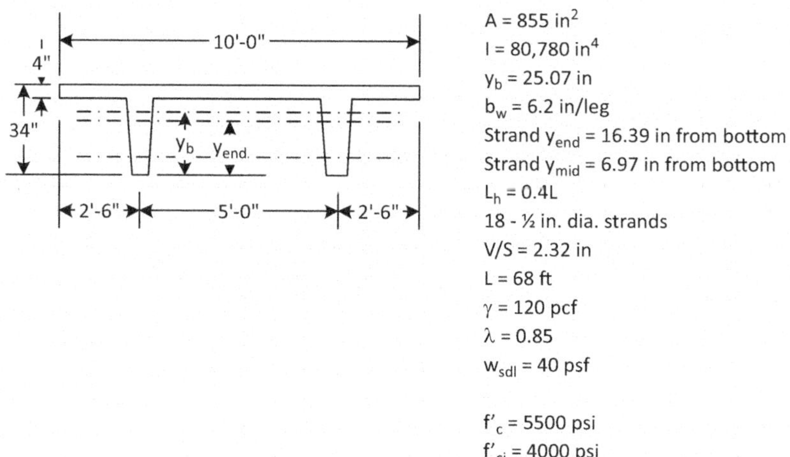

Fig. 16.8   34 in. deep double-T beam with lightweight concrete

16.7.5   10DT34-80

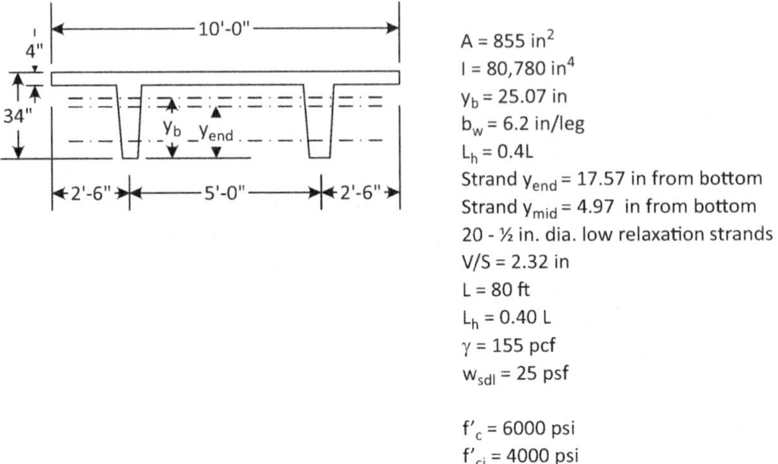

Fig. 16.9   34 in. deep double-T beam with harped strand

16.7.6   10DT34-60

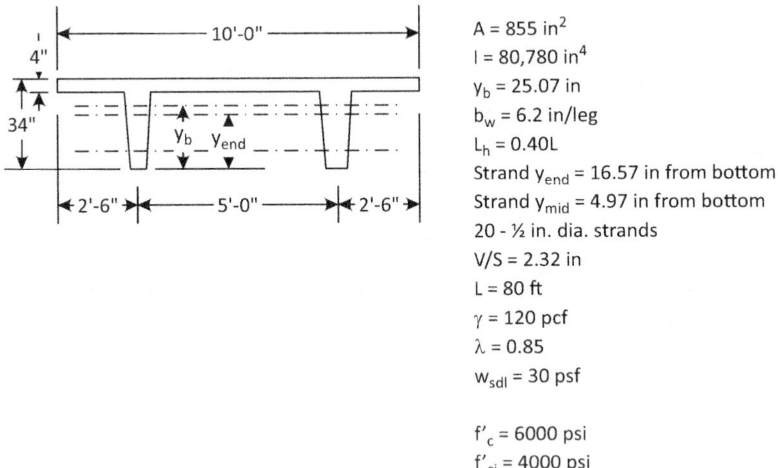

Fig. 16.10   34 in. deep double-T beam, harped strand and lightweight concrete

Appendixes

Appendix A: Properties of Prestressing and Reinforcement Steel

Table A.1 Properties of 270 ksi prestressing strand, $f_{pu} = 270$ ksi

Nominal diameter (in.)	Area, A_{ps} (in.^{2})	Weight (plf)	$0.7 f_{pu} A_{ps}$ (kips)	$0.75 f_{pu} A_{ps}$ (kips)	$0.8 f_{pu} A_{ps}$ (kips)	$f_{pu} A_{ps}$ (kips)
3/8	0.085	0.29	16.1	17.2	18.4	23.0
7/16	0.115	0.40	21.9	23.3	24.8	31.0
½	0.153	0.52	28.9	31.0	33.0	41.3
½ special	0.167	0.53	31.6	33.8	36.1	45.1
0.6	0.217	0.74	41.0	43.0	46.9	58.6

Adapted from PCI Design Handbook, 8th ed. 2017

Table A.2 Properties of plain prestressing bars, $f_{pu} = 145$ ksi

Nominal diameter (in.)	Area, A_{ps} (in.^{2})	Weight (plf)	$0.7 f_{pu} A_{ps}$ (kips)	$0.8 f_{pu} A_{ps}$ (kips)	$f_{pu} A_{ps}$ (kips)
3/8	0.442	1.50	44.9	51.3	64.1
7/8	0.601	2.04	61.0	69.7	87.1
1	0.785	2.67	79.7	91,0	113,8
1 1/8	0.994	3.38	100.9	115.3	144.1
1 ¼	1.227	4.17	124.5	142.3	177.9
1 ½	1.485	5.05	150.7	172.2	215.3

Adapted from PCI Design Handbook, 8th ed. 2017

Table A.3 Properties of prestressing bars, $f_{pu} = 160$ ksi

Nominal diameter (in.)	Area, A_{ps} (in.^{2})	Weight (plf)	$0.7 f_{pu} A_{ps}$ (kips)	$0.8 f_{pu} A_{ps}$ (kips)	$f_{pu} A_{ps}$ (kips)
3/8	0.442	1.50	49.5	56.6	70.7

(continued)

© Springer Nature Switzerland AG 2019
C. W. Dolan, H. R. Hamilton, *Prestressed Concrete*,
https://doi.org/10.1007/978-3-319-97882-6

Table A.3 (continued)

Nominal diameter (in.)	Area, A_{ps} (in.^{2})	Weight (plf)	$0.7\,f_{pu}A_{ps}$ (kips)	$0.8\,f_{pu}A_{ps}$ (kips)	$f_{pu}A_{ps}$ (kips)
7/8	0.601	2.04	67.3	77.0	96.2
1	0.785	2.67	87.9	100.5	125.6
1 1/8	0.994	3.38	111.3	127.2	159.0
1 ¼	1.227	4.17	137.4	157.0	196.3
1 ½	1.485	5.05	166.3	190.1	237.6

Adapted from PCI Design Handbook, 8th ed. 2017

Table A.4 Deformed prestressing bars, $f_{pu} = 150$ ksi

Nominal diameter (in.)	Area in^2	Weight plf	$0.7\,f_{pu}A_{ps}$ kips	$0.8\,f_{pu}A_{ps}$ kips	$f_{pu}A_{ps}$ kips
5/8	0.28	0.98	29.4	33.6	42
3/8	0.42	1.49	44.1	50.4	63
1	0.85	3.01	89.3	102.0	128
1-1/4	1.25	4.39	131.3	150.0	188
1-3/8	1.58	5.56	165.9	159.6	237
1-3/4	2.58	9.10	270.9	309.6	387
2-1/2	5.16	18.20	541.8	619.2	774
3	6.85	24.09	719.3	822.0	1028

Table A.5 Properties of prestressing wire

Diameter	Area in^2	Weight plf	f_{pu} ksi	$0.7\,f_{pu}A_{ps}$ kips	$0.8\,f_{pu}A_{ps}$ kips	$f_{pu}A_{ps}$ kips
0.192	0.0290	0.098	250	5.07	5.80	7.25
0.196	0.0302	0.100	250	5.28	6.04	7.55
0.250	0.0491	0.170	240	8.25	9.43	11.78
0.276	0.0598	0.200	235	9.84	11.24	14.05

Table A.6 Properties of ASTM Standard reinforcement

Dar sizedesignation		Nominal diameter in.	Area, A_p in.^{2}	Weight plf
US	SI			
#3	#10	0.375	0.11	0.376
#4	#13	0.50	0.20	0.668
#5	#16	0.625	0.31	1.043
#6	#19	0.75	0.44	1.502
#7	#22	0.875	0.60	2.044
#8	#25	1.00	0.79	2.67
#9	#29	1.127	1.00	3.40
#10	#32	1.27	1.27	4.30
#11	#36	1.41	1.56	5.31
#14	#43	1.69	2.25	7.65

(continued)

Table A.6 (continued)

Dar sizedesignation		Nominal diameter	Area, A_p	Weight
US	SI	in.	in.^{2}	plf
#18	#57	2.26	4.00	13.60

Adapted from PCI Design Handbook, 8th ed. 2017
It is common for many mills to mark bars with a SI designation. In such cases, the bar properties are identical to the US customary designation

Table A.7 Properties of plain and deformed welded-wire reinforcement (ASTM A1064)

Wire designation[a]		Nominal diameterin. [mm]	Nominal areain.^{2} [mm]
Plain	Deformed		
W1.4		0.134 [3.39]	0.014 [9.03]
W2	D2	0.160 [4.05]	0.020 [12.9]
W2.5		0.178 [4.53]	0.025 [16.1]
W2.9		0.192 [4.88]	0.029 [18.7]
W3.5		0.211 [5.36]	0.035 [22.6]
W4	D4	0.226 [5.73]	0.040 [25.8]
W4.5		0.239 [6.08]	0.045 [29.0]
W5	D5	0.252 [6.41]	0.050 [32.3]
W5.5		0.265 [6.72]	0.055 [35.5]
W6	D6	0.276 [7.02]	0.060 [38.7]
W8	D8	0.319 [8.11]	0.080 [51.6]
W10	D10	0.357 [9.06]	0.100 [64.5]
W11	D11	0.374 [9.50]	0.110 [71.0]
W12	C12	0.391 [9.93]	0.120 [77.4]
W14	D14	0.422 [10.7]	0.140 [90.3]
W16	D16	0.451 [11.5]	0.160 [103]
W18	D18	0.479 [12.2]	0.180 [116]
W20	D20	0.505 [12.8]	0.200 [129]
W22	D22	0.529 [13.4]	0.220 [142]
W24	D24	0.553 [14.0]	0.240 [155]
W26	D26	0.575 [14.6]	0.260 [168]
W28	D28	0.597 [15.2]	0.280 [181]
W30	D30	0.618 [15.7]	0.300 [194]

[a]The number following the prefix indicates the nominal cross-sectional area of the deformed wire in square inches multiplied by 100

Appendix B: Beam Equations

Appendix B.1: Camber and Equivalent Load Equations

Table B.1 Tendon configurations and their equivalent loads

Tendon configuration	Δ	Equiv. load
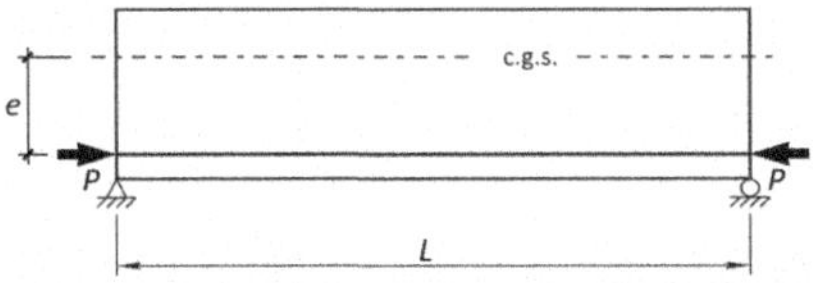	$\dfrac{PeL^2}{8EI}$	
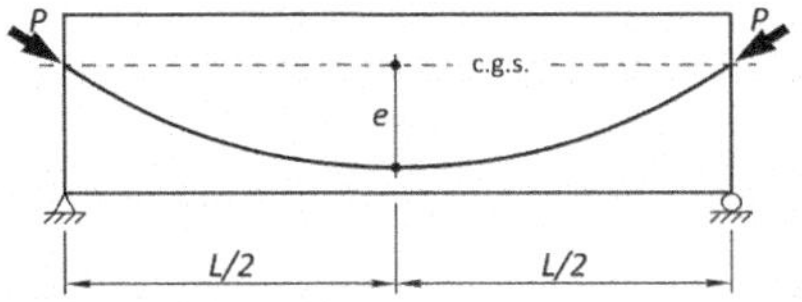	$\dfrac{5PeL^2}{48EI}$	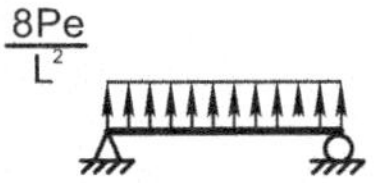
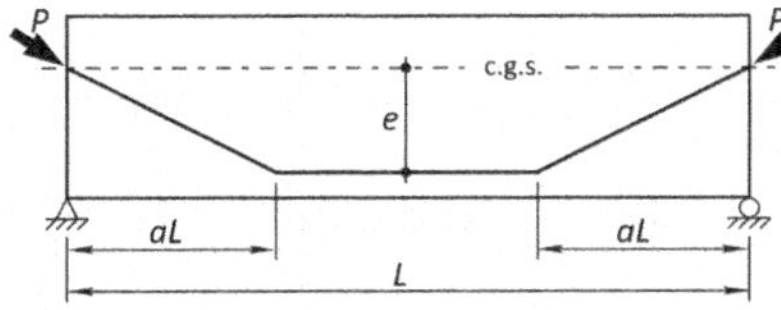	$PeL^2\left(\dfrac{1-\frac{4}{3}a^2}{8EI}\right)$	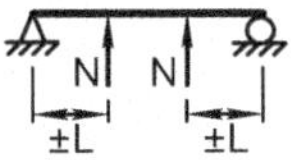
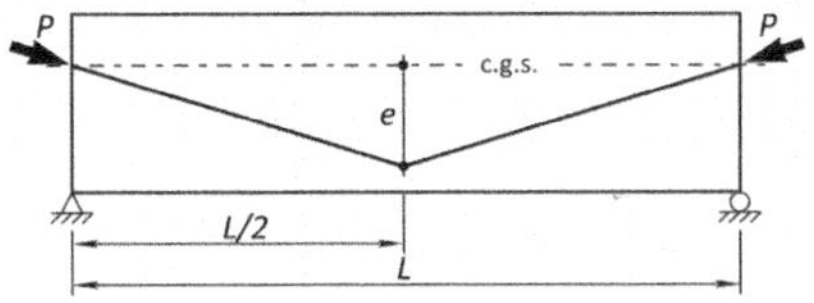	$\dfrac{PeL^2}{12EI}$	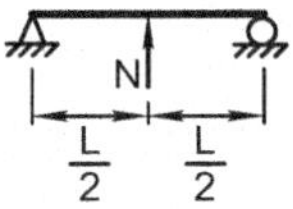

Appendix C: Representative Section Properties

Appendix C.1: AASHTO Bridge Girders

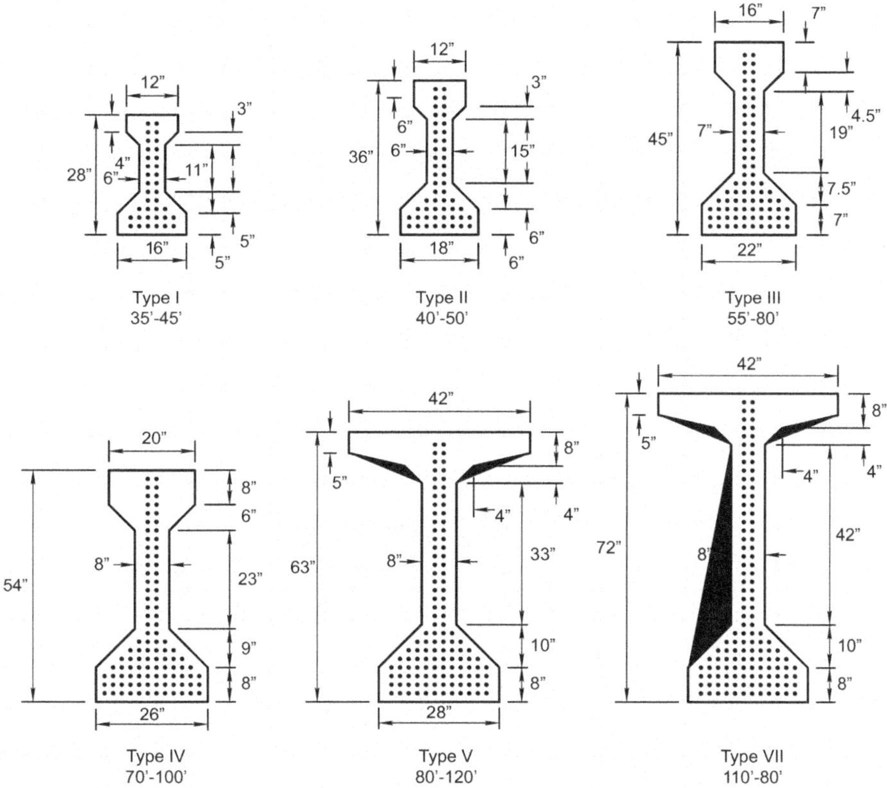

Fig. C.1 AASHTO bridge girder sections and strand patterns

Table C.1 AASHTO Bridge Girder properties

Type	h in.	A_g in.^{2}	I_g in.^{4}	c_{top} in.	c_{bottom} in.	w_o plf	r^2 in.^{2}
I	28	276	22,750	15.41	12.59	288	82
II	36	369	50,979	20.17	15.83	384	138
III	45	560	125,390	24.73	20.27	583	224
IV	54	789	260,741	29.27	24.73	822	330
V	63	1013	521,180	31.04	31.96	1055	514
VI	72	1085	733,320	35.62	36.38	1130	676

Appendix C.2: PCI/AASHTO Bridge Girders

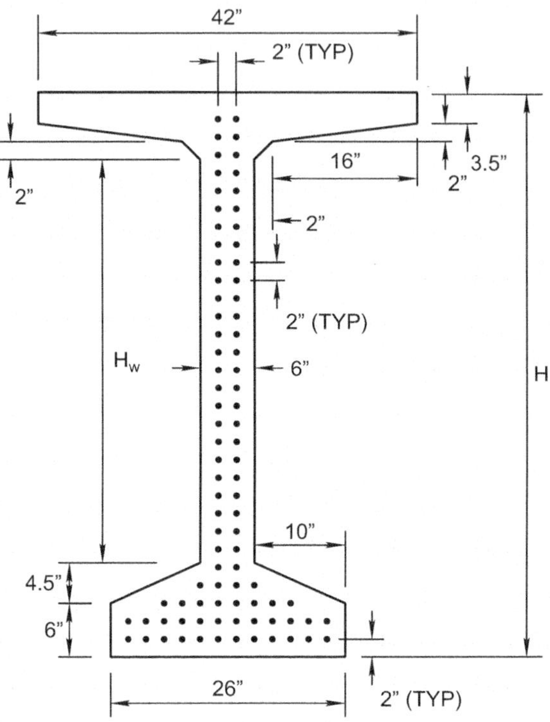

Fig. C.2 PCI bridge girder section and strand pattern

Table C.2 PCI bridge girder properties

Type	H in.	H_w in.	A_g in.^{2}	I_g in.^{4}	c_{bottom} in.	w_o plf
BT-54	28	36	659	268,077	27.63	686
BT-63	36	45	713	392,638	32.12	743
BT-72	45	54	767	545,894	36.60	799

Appendix C.3: Double-T Sections

Double-T sections are one of the more common precast prestressed concrete elements. The exact section properties vary by manufacturer. Typical sections are shown in Fig. C.2 and selected properties are given in Table C.2. Extensive tables and span capacities are given in the PCI Design Handbook.

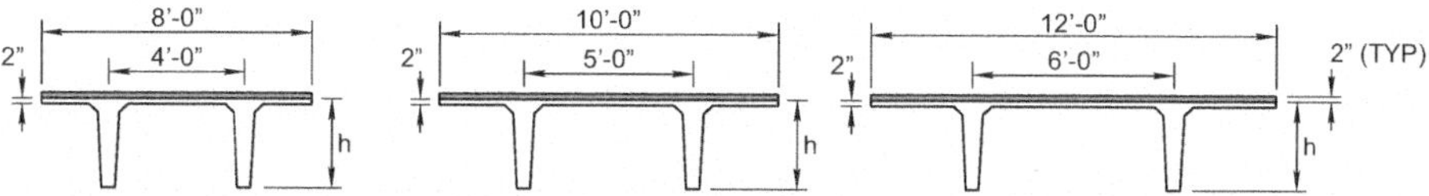

Fig. C.3 Selected composite double-T sections and properties

Table C.3 Selected composite double-tee properties for normal weight concrete

Type	No topping						With 2-in. topping			
Width × height	A_g in.^{2}	I_g in.^{4}	c_{top} in.	c_{bottom} in.	w_o plf	V/S in.	I_g in.^{4}	c_{top} in.	c_{bottom} in.	w_o plf
$8'\text{-}0'' \times 24''$	401	20,985	6.85	17.15	418	1.41	27,720	6.73	19.27	618
$8'\text{-}0'' \times 32''$	567	55,646	10.79	21.21	591	1.79	71,866	10.34	23.66	791
$10'\text{-}0'' \times 24''$	449	22,469	6.23	17.11	468	1.35	29,396	6.11	19.89	718
$10'\text{-}0'' \times 32''$	615	59,720	10.02	21.98	641	1.69	77,118	9.46	24.54	891
$12'\text{-}0'' \times 28''$	640	44,563	7.99	20.01	677	1.62	57,323	7.53	22.47	977
$12'\text{-}0'' \times 32''$	690	64,620	9.25	22.75	551	1.70	88,305	7.92	26.08	851

Adapted from PCI Design Handbook, 7th ed. 2010

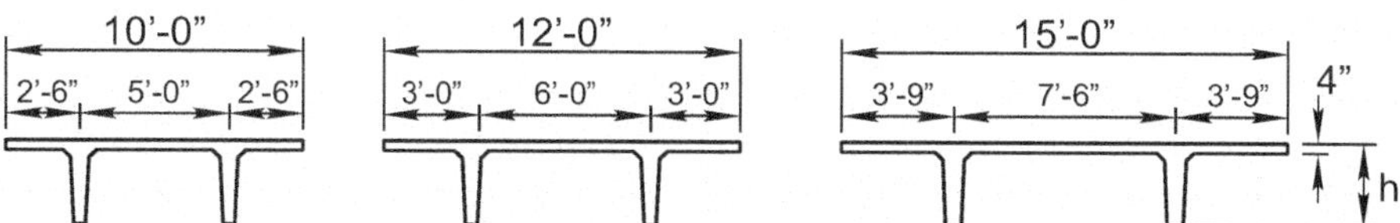

Fig. C.4 Selected composite double-T sections

Table C.4 Selected pretopped double-tee properties for normal weight concrete

Width × height	A_g in.^{2}	I_g in.^{4}	c_{top} in.	c_{bottom} in.	w_o plf	V/S in.
$10'\text{-}0'' \times 26''$	689	30,716	5.71	20.29	718	2.05
$10'\text{-}0'' \times 34''$	855	80,780	8.93	25.07	891	2.32
$12'\text{-}0'' \times 30''$	928	59,997	7.06	22.94	967	2.30
$12'\text{-}0'' \times 34''$	978	86,072	8.23	25.77	1019	2.39
$15'\text{-}0'' \times 30''$	1133	78,625	7.25	22.75	1180	2.42

(continued)

Table C.4 (continued)

Width × height	A_g in.^{2}	I_g in.^{4}	c_{top} in.	c_{bottom} in.	w_o plf	V/S in.
15′-0″ × 34″	1185	109,621	8.35	25.65	1234	2.45

Adapted from PCI Design Handbook, 8th ed. 2017

Appendix C.4: Hollowcore Slabs

Hollowcore slabs or planks come in a number of varying configurations depending on the manufactured. The cores range from circular to oval to more complex shapes as seen in Fig. C.5. The slabs are extruded with a continuously running slipform machine using a zero slump concrete. Prestressing strands are centered between the cores with ¾ in. clearance. Table C.5 summarizes several hollowcore sections and properties.

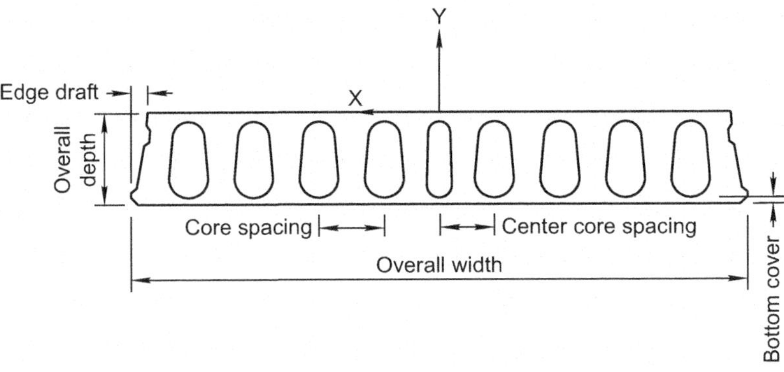

Fig. C.5 Representative hollowcore section (Courtesy Elematic Inc.)

Table C.5 Selected properties of hollowcore slabs

Manufacturer	Untopped					With 2″ topping		
	Width × depth	A_g in.^{2}	I_g in.^{4}	c_{bottom} in.	w_o psf	I_g in.^{4}	c_{bottom} in.	w_o plf
Flexicore	2′-0″ × 6″	86	366	3.00	45	793	4.20	70
Spancrete	4′-0″ × 8″	258	1806	3.98	63	5787	5.22	88
Spancrete	4′-0″ × 12″	355	5784	6.28	86	8904	7.58	111
Elematic	8′-0″ × 10″	549	6642	5.00	73	11,827	6.50	98

Adapted from PCI Design Handbook, 8th ed. 2017

Appendix D: Equivalent Column Stiffness

In the equivalent frame approach to determining moments in a slab structure, the equivalent column recognizes that the slab rotation is minimum at the column and greater as the slab moves away from the column, Fig. D.1. The **equivalent column** replaces the real column by accounting for the torsional flexibility of the slab.

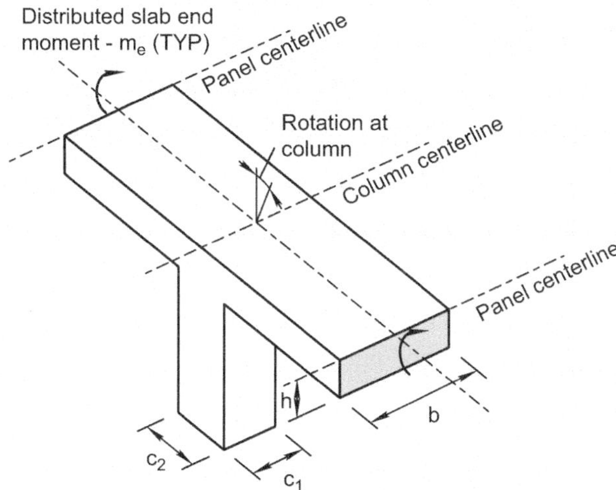

Fig. D.1 Equivalent column

The slab in the column strip connecting to the column is subject to an end moment m_e which in turn becomes a torque at the end of the slab. This torque is resisted by the column and the slab beam strip. Figure D.1 indicates the slab beam strip is a rectangle b wide and h deep, where b is the width of the column plus half the depth of the slab on either side of the column and h is the thickness of the slab. If the two-way slab rested on a support beam, that beam would be included. The equivalent column then replaces the actual column and the end beam and is defined so that the total flexibility (the inverse of stiffness) of the combined column is the sum of the flexibilities of the components. Therefore:

$$\frac{1}{K_{ec}} = \frac{1}{\Sigma K_c} + \frac{1}{K_t} \tag{D.1}$$

where:

K_{ec} is the stiffness of the equivalent column
K_c is the flexural stiffness of the real column
K_t is the torsional stiffness of the transverse slab strip

and all stiffnesses have units of moment per unit rotation. The flexural stiffness of the real column is $4E_c I/l$, and the summation indicates that there is a column below and there may be a column above the slab. The torsional stiffness of the transverse slab strip is:

$$K_\mathrm{t} = \sum \frac{9E_\mathrm{c}C}{l_2\left(1 - \frac{c_2}{l_2}\right)^3} \tag{D.2}$$

where:

C is the cross sectional constant for the transverse slab strip
l_2 is the width of the slab in the direction of the transverse slab strip
c_2 is the width of the column in the transverse direction

The summation applies to the typical case where the slab strip is on both sides of the column. The torsional cross section constant is

$$C = \left(1 - 0.63\frac{x}{y}\right)\frac{x^3 y}{3} \tag{D.3}$$

where x is the smaller dimension of the transverse slab strip section and y is the larger dimension, corresponding to h and b respectively in Fig. D.1.

Author Index

© Springer Nature Switzerland AG 2019
C. W. Dolan, H. R. Hamilton, *Prestressed Concrete*,
https://doi.org/10.1007/978-3-319-97882-6

443

Subject Index

Printed by Printforce, United Kingdom